Dynamic Modeling and Econometrics in Economics and Finance

Volume 25

More information about this series at http://www.springer.com/series/5859

Josef L. Haunschmied ·
Raimund M. Kovacevic ·
Willi Semmler · Vladimir M. Veliov
Editors

Dynamic Economic Problems with Regime Switches

Springer

Editors
Josef L. Haunschmied
Operations Research and Control Systems, Institute of Statistics and Mathematical Methods in Economics
Vienna University of Technology
Wien, Austria

Raimund M. Kovacevic
Operations Research and Control Systems, Institute of Statistics and Mathematical Methods in Economics
Vienna University of Technology
Wien, Austria

Willi Semmler
Department of Economics
The New School for Social Research
New York, NY, USA

Vladimir M. Veliov
Operations Research and Control Systems, Institute of Statistics and Mathematical Methods in Economics
Vienna University of Technology
Wien, Austria

ISSN 1566-0419 ISSN 2363-8370 (electronic)
Dynamic Modeling and Econometrics in Economics and Finance
ISBN 978-3-030-54578-9 ISBN 978-3-030-54576-5 (eBook)
https://doi.org/10.1007/978-3-030-54576-5

Mathematics Subject Classification: 91-06, 91Bxx, 90-06, 93Exx, 49-06

This Springer imprint is published by the registered company Springer Nature Switzerland AG
The registered company address is: Gewerbestrasse 11, 6330 Cham, Switzerland

Preface

This book is about instantaneous (or relatively fast) changes of the regime of operation of controlled dynamical systems with a focus on models arising in economics. The dynamics is described by deterministic or stochastic ordinary differential equations and regime changes mean switches from one to another dynamics with possible change of the state space as well as of the objective function. Regime switches may be caused by several reasons:

(i) Exogenous changes in the dynamics (for example, due to sudden environmental disasters or social/political reform);
(ii) Unintended internally driven changes in the dynamics (e.g., disasters caused by human activities or bankruptcy of firms);
(iii) Intended (controlled) shifts to new dynamics (technological innovations, merge of firms, immergence of backstop technologies, etc.);
(iv) Changes of preferences/objectives (involving environmental concerns, shifting from individual to cooperative objective, etc.).

Several of the above triggers of switches may be present in a single model. A typical example is provided by the so-called multistage (or multi-phase) models, where every stage may have its own dynamics and its own objectives.

While exogenous changes are an important subject even without considering optimal decisions—which is, e.g., shown by many econometric papers on Markov– and hidden Markov models, Threshold Autoregressive (TAR) models, and Smooth Transition Autoregressive (STAR) models—the focus of this book remains on optimization and control of dynamics. Still, type (i) plays an important role in control theory. This is particularly true in stochastic control and related financial applications, where the term "regime switch" usually refers exactly to exogenous changes.

The development of the theory of regime changes in control problems was originally driven by numerous engineering applications. This theory addresses the so-called *switching (or switched) systems*, and is a part of the *hybrid system* theory, with the specific feature that the discrete event sub-models reduce to relatively simple laws for switching from one dynamics to another. In parallel to points (i)–(iii) above, the times of switch and the law of instantaneous state transition at the switching times can be internally or externally driven. In the former, the switch of the dynamics occurs “automatically” when the trajectory of the system crosses the boundary between exogenously given domains in the time-state space. In the latter case, the switching times and the transition laws are partly or entirely controlled, that is, they depend on external decisions.

The theory of switched systems related to engineering applications is developed from many perspectives: controllability, stabilization, optimality conditions, dynamic programming, Hamilton–Jacobi–Bellman equations, etc. On the other hand, models arising in economics, utilization of resources, population dynamics, and social sciences pose a number of problems involving regime switches which go beyond the traditional engineering applications. Many models in these areas require investigation of infinite horizon, multiple solutions, periodic optimal behavior, chattering, etc. The respective theoretical tools are still under development, often restricted to particular narrow classes of problems with switches. Even more, in all scientific areas mentioned above, consideration with more than one decision-maker naturally arise. Thus, the field of differential games for systems with switches in the dynamics becomes a hot topic, in particular in economics. As also seen in this book, even switches from a differential game to optimal control and vice versa are of interest.

With the aim of promoting and facilitating the development of optimal control and dynamic games and their applications in economics, the Vienna University of Technology initiated a conference series originally named “Viennese Workshop on Optimal Control, Dynamic Games and Nonlinear Dynamics” called nowadays “Viennese Conference on Optimal Control and Dynamic Games” (VC). The purpose was to bring together specialists in optimal control, dynamic games, and dynamical systems theory, with economists, demographers, and social scientists. The research presented at the 14 VCs organized to date covers all areas of the modern theory of optimal control and differential games with applications that range from “optimal wine consumption” and “dynamics of extra-marital affairs” to mean field games and PDE constrained optimization. The decision to publish a book on dynamic economic problems with regime switches was made at the 14th VC held in Vienna in July 2018, in which many contributors to this book participated. Additional chapters are included here by invited distinguished specialists in the area.

About the Content of the Book

Each chapter of the book has its own abstract, therefore the present editorial notes are structured in accordance with subjects, drivers of the switches, and mathematical techniques, rather than chapters. Referring to chapters we use the chapter number and the name of the first author.

The book begins with a review paper by N. Van Long, which gives a broad view on the problematic of regime switch in economic models and provides a survey in the area, emphasizing the techniques of optimal control and dynamic games. Thematically, the contributions in the book address a variety of subjects: capital accumulation, Chaps. 3 (Kovacevic), 10 (Dawid), 11 (Lambertini), 12 (Palokangas); innovations, Chaps. 7 (Yegorov), 10 (Dawid); cash flow optimization in finance, Chap. 13 (Savku); population economics, Chaps. 2 (Feichtinger), 5 (Orlov), 12 (Palokangas); economics of pollution and climate change, Chaps. 6 (Semmler), 9 (Boucekkine), 12 (Palokangas); institutional change, Chap. 2 (Feichtinger); dynamics of addiction, Chap. 4 (Kuhn).

The chapters of the book cover most of the drivers for switches as described at the beginning:

(i) *Exogenous switches*: switches driven by sudden environmental changes appear in Chap. 3 (Kovacevic); exogenous change of parameters may lead to abrupt change of strategies for maximizing biological fitness, Chap. 5 (Orlov); introduction of new control mechanisms (involving green bonds) at exogenously given times in a climate change integrated assessment model, Chap. 6 (Semmler); regime switches are generated by a Markov chain in a general stochastic optimal control model with applications for maximization of utility from cash flow, Chap. 13 (Savku).

(ii) *Unintended internally driven changes in the dynamics:* shift from non-addicted to addicted use of drugs, depending on the accumulated addictive stock, Chap. 4 (Kuhn); jump of model parameters due to reaching a threshold manifold in the state space, Chap. 8 (Bondarev); population growth and capital accumulation, as state variables, may trigger environmental disasters, Chap. 12 (Palokangas).

(iii) *Controlled shifts of the dynamics*: in Chap. 2 (Feichtinger) the time of switch from dictatorship economy to emigration of the dictator's elite is a decision variable; switch in climate policies, Chap. 6 (Semmler); switching from one to another type of scientific carrier aiming at maximal scientific output, Chap. 7 (Yegorov); switching from competition to cooperation in a polluting industry, Chap. 9 (Boucekkinne); shift to a new dynamics in a game between two firms as a result of the decision of one of the players to introduce a new product, Chap. 10 (Dawid).

(iv) *Change of objectives*: multistage models in which the driver of the change is the objective appear in Chap. 9 (Boucekkinne), where the objectives of two players (polluting firms) change from noncooperative to cooperative. Change of objectives is present in almost all chapters, but there it is more a result of the changed dynamics than a driving force of the switch.

The models investigated in this book are mostly deterministic. Stochastic considerations are involved in Chaps. 3 (Kovacevic) and 12 (Palokangas), where exogenous stochasticity only influences the switches and not the dynamics of otherwise deterministic models. Only Chap. 13 (Savku) involves fully stochastic dynamics with exogenous switches generated by Markov chains and belongs to the extensive literature on stochastic regime switching in finance and econometrics.

About the Order of Chapters in the Book

Half of the chapters after the survey paper by Van Long in Chap. 1, namely Chaps. 2–7, concern problems of optimal control, the remaining Chaps. 8–13 deal with differential games. In each of the two groups, the chapters are alphabetically ordered with respect to the first author's name.

Vienna, Austria
April 2020

Josef L. Haunschmied
Raimund M. Kovacevic
Willi Semmler
Vladimir M. Veliov

Contents

Contributors

Anton Bondarev International Business School Suzhou, Xi'an Jiaotong-Liverpool University, Suzhou, People's Republic of China

Tony Bonen Labour Market Information Council, Ottawa, Canada

Raouf Boucekkine Aix-Marseille University (Aix-Marseille School of Economics), CNRS, and EHESS, Marseille, France

Carmen Camacho Paris Sciences et Lettres, Paris School of Economics and CNRS, Paris, France

Matthew Cantele International Institute for Applied Systems Analysis, Laxenburg, Austria;
School of BioSciences, University of Melbourne, Parkville, Australia

Herbert Dawid Department of Business Administration and Economics, Institute of Mathematical Economics, Bielefeld University, Bielefeld, Germany

Gustav Feichtinger Institute of Statistics and Mathematical Methods in Economics, Operations Research and Control Systems, Vienna University of Technology, Wien, Austria

Serhat Gezer Department of Business Administration and Economics, Institute of Mathematical Economics, Bielefeld University, Bielefeld, Germany

Dmitry Gromov Krasovskii Institute of Mathematics and Mechanics, Ural Branch of the Russian Academy of Sciences, Yekaterinburg, Russia;
Faculty of Applied Mathematics and Control Processes, St. Petersburg State University, Saint Petersburg, Russia

Raimund M. Kovacevic Vienna University of Technology, Vienna, Austria

Michael Kuhn Wittgenstein Centre (IIASA, OeAW, University of Vienna), Vienna Institute of Demography/OeAW, Vienna, Austria

Luca Lambertini Department of Economics, University of Bologna, Bologna, Italy

George Leitmann College of Engineering, University of California at Berkeley, Berkeley, CA, USA

Ngo Van Long McGill University, Montreal, Canada

Helmut Maurer Institute for Analysis and Numerics, University of Münster, Müntser, Germany

Sergey Orlov International Institute for Applied Systems Analysis, Laxenburg, Austria;
Faculty of Computational Mathematics and Cybernetics, Lomonosov Moscow State University, Moscow, Russia

Andreas Novak Faculty of Business, Economics and Statistics, University of Vienna, Wien, Austria

Tapio Palokangas University of Helsinki, Helsinki, Finland

Elena Rovenskaya International Institute for Applied Systems Analysis, Laxenburg, Austria;
Faculty of Computational Mathematics and Cybernetics, Lomonosov Moscow State University, Moscow, Russia

E. Savku Ecole Polytechnique, CMAP, Palaiseau, France;
Middle East Technical University, IAM, Dumlupinar Bulvari, Universiteler Mahallesi, Cankaya, Ankara, Turkey

Willi Semmler New School of Social Research, Newyork, USA;
Bielefeld University, Bielefeld, Germany;
IIASA, Laxenburg, Austria

Vegard Skirbekk Centre for Fertility and Health, Norwegian Institute of Public Health, Oslo, Norway;
Columbia Aging Center, Mailman School of Public Health, Columbia University, New York, USA

Marcin Stonawski Center for Advanced Studies of Population and Religion (CASPAR), Cracow University of Economics, Cracow, Poland

G.-W. Weber Ecole Polytechnique, CMAP, Palaiseau, France;
Middle East Technical University, IAM, Dumlupinar Bulvari, Universiteler Mahallesi, Cankaya, Ankara, Turkey

Franz Wirl Faculty of Business, Economics and Statistics, University of Vienna, Wien, Austria

Stefan Wrzaczek Wittgenstein Centre (IIASA, OeAW, University of Vienna), Vienna Institute of Demography/OeAW, Vienna, Austria

Yuri Yegorov University of Vienna, Vienna, Austria

Benteng Zou University of Luxembourg, Luxembourg City, Luxembourg

Chapter 1
Managing, Inducing, and Preventing Regime Shifts: A Review of the Literature

Ngo Van Long

JEL Codes: P48 · Q28 · Q58

1.1 Introduction

Many significant changes that occur to human societies, both at the macro-level and at the micro-level, are often associated with "sudden" shifts in the regimes or the modes of operations.[1] Examples of regime shifts in economics include the introduction of a new technology which makes the old mode of production obsolete [37], changes in the property rights regime (such as the enclosure process which created a landless working class in England), the emancipation of slave labor, revolutions (see, e.g., Campante and Chor [21], Lang and De Sterck [65], Boucekkine et al. [16], Michaeli and Spiro [92], for models related to the Arab Spring), the transfers of power from a colonial regime to a democratic regime, and human-induced climatic changes that can wipe out a large number of species. At the individual level, regime shifts include sudden events which change one's activities and consumption patterns, such as retirement, divorce, serious illness, or conversion to a new faith.

[1] From an historical perspective, what is termed "sudden" can correspond to hundred of years. For example, in England, the change in property right regime brought about by the "enclosure" movement took more than 300 years. Between 1605 and 1914, over 5000 "inclosure acts" were passed by Parliament, which transferred to private owners' land that was previously common properties. The general question as to whether most changes occur as discrete jumps or in a continuous fashion is a matter of debate, which to some extent hinges on what one means when words such as continuity and suddenness are used. For example, the theory of punctuated equilibrium, put forward by Eldredge and Gould [38] as a "better description" of the evolutionary process than Darwin's gradualism, has been opposed by Dawkins [32] on the ground that it was wrong to interpret gradualism as "constant speedism".

N. V. Long (✉)
McGill University, Montreal, Canada
e-mail: ngo.long@mcgill.ca

J. L. Haunschmied et al. (eds.), *Dynamic Economic Problems with Regime Switches*, Dynamic Modeling and Econometrics in Economics and Finance 25,
https://doi.org/10.1007/978-3-030-54576-5_1

Another source of regime shift is changes in preferences. Kemp and Long [57] analyzed the consequences of a shift in preferences of a decision-making body: a peaceful transfer of power anticipated by a colonial administration that plans to hand over the administration of a colony to a democratic government to be elected by local residents. Nkuiya and Costello [94] argued that a society's environmental preferences may change in the future when the citizens become more acutely aware of costs and benefits of conservation. These preference changes may themselves be triggered by a series of events. The authors wrote that "The modern environmental movement in the United States, where the Environmental Protection Agency, the Clean Water Act, and the National Environmental Policy Act were all formed over a relatively short period of time, is thought to have been triggered by a series of environmental disasters that raised the environmental profile sufficiently to incite public action" (p. 194). Related works on change in preferences or uncertain future preferences include Le Kama [69], Beltrati et al. [10], Le Kama and Schubert [70], Leonard and Long [75], and Itaya and Tsoukis [54].

Regime shifts can occur in the natural environment even in the absence of human activities. For example, lakes may shift from oligotrophic conditions (i.e., exhibiting a deficiency of plant nutrients, such that the water is very clear) to eutrophic conditions (displaying an abundance of nutrients), impacting fish populations and water quality [20, 22, 23, 104]. Coral reef systems can undergo changes from coral-dominated state to algal-dominated states. Forested land can become grassland. A biological invasion can wipe out wild and domestic animals and plants [96]. A disease can spread and become persistent after crossing an epidemiological threshold. For analyses of thresholds in epidemic diseases, see Veliov [124], Sims et al. [106], among others.

From an economic view point, regime shifts are often caused by a desire for changes on the part of some powerful coalitions of economic agents in order to further their interests. Throughout human history, many conflicts between nations or between social classes within a nation (e.g., the "elite" versus the "citizens") are attributable to attempts of possession or expropriation of natural resources. (See, for example, Long [80] on the nationalization of mines; Acemoglu and Robinson [2, 3] on class conflicts; van der Ploeg [120, 122] on resource wars; and Long [84] for a review of the theory of contests).[2] Smith [109] pointed out that the desire to possess more natural resources was one of the motives behind the European conquest of the New World and the establishment of colonies around the globe, some of which thrived on the systematic large-scaled exploitation of slave labor. Many changes that occur in our natural environment (such as climate change, with possible tipping points) can be attributed to the race among industrialized nations to become a dominant actor in the world scene.[3] Conflicts often arise because of lack of well-defined property rights in the exploitation of resources. In fact, the word "rivals" were derived from the

[2]The Arab Spring, which undoubtedly has many facets, is not unrelated to the contests for rents between the elite and the citizens.

[3]To be fair, humans are also one of nature's most cooperative species. See, for example, Seabright [105], Grafton et al. [42], and Roemer's book, "How We Cooperate: A Theory of Kantian Optimization", (2019, Yale University Press).

Latin word "rivales" which designated people who drew water from the same stream (rivus).[4] Indeed, Couttenier and Soubeyran [26, 27] found that natural resources played a key role in causing civil conflicts and documented the empirical relationship between water shortage on civil wars in Sub-Saharan Africa.

How do economic agents manage expected shifts in regimes? How do they try to influence or prevent the arrival of such shifts? This chapter provides a selective survey of the analysis of regime shifts from an economic view point, with particular emphasis on the use of the techniques of optimal control theory and differential games.

This paper is organized as follows. Section 1.2 gives an overview of the concepts of regime shifts, thresholds, and tipping points. Section 1.3 shows how unknown tipping points affect the optimal current policy of decision-makers, with or without ambiguity aversion. Focus of Section 1.4 is on political regime shifts in a two-class economy, where we review models of revolution and of how the elite may try to prevent revolution by using policy instruments such as repression, redistribution, and gradual democratization. Section 1.5 reviews models of dynamic games in resource exploitation involving regime shifts and thresholds. Section 1.6 reviews some studies of regime shifts in industrial organization theory, with focus on R&D races, including efforts to sabotage rivals in order to prevent entry. Section 1.7 reviews games of regime shifts when players can manage a Big Push. Section 1.8 discusses some directions for future research.

1.2 Regime Shifts, Thresholds, and Tipping Points

In this section, we briefly introduce the concepts of regime shift, threshold, and tipping point and give a brief overview of the literature on these topics. More detailed discussions will be provided in later sections.

1.2.1 Regime Shifts

A regime shift is a discrete break in a dynamic system: at the regime-switching time, there is a discrete change in either the objective function or the transition dynamics. Regime shifts can be anticipated to some extent, and such anticipation affects the behavior of economic agents prior to the actual occurrence of the shifts. A prototype model of optimal response to anticipated regime shifts (in the form of an anticipated machine failure) was developed by Kamien and Schwartz [55]. This model predates models of responses to threat of environmental collapses [28, 100]. Along the same vein, Long [80] showed that a monopolist mine owner that expects the nationalization of the mine to occur at some unknown date in the future would

[4] Dictionnaire LE ROBERT, Société du Nouveau Littré, Paris: 1979.

hasten his extraction rate. Another early contribution to the modelization of responses to anticipated regime shifts was done by Kemp and Long [57]. They formulated an optimal control problem with several state variables to show how economic agents would alter their optimal plans if a regime shift is anticipated to take place at some known future date. They mentioned two classes of shifts: a shift in preferences and a shift in technology. In the case of preference shifts, they supposed that a first economic actor (an individual or a group) cares not only about its own present happiness but also about the future happiness of the second individual or groups whose preferences may differ substantially from those of the first actor. As an example of preference shift, they considered the task of an imperial power committed to the eventual independence of its colony. Another example of optimal actions under an anticipated technology shift is that a firm knows that at some future point in time a patent will expire and a new process becomes available to it.

In Kemp and Long [57], at each regime-shift time t_i, there are H inequality constraints involving the n state variables:

$$S_h^i(x_1(t_i), \ldots, x_n(t_i)) \geq 0, h = 1, 2, \ldots, H.$$

They proved that the co-state variables are continuous at the time of the shift, unless one of the constraints (say $S_{h^*}^i$) is binding, in which case the co-state associated with the state variable x_j will jump downward if and only if $\partial S_{h^*}^i / \partial x_j(t_i) \geq 0$. In the case of a parent who expects to transfer the family business to an offspring, Kemp and Long [57] showed that prior to the shift time, the consumption path may be non-monotone. They also considered the case of a mining firm that expects a discrete shift in the price. Anticipation of the price shift makes the firm modify its current extraction plan. This result has implications to what is now known as the Green Paradox: announcing policies that are intended to mitigate climate changes (e.g., policies that institute future sharp rises in carbon tax) may worsen the climate, as the anticipation of the future tax hike induces coal and oil producers to quickly exhaust their stocks of fossil fuels, resulting in the unintended consequence of hastening the rise in global temperature [87, 107].

A related work on regime shifts is Hillman and Long [52], who considered an economy that owns a small stock of exhaustible resource (say oil) which competes in the domestic market with imported perfect substitutes. The shift in question is change from the free-trade regime to an embargo regime, in which the economy is subject to a trade embargo by, say, a foreign oil cartel. Unlike Kemp and Long [57] who assumed a known date of regime shift, Hillman and Long [52] supposed that the date when a foreign embargo will be imposed on the home country is a stochastic variable. They proved that the planner of the economy that anticipates the threat of embargo will extract its oil more conservatively. Interestingly, if the economy's resource stock is exploited by perfectly competitive domestic firms, these firms will replicate the planner's conservationist solution, because they anticipate an upward jump in domestic oil price as soon as the embargo occurs. A striking result is that if the domestic stock is exploited by a single firm, then the embargo threat will cause this firm to overextract the resource in the pre-embargo phase, in anticipation that it will become a monopolist in the domestic market immediately after the embargo takes place. This overextraction prior to the embargo can be explained intuitively: the monopolist's profit under the embargo regime is higher, the lower is its stock at the time of the regime shift. Interestingly, in this scenario, Robert Solow's dictum (that the monopolist is the conservationist's best friend) fails to hold.

In both Kemp and Long [57] and Hillman and Long [52], it was assumed that the date of regime shift, whether known or stochastic, is outside the control of the planner. The case of where a regime shift date can be chosen by the planner was considered by Hung et al. [53] using a model of transition from fossil energy to a non-exhaustible substitute (such as solar energy). The date of transition, t_T, is optimally chosen to maximize the discounted stream of social welfare, by balancing the cost of investment with the discounted future stream of benefits derived from consuming the clean substitute. At the date of transition, a lumpy cost K must be incurred. The authors show that the transition date is determined by the condition that the current-value Hamiltonian immediately before the transition, $H(t_T^-)$, is smaller than that the Hamiltonian immediately after the transition, $H(t_T^+)$, by the amount rK, where r is the interest rate. This implies that the equilibrium price path of energy has a discrete downward jump at the time of regime shift. As expected, the regime-shift decision involves a trade-off, since adopting a new regime brings immediate costs as well as future benefits. The result of the paper by Hung et al. [53] is consistent with the multi-stage optimization analysis of regime switching by Tomiyama [110] and Amit [4] which also endogenously determine switching times.

The analysis of dynamic responses to regime shifts can be conducted using two approaches: the optimal control theory/dynamic programming approach (where a single decision-maker decides how to cope with an anticipated regime shift), and the dynamic games approach, where multiple agents plan their responses in a non-cooperative way, while strategically reacting to one another. The first approach is clearly simpler, but it misses out some important strategic considerations. Representative papers using the first approach include Tsur and Zemel [114, 115], Ren and Polasky [98], Nkuiya and Costello [94], and Lemoine and Traeger [73], among others. Papers using the dynamic games approach include Tornell [111], Maler et al. [90], Doraszelski [29, 37, 86]. A paper that does not include differential games but does take into account game-theoretic considerations is Nakuiya et al. [93], where the threat of regime shift occurs only in the first period. They found that countries are more likely to ratify a climate change today when they face endogenous uncertainty about a possible future upward shift in damage costs.

1.2.2 *Thresholds and Tipping Points*

Quite often, a regime shift occurs when a certain endogenous variable crosses some thresholds, the exact value of which may, or may not, be known to a decision-maker. One may make a distinction between two types of thresholds in the state variable of a dynamic system. The first type of threshold, once crossed, induces a *discrete* change in the differential equation (or difference equation) that describes the state dynamics, or in the preferences of the decision-makers. The second type of threshold (as in the case of the shallow lake problem that will be discussed in Sect. 1.5.2) only involves, upon crossing it, a change in the basin of attraction, or a substantial change in the qualitative properties of the optimal policy. Thresholds play an important role in economic models of social change. The proverbial "last straw that breaks the back of the donkey" is a case in point. In an interesting article on racial segregation, Schelling [103] showed that a small change in the initial mixture of blacks and whites in a neighborhood may eventually lead to a complete segregation. If there is a limit to how small a minority the members of either color are willing to be, for example, a 25% minority, then "initial mixtures ranging from 25 to 75% will survive but initial mixtures more extreme than that will lose their minority members and become all of one color" (p. 148). His models contributed to the explanation of the phenomenon called " neighborhood tipping", which occurs "when a recognizable new minority enters a neighborhood in sufficient numbers to cause the earlier residents to begin evacuating" (p. 181).

Another early interesting work on threshold is that of Azariadis and Drazen [7]. They show that the success or failure of a developing economy depends on whether it manages to pass a certain threshold level of externalities. Similarly, in the context of the tragedy of the commons, Lasserre and Soubeyran [67] found that a small amelioration of institutions can move an economy to a superior equilibrium. Along the same vein, Leonard and Long [74] demonstrated how a strengthening of the enforcement of property rights, financed by taxation supported by a self-interested electorate, could move the economy to an efficient steady state. These papers assume that economic agents care only about their material well-being. As a counterpoint, Long [85] offers a model where economic agents care also about their self-image. Long assumes that economic agents feel bad if their action falls short of the Kantian ideal. Using an overlapping generations model in which pro-social attitudes evolve across generations Long [85] shows that there is a threshold level of pro-socialness beyond which the economy will converge to a steady state with a high level of both pro-socialness and material prosperity, while below the threshold, society's level of pro-socialness will eventually vanish, and the economy will end up in poverty.

Quite often while the decision-maker is aware of the possibility of thresholds and tipping points, there is considerable uncertainty as to the exact location of the tipping points. This is a particularly relevant issue in the analysis of optimal responses to climate change. Heal [51] and Tsur and Zemel [114] assume that the decision-maker has imperfect knowledge of the underlying climate threshold. Keller et al. [56] study optimal economic growth under uncertain climate thresholds. While Keller et al. [56], Gjerde et al. [41], and Lontzek et al. [88] model climate tipping points as directly reducing output or utility, Lemoine and Traeger [72, 73] and van der Ploeg [121] model tipping points as a shift in the dynamics of the climate system.

The use of optimal control theory enriches the analysis of thresholds. Skiba [108] showed that if an optimal control problem exhibits two steady states that are locally stable in the saddle-point sense, then there exists in the state space a threshold that separates the two basins of attraction. Later authors call such a threshold a "Skiba point."There is a large literature on Skiba points (see, e.g., Feichtinger and Wirl [39], Wagener [125], Hartl et al. [49], Wirl and Feichtinger [126], Wirl [127], Yanase and Long [129]).

A key feature of a Skiba point is that at such a point, the decision-maker is indifferent between two trajectories, each converging to a different steady state. For example, Hartl and Kort [48] show that a firm facing an emission tax may choose between achieving a steady state with a high capital stock which is compatible with efficient abatement efforts or a low capital stock with no abatement efforts. The firm has to invest more to reach the high capital stock equilibrium. This implies that there is a discontinuity of the policy function at the Skiba point k^S. Immediately to the right of k^S, the firm invests a great deal more than to the left of k^S. Discontinuity, however, is not a generic feature of Skiba point. Wirl and Feichtinger [126] show the existence of a Skiba point with a continuous policy function. This requires that the unstable steady state is a node (rather than a focus). Hartl et al. [49] give a complete classification of Skiba points near unstable steady states: focus, continuous node, and discontinuous node. These papers assume that there is a single decision-maker. When there are several decision-makers interacting in a dynamic game, the study of Skiba points becomes much more complicated. See Sect. 1.6 for details.

1.3 Unknown Tipping Points: The Hazard Rate Function Approach

The precise points at which tipping may occur are typically unknown, because of lack of scientific information [72]. A standard approach to model unknown tipping points is to use the hazard function approach [25, 33, 41, 115]. For example, in the context of risks of abrupt climate change that are associated with the stock of green house gases (GHG), one could imagine that an adverse climatic event may occur at some unknown time T in the future that would inflict severe economic damages. A convenient formulation is to suppose that the distribution of the random occurrence date T is related to a hazard rate function $h(X)$ where $X(s) \geq 0$ is the stock of GHG at date s, with $h(X) > 0$ for $X > 0$. Given that the adverse climatic event has not occurred at or before time t_0, for any given $t > t_0$ the probability that T will occur after time t is specified as follows:

$$\Pr(T > t \mid T > t_0) = e^{-\int_{t_0}^{t} h(X(s))ds}.$$

The conditional probability that the adverse event occurs before some time $t > t_0$ is

$$F(t \mid t_0) = 1 - e^{-\int_{t_0}^{t} h(X(s))ds}$$

. Notice that this formulation implies that, assuming that $h(X)$ is strictly positive, the event is definitely going to take place at some time in the future:

$$\lim_{t \to \infty} F(t|\, t_0) = 1.$$

The corresponding conditional density function is

$$f(t|\, t_0) = F'(t|\, t_0) = h(X(t))e^{-\int_{t_0}^{t} h(X(s))ds}.$$

Thus, at time t_0, the conditional probability that the adverse event will occur at some time during the time interval $(t_0, t_0 + \Delta t)$ is approximately

$$h(X(t_0)) \times \Delta t$$

provided that Δt is sufficiently small.

In general, the event need not be a climatic event, and the state variable X need not refer to the stock of GHG. Thus, X could refer to, say, the stock of fish in a fishing ground, and the event could be a collapse of the fish stock or a change in its growth function, $G(X)$. Or X could simply be time, as in the nationalization model of Long [80]. While it is usually assumed that the hazard rate depends only on the state variable, some authors have allowed the hazard rate to depend on both a state variable and a control variable, under the assumption that the feedback control rule is continuous in the state variable. See, for example, Doraszelski [37, p. 22], van der Ploeg [122] and Haurie et al. [50].

The hazard rate approach can be applied to a single occurence or to recurrent events. See Tsur and Zemel [115] for the distinction. For an analysis of recurrent environmental catastrophes, see Tsur and Zemel [118], where increased GHG concentration implies higher frequency of occurrence. They focus on long-run properties, using techniques developed in Tsur and Zemel [119].

1.3.1 The Ambiguous Effect of Anticipation of Regime Shifts

How does the possibility of a regime shift influence the behavior of the decision-maker? In general, the answer to this question is ambiguous. We can illustrate this ambiguity by considering a model of a fishery where the regime shift takes the form of a change in the natural growth rate of the stock, from $G_1(.)$ to $G_2(.)$, where $G_2(X) < G_1(X)$ for all $X \geq 0$. The special case where $G_2(X)$ is identically zero corresponds to a stock collapse (i.e., the fish stock X suddenly becomes zero at a random date T). Let us consider the fishery model of Polasky et al. [97], where an analytical solution can be obtained thanks to the authors' assumption that the instantaneous payoff function is linear in the harvesting rate, y. Before the regime shift, taking into account human's exploitation of the fish stock, the net rate of growth of the fish stock is

$$\dot{X} = G_1(X) - y.$$

Let us first consider the optimal harvest policy if the decision-maker is certain that there will never be a regime change. Let $r > 0$ be the rate of discount. The decision-maker's objective is to maximize

$$\int_0^{\infty} e^{-rt} py(t)dt,$$

where $p > 0$ is a constant. Assume that $G(.)$ is hump-shaped, with $G(0) = 0$, $G'(0) > r, G'' < 0$, and $G'\left(\overline{X}\right) = 0$ for some $\overline{X} > 0$. Assume that $0 \leq y \leq y_m$ where y_m is an exogenous upper bound on y, with $y_m > G\left(\overline{X}\right)$. As an example, one can assume that

$$G_i(X) = X\left(1 - \frac{X}{K_i}\right),$$

where $K_i > 0$ is called the carrying capacity. Under these assumptions, it is well known that, with no threat of regime shift, the decision-maker will aim at a steady-state stock X^* such that $G'(X^*) = r$, and that the optimal y is equal to zero for $X < X^*$, and is equal to y_m for $X > X^*$ (see Clark [24]).

What happens if there is a threat of a regime shift from $G_1(.)$ to $G_2(.)$ as specified above? Assume that the hazard rate function is $h(X) \geq 0$, with $h'(X) \leq 0$ (i.e., the risk is lower when the stock is larger).[5] What would be the steady-state stock that the decision-maker aims at? Let us call this stock X_1. Is X_1 greater than or smaller than X^*? Polasky et al. [97] show that the answer is ambiguous if $G_2(X) \equiv 0$, unless additional assumptions are made. To understand this ambiguity, recall that the HJB equation when the system is in regime 1 is given by

$$rV_1(X) = \max_{0 \leq y \leq y_m} \left[py + V_1'(X)(G_1(X) - y)\right] + h(X)\left[V_2(X) - V_1(X)\right], \quad (1.1)$$

where $V_i(.)$ is the value function under the ith regime.[6] (See, e.g., Dockner et al. [35] for a general formulation of this type of regime shifts).

If the system is in regime 1, when one maximizes the right-hand side of the HJB equation with respect to y, the optimal harvesting effort is $y = 0$ for all X such that $V_1'(X) > p$ and $y = y_m$ if $V_1'(X) < p$ (with y indeterminate if $V'(X) = p$). One searches for a value $X_1 < K_1$ such that $y = 0$ for $X < X_1$ and $y = y_m$ for $X > X_1$. Then Eq. (1.1) yields

$$0 = V_1'(X)G_1(X) + h(X)V_2'(X) - [r + h(X)]\, V_1(X) \text{ for } X < X_1 \quad (1.2)$$

$$0 = py_m + V_1'(X)(G_1(X) - y_m) + h(X)V_2'(X) - [r + h(X)]\, V_1(X) \text{ for } X > X_1. \quad (1.3)$$

Assuming that $V_1'(X)$ is continuous at X_1, the two Eqs. (1.2) and (1.3) yield

[5] Polasky et al. [97, p. 233] assume that $h'(X) = 0$ at $X = K_1$.

[6] Note that in writing the above HJB equation, it is assumed that $V_1'(X)$ is defined for all $X > 0$.

$$V_1'(X_1) = p \text{ and } V_1(X_1) = \frac{pG(X_1) + h(X_1)V_2(X_1)}{r + h(X_1)}. \tag{1.4}$$

Furthermore, assume that $V_1''(X)$ exists. Then differentiating Eqs. (1.2) and (1.3) with respect X, one obtains

$$G_1(X)V_1''(X) = \phi(X) \text{ if } X < X_1$$

and

$$\left[G_1(X) - y_m\right] V_1''(X) = \phi(X) \text{ if } X > X_1,$$

where

$$\phi(X) \equiv \left[r + h(X) - G_1'(X)\right] V_1'(X) - h(X)V_2'(X) + h'(X)\left[V_1(X) - V_2(X)\right].$$

Under the assumption that $V_1''(X_1) \leq 0$, one can see that $\phi(X)$ is negative to the left of X_1 and positive to the right. The assumed continuity of V_1' and V_2' then implies that $\phi(X_1) = 0$. This equation and (1.4) taken together imply that

$$G_1'(X_1) = r + h(X_1)\left[1 - \frac{V_2'(X_1)}{p}\right] + \frac{h'(X_1)}{r + h(X_1)}\left[G_1(X_1) - \frac{r}{p}V_2(X_1)\right]. \tag{1.5}$$

Equation (1.5) shows that, in general, one cannot determine whether the (regime 1) steady-state stock X_1 exceeds or falls short of the steady-state stock X^* (which applies if there is no threat of regime shift). To see this ambiguity, consider the case where $G_2(X) = 0$ identically (i.e., the stock collapses immediately after the adverse event occurs), so that $V_2(X) = 0$ identically. Consider two benchmark subcases. First, the subcase where $h(X) = \lambda$, a positive constant. Then Eq. (1.5) gives $G_1'(X_1) = r + \lambda > r = G_1'(X^*)$. That is, under the threat of an exogenous regime shift, the planner's exploitation is less conservationist than under the no-threat scenario. (This is reminiscent of the result of Long [80]: the threat of nationalization leads to more aggressive extraction of the mine.) Second, the subcase where $h'(X) < 0$. Then Eq. (1.5) gives

$$G_1'(X_1) = r + h(X_1) + \left\{\frac{h'(X_1)G_1(X_1)}{r + h(X_1)}\right\}.$$

Since the term inside the curly bracket is negative for $X_1 < \overline{X}$, we can no longer be sure that $G_1'(X_1) > r$. Thus, it is possible that $X_1 > X^*$, i.e., the decision-maker's exploitation is more conservationist, because she wants to achieve a lower hazard rate at the steady state of regime 1.

The above "ambiguity result" is in line with previous works for the cases of threats of forest fire and fishery collapse [99, 100], nuclear power risks [8, 91], and environmental threats [25, 116, 117].

1.3.2 Knightian Uncertainty: Decision-Making Under Ambiguity About Tipping Points

In many real-world problems, such as climate change, our knowledge is so thin that it may not be appropriate to use models that assume a known distribution of stochastic shocks. The terms "Knightian uncertainty" or "deep uncertainty" and "ambiguity" have been used interchangeably to refer to situations in which the underlying probabilities are not known. A number of studies have explored the implications of Knightian uncertainty in the context of climate change. Lange and Treich [66] use a two-period model to show that ambiguity aversion about damages induces the decision-maker to opt for lower emissions. A number of authors use aversion to Knightian uncertainty to motivate the robust control approach to abatement policies [5, 76].

Lemoine and Traeger [73] analyze the effect of ambiguity aversion on optimal policy in the face of an unknown tipping point. Their point of departure is a model of rational behavior under deep uncertainty that was axiomatized in Traeger [113], which is closely related to the recursive smooth ambiguity model of Klibano et al. [59, 60]. In Lemoine and Traeger [73], the vector of state variables is denoted by S_t. This vector can include the capital stock, temperature level, carbon dioxide, and time. The vector of control variables is denoted by x_t. In each period, there is a deterministic utility flow $u_t = u(x_t, S_t)$. The decision-maker maximizes the expected intertemporal payoff over the infinite horizon. There are two value functions, $V_0(S)$ and $V_1(S)$, which apply to the pre-tipping world and the post-tipping world. The system dynamics are described by $S_{t+1} = G_0(x_t, S_t)$ for the pre-tipping world and $S_{t+1} = G_1(x_t, S_t)$ for the post-tipping world.[7] In the absence of ambiguity, $V_0(S)$ is related to $V_1(S)$ through the Bellman equation:

$$V_0(S_t) = \max_{x_t} \left\{ u(x_t, S_t) + \beta \left[(1 - h(S_t, S_{t+1}))V_0(S_{t+1}) + h(S_t, S_{t+1})V_1(S_{t+1}) \right] \right\} \tag{1.6}$$

subject to

$$S_{t+1} = G_0(x_t, S_t),$$

where $\beta \in (0, 1)$ is the discount factor and $h(S_t, S_{t+1})$ is the hazard rate function which gives the probability of the tipping that occurs in period $t + 1$.

In the context of climate change, Lemoine and Traeger [73] define ambiguity as the decision-maker's lack of confidence in the hazard rate function $h(S_t, S_{t+1})$. They propose to capture this lack of confidence by introducing into the recursive utility model a concave function, which I denote by $\Phi(.)$, such that the Bellman equation is modified as follows:

[7] In their formulation, they also add a random variable ε_t that represents stochastic shocks with a known distribution (these shocks are not ambiguous). For simplicity of exposition, I have omitted this variable.

$$V_0(S_t) = \max_{x_t} \left\{ u(x_t, S_t) + \beta\Phi^{-1}\left[(1 - h(S_t, S_{t+1}))\Phi(V_0(S_{t+1})) + h(S_t, S_{t+1})\Phi(V_1(S_{t+1}))\right]\right\}. \quad (1.7)$$

Since the model cannot be solved analytically, Lemoine and Traeger resort to numerical simulations. To facilitate the simulations, they assume that the function Φ contains two parameters, $\eta \geq 0$ and $\gamma \geq 0$, where η is a measure of aversion to risk and γ takes into account the decision-maker's aversion to ambiguity.

They consider two different classses of models. In the first class of models, when a tipping point is crossed, there is a sudden increase in the climate feedbacks that amplify global warming. This type of tipping points increases the effects of emissions on temperature.[8] In the second class of models, a tipping point triggers an increase in the decay rate of CO_2, i.e., a weakening of the carbon sinks.[9] Numerical simulations show that in either class of model, an increase in ambiguity aversion (an increase in γ) will increase the optimal carbon tax and reduce the peak level of CO_2 along an optimal path. The authors decompose the total effect of an increase in γ into two effects: (a) the marginal hazard rate effect (MHE), which reflects the awareness that present policies influence on the chance of tipping, and (b) the differential welfare impact (DWI), which compares the effects of abatement on pre-tipping welfare and on post-tipping welfare. The sign of DWI is ambiguous. In the numerical calculations, aversion to Knightian uncertainty increases the contribution of MHE to the carbon tax, but tends to reduce the carbon tax via the DWI effect. However, the overall effect of aversion to Knightian uncertainty is to increase the carbon tax.

1.4 Preventing Regime Shifts: The Role of Repression, Redistribution, and Education in a Two-Class Economy

There is a large literature on the threat of revolution that an autocratic regime faces. The early theoretical models of revolutions [43, 64] abstract from strategic considerations.[10] More recent works, such as Acemoglu and Robinson [1–3], offer models on interaction between the ruling elite and the citizens, where coups and revolutions can occur in response to exogenous economic shocks. In Acemoglu and Robinson [2], there are two groups of agents, the poor and the elite. Each group consists of infinitely lived individuals. The elite has more capital than the poor. The majority of people are poor, and initially it is the elite that has the political power. The poor can attempt a revolution at any time, but revolution is costly (a fraction of national

[8] In the standard DICE model of Nordhaus [95] without tipping points, there is a parameter called "climate sensitivity,"defined as the equilibrium warming from doubling the stock of GHGs. Lemoine and Traeger [73] introduce Knightian uncertainty about a climate-feedback tipping point that increases this parameter from its pre-tipping value of 3 °C to, say, 5 °C.

[9] The authors assume that when the unknown tipping point is crossed, there is a sudden decrease in the rate of transfer of CO_2 out of the atmosphere.

[10] The non-strategic approach is also taken in a recent interesting paper by Michaeli and Spiro [92], where they show a number of interesting results, including a demonstration of how the implementation of popular policies, such as Perestroika, can trigger a revolution.

income is destroyed). If a revolution is successful, a fraction of assets of the elite is expropriated. The elite can avoid a revolution by embarking on a process of democratization. The productivity of capital is a random variable, which is revealed at the beginning of each period: it can be low or high. This random variable affects the opportunity costs of revolution in a nondemocracy as well as the elite's opportunity costs of mounting a coup to overthrow a democracy. These models typically assume that the elites are killed or evicted during or after a successful revolution.

In contrast, Boucekkine et al. [16] assume that after the elites are removed from power, they co-exist with the citizens and share access to the country's stock of resources. Boucekkine et al. [16, p. 189] argue that this is consistent with what happened in countries such as Tunisia, where "the Arab Spring events have successfully overthrown the ruling dynasty but have failed to renew the political and economic life to a large extent." Their paper models the efforts of the elite to prolong their regime as much as possible. The elite has two policy instruments: repression and redistribution.[11] The model is solved in two stages. In the post-revolution stage, the elite and the citizens have equal access to the country's resources, and they solve a differential game of resource exploitation in the manner postulated by Tornell and Lane [112]. In the pre-revolution stage, the authors assume a Stackelberg model of differential game. In this game, the elite (the Stackelberg leader) is able to commit to a redistribution parameter, $1 - u_E$, and a repression parameter r_E, while the citizens, taking these parameters as given, choose the date T at which they start a revolution.[12] Revolution is costly: it destroys a fixed amount, χ, of the country's capital stock, and on top of that, the citizens must incur a direct switching cost (DSC), ψ. This cost is an increasing and concave function of the level of repression, r_E. It is found that the date T is increasing in $1 - u_E$ and in χ, and decreasing in the economy's initial resource stock. Knowing how the citizens' choice of revolution time depends on the redistribution parameter and the repression parameter, the elite sets these parameters to maximize their own payoffs. This is a deterministic optimal control problem. The authors show that if the vulnerability of the economy is high, the revolution will occur in finite time. However, if the vulnerability is intermediate, in equilibrium the dictatorship survives.

The model by Boucekkine et al. [16] allows the ruling class to resort only to two policy instruments: repression and redistribution. In a follow-up paper, Boucekkine et al. [17] consider a third policy instrument that can help the elite prevent a violent revolution: education of the mass that eventually leads to a peaceful handover of power. They develop a dynamic optimization model that portrays the ruling class's policy choice to cope with the threat of revolution. In this model, the elite may choose between (a) keeping the population largely uneducated, while redistributing income just enough to avert a revolution, and (b) embarking on a path of education and development and eventually relinquishing autocratic power, ensuring a smooth democratic

[11] In the model of Boucekkline et al. [16], these are parameters that the elite chooses at the beginning of the program; they are not control variables in the standard sense of being piece-wise continuous functions of time.

[12] The fraction of national income that is distributed to the citizens is $1 - u_E$.

transition. In their model, from the point of view of the elite ruling class, education of the oppressed class has two opposing effects. On the one hand, the life-satisfaction threshold above which the population would not revolt is increasing in education, i.e., a more educated mass tends to demand higher life prospects at the expense of the elite. On the other hand, education contributes to economic development and is conducive to a political culture of negotiation and a recognition of the merit of trying to achieve a compromise [9, 18, 78, 79].

In Boucekkine et al. [17], the ruling class derives income from natural resources (available in a fixed quantity, R, per period). The working class's income consists of wage income, wH, and a transfer Θ from the government. Here, H is the level of human capital, which accumulates as a result of education, E, that the ruling class provides. If the sum $wH + \Theta$ is below a certain threshold, a revolution will take place. This threshold is increasing in the level of human capital. By choosing Θ and by influencing H through education expenditure, the elite can avoid a revolution and stay in power forever. However, it may be to the elite's advantage to relinquish power at some planned date T through a process of democratization, if the anticipated payoff to the elite at the time of handover, $S(H(T))$, is sufficiently attractive. Boucekkine et al. [17] assume that this payoff is increasing in $H(T)$. The elite class chooses consumption, $C(t)$, redistribution, $\Theta(t)$, education expenditure, $E(t)$, and a terminal time, T, to maximize its intertemporal welfare, subject to the no-revolt constraint. Their intertemporal welfare is

$$U = \int_0^T e^{-\rho t} u(C(t))dt + e^{-\rho T} S(H(T)).$$

This is a standard optimal control problem. The authors show that, depending on parameter values, the optimal solution may be one of three possible varieties: (i) permanent dictatorship with no education, regardless of the initial H_0, (ii) education-driven democratization, with a finite time for power handover, also regardless of the initial H_0, and (iii) human-capital poverty trap: there exists a threshold level $\overline{H}$ such that if $H_0 < \overline{H}$ then permanent dictatorship is optimal for the elite, and if $H_0 > \overline{H}$ then democratization through education is optimal.

1.5 Dynamic Games Involving Natural Resources with Threat of Regime Shifts and Thresholds

In this section, we review some dynamic game models of natural resource exploitation that feature either a threat of regime shift or a threshold.

1.5.1 *Extraction of an Exhaustible Resource Under Threat of Regime Shift*

Laurent-Lucchetti and Santugini [68] study a dynamic game model of common property exhaustible resources under uncertainty about full or partial expropriation, generalizing the nationalization model of Long [80]. Consider a host country that allows two firms to exploit a common resource stock under a contract that requires each firm to pay the host country a fraction τ of its profit. Under the initial agreement, $\tau = \tau_L$. However, there is uncertainty about how long the agreement will last. The host country can legislate a change in τ to a higher value, τ_H. It can also evict one of the firm. The probability that these changes occur is exogenous. Formulating the problem as a dynamic game between the two firms, in which the risk of expropriation is exogenous and the identity of the firm to be expropriated is unknown *ex ante*, the authors find that weak property rights have an ambiguous effect on present extraction. Their theoretical finding is consistent with the empirical evidence provided by in Bohn and Deacon [15].

How does the threat of being removed from office influence a government's extraction path of an exhaustible resource stock and its exploration efforts? A recent paper by van der Ploeg [122] offers three related models that shed light on this question. In Model 1, an incumbent faces the threat of removal from office (for ever) by a rival faction. This model is related to the model of the effects of political uncertainty about nationalization [15, 61, 68, 80]. The author assumes that the incumbent government (player A) faces the risk of being overthrown by a rival faction (player B). The hazard rate is a constant, $h > 0$. Once player A is removed from office, it receives a smaller share of the resource rent. Under this scenario, it is found that resource extraction by the incumbent is more voracious. Furthermore, the incumbent tends to invest less in the exploration for the resources, because of the holdup problem.

In Model 2, van der Ploeg [122] considers the scenario of ongoing political resource conflict cycles between two political factions. Once a faction is in office, it faces a hazard rate h of being removed by the other factions. After being removed, the faction can regain office, also with the hazard rate h. The author assumes that both factions are obliged to share equally the resource rents, but the faction that is in office enjoys utility more. This is captured by introducing a multiplicative partisan in-office bias, $\beta > 1$, in line with Aguiar and Amador [6]. The author shows that with perennial ongoing political cycles, resource depletion is rapacious especially if the partisan in-office bias is large (high β) and there are frequent changes of government (high h).

In Model 3, the author endogenizes the hazard rate. Again, there are two factions, A and B. If faction A is the incumbent, it faces the hazard rate h^A of being removed from office. Being in office, it can choose the resource extraction rate R^A, and obtains the resource rents $\pi(R^A)$, of which a fraction $\tau < 0.5$ must be transferred to the other faction (according to some constitutional convention).

Assume that h^A is a function of A's defence effort, f^A, and of B's attack effort, f^{B*}. Using the common formulation of the rent-seeking literature, assume that

$$h^A = H\frac{(f^{B^*})^\phi}{(f^A)^\phi + (f^{B^*})^\phi} \text{ and } h^B = H\frac{(f^{A^*})^\phi}{(f^B)^\phi + (f^{A^*})^\phi}$$

, where H is a constant and $\phi \in (0, 1)$. Each faction has a maximum of N units of efforts, and the income derived from $N - f$ is $w(N - f)$, where $w > 0$ is the wage rate. Let S denote the stock of the exhaustible resource. Let $V^A(S)$ and $V^{A*}(S)$ denote, respectively, faction A's value function when it is the incumbent and when it is not the incumbent. Then the HJB equations for A are

$$rV^A(S) = \max_{f^A, R^A} \left\{\beta(1-\tau)\pi(R^A) + w(N - f^A) - V_S^A(S)R^A + h^A\left[V^{A^*}(S) - V^A(S)\right]\right\}$$

$$rV^{A^*}(S) = \max_{f^{A*}} \left\{\tau\pi(R^B) + w(N - f^{A^*}) - V_S^{A*}(S)R^B - h^B\left[V^{A*}(S) - V^A(S)\right]\right\}.$$

Faction B is in a similar situation. Assuming that the function $\pi(R)$ is iso-elastic, the value functions can be solved analytically. It is found that dynamic resource wars are more intense if S is high and w is low. Depletion of the reserves is less rapid if τ is closer to 0.5, and if the government's stability is high (a low H). An increase in the partisan in-office bias parameter β leads to more rapacious extraction.

1.5.2 Dynamic Games Involving Natural Resources with Thresholds and Non-linear Dynamics

Examples of dynamic games involving natural resource stocks with non-linear dynamics include fishery games and lake-pollution games. Most fishery models assume that the transition equation is concave in the state variable. Even so, multiple steady-state equilibria can exist in concave optimal control fishery problems (see Long [81], where it is found that there are three steady-state equilibria, of which the middle one is unstable). Limit cycles can also be optimal [82, 83], pp. 294–295; [58]. The lake-pollution game model is another interesting example of multiple equilibria, where the transition equation is neither concave nor convex in the state variable. This implies that there are potentially several steady states. We describe below a lake-pollution model based on Maler et al. [90].

The state variable, $s(t)$, denotes the amount of phosphorus sequestered in algae. There are n players. Player i discharges $c_i(t) \geq 0$ units of phosphorus to the lake. The transition equation is

$$\frac{ds}{dt} = -\delta s(t) + \left[\frac{s^2(t)}{s^2(t)+1}\right] + \sum_{i=1}^{n} c_i(t), \qquad x(0) = x_0 \geq 0,$$

where $\delta > 0$ and $s(t) \geq 0$. The term inside the square brackets is the internal release of phosphorus that has been sequestered in sediments and in submerged vegetation;

this term is bounded above by 1. Thus, for any given constant aggregate discharge $C \equiv \sum_{i=1}^{n} c_i$, the steady-state stock of pollution is bounded above by $(C+1)/\delta$. The transition equation can be re-arranged to yield

$$(s^2+1)\frac{ds}{dt} = -\delta s^3 + (1+C)s^2 - \delta s + C \equiv h(s; C, \delta).$$

Since $h(0; C, \delta) = C > 0$ and $h(\infty; C, \delta) = -\infty$, there exists at least one positive steady state.

Suppose c_i is constant. Then it can be shown that, provided $0 < \delta < 3\sqrt{0.375}$, there exists a certain range of c_i such that there are three steady states, denoted by s_L, s_M, and s_H where $s_L < s_M < s_H$, where s_M is unstable, and s_L and s_H are locally stable. (In the lake-pollution literature, s_L is usually referred to as the oligotrophic state, and s_H is the eutrophic state.)

Suppose initially the system is at the low steady state s_L. Consider a temporarily sustained increase in c_i. If this increase crosses a threshold level, there will be a sudden flip to s_H. This is called a tipping point. If $\delta \leq 1/2$, the flip is irreversible, since c_i cannot be negative.[13] In what follows, we assume $1/2 < \delta < 3\sqrt{0.375}$,

Suppose that the net benefit function of player i is

$$B_i = \ln c_i - \omega s^2,$$

where $\omega > 0$. Player i's overall payoff is

$$\int_0^{\infty} e^{-\rho t}\left[\ln c_i - \omega s^2\right] dt.$$

Let us compare the open-loop Nash equilibrium with the Markov-perfect Nash equilibrium of this game.

Under open-loop behavior, the Hamiltonian for player i is

$$H_i = \ln c_i - \omega s^2 + \psi_i \left[\frac{s^2(t)}{s^2(t)+1} - \delta s + c_i + (n-1)c_j\right].$$

Assuming a symmetric Nash equilibrium, so that $c_i = c_j = c$, and defining $C = nc$, the necessary conditions are

$$\frac{1}{c_i} + \psi_i = 0$$

$$\dot{s} = \frac{s^2(t)}{s^2(t)+1} - \delta s + C, \qquad s(0) = s_0,$$

[13] With $\delta = 1/2$ and $C = 0$, one obtains $h(s; 0, 1/2) = -(s/2)(s^2 - 2s + 1)$. Then $h = 0$ at $s_L = 0$ and $s_M = s_H = 1$.

$$\dot{\psi}_i = \left[\delta + \rho - \frac{2s}{\left(s^2+1\right)^2}\right]\psi_i + 2\omega s.$$

The transversality condition is $\lim_{t\to\infty} e^{-\rho t}\psi_i(t) = 0$.

The symmetric open-loop Nash equilibrium is the solution of the following system of differential equations:

$$\dot{s} = \frac{s^2(t)}{s^2(t)+1} - \delta s + C, \qquad s(0) = s_0,$$

$$\frac{\dot{C}}{C} = -\left[\delta + \rho - \frac{2s}{\left(s^2+1\right)^2}\right] + \frac{2\omega s C}{n},$$

with the transversality condition $\lim_{t\to\infty} e^{-\rho t}(n/C) = 0$. This system may possess multiple steady states, depending on parameter values.

It is useful to compare the open-loop Nash equilibrium and the social optimum. In the latter case, assume that a social planner maximizes the sum of the welfare of the n regions. This leads to the a different system of differential equations:

$$\dot{s} = \frac{s^2(t)}{s^2(t)+1} - \delta s + C, \qquad s(0) = s_0,$$

$$\frac{\dot{C}}{C} = -\left[\delta + \rho - \frac{2s}{\left(s^2+1\right)^2}\right] + 2\omega s C,$$

where we can see that the evolution of aggregate discharge is independent of the number of regions, n.

Comparing the two sets of differential equations, we notice that an open-loop Nash equilibrium with pollution damage parameter ω can be found by solving the optimization problem of a social planner who happens to have a lower damage parameter, say ω', where $\omega' = \omega/n$.[14]

To illustrate, consider the following parameter values: $\delta = 0.6$, $\omega = 1$, $\rho = 0.03$. Then the social planner's solution has a unique steady state, $s = 0.353$. It is stable in the saddle-point sense. It can be shown that the social planner's optimal path of C is non-monotone when the initial level of pollution is large: C at first declines, then increases again, approaching the steady-state level of discharge from below. On the other hand, when players do not cooperate, the open-loop Nash equilibrium has three steady states: an unstable one with a medium level of pollution, situated in between two saddle-point stable ones, $s_L = 0.398$ and $s_H = 1.58$.

[14]This property is crucially dependent on the special structure of the model, and on the assumed functional forms, e.g., logarithmic utility.

Suppose that it is socially desirable to achieve the oligotrophic steady state. Then, as expected, a time-dependent tax rate per unit of discharge can guide the open-loop players to achieve the socially optimal rate of discharge. Maler et al. [90] considered the restriction that the tax rate must be time-independent. They found for $n \leq 7$, under a suitably chosen time-independent tax rate, the phase diagram for the open-loop Nash equilibrium is qualitatively similar to the phase diagram under a social planner, and the optimal steady state can be achieved, though welfare along the path toward the steady state will generally fall short of the welfare level that would be achieved under central control. However, for $n > 7$, the phase diagram for the open-loop Nash equilibrium under a fixed tax rate can be quite irregular, such that it may not be possible to guide the system to the socially optimal steady state.

We now turn to the symmetric Markov-perfect Nash equilibrium. The HJB equation for player i is

$$\rho V_i(s) = \max_{c_i} \left\{ \ln c_i - \omega s^2 + V_i'(s) \left[\frac{s^2}{s^2+1} - \delta s + c_i - (n-1)c_j(s) \right] \right\}.$$

Using symmetry, the equilibrium feedback strategy must satisfy

$$c_i(s) = -\frac{1}{V'(s)} \equiv c(s).$$

Then, the HJB equation yields the identity

$$\rho V(s) = \ln c(s) - \omega s^2 - \frac{1}{c(s)} \left[\frac{s^2}{s^2+1} - \delta s + nc(s) \right] \text{ for all } s.$$

Differentiating this identity, we obtain the following differential equation:

$$\left[\delta s - c(s) - \frac{s^2}{s^2+1} \right] c'(s) = \left(\rho + \delta - 2\omega s - \frac{2s}{\left(s^2+1\right)^2} \right) c(s).$$

Since a closed-form solution cannot be obtained, numerical solutions can be computed after specifying parameter values.[15] With $\delta = 0.6$, $\omega = 1$ and $\rho = 0.03$, it is found that the locus of possible steady states (in the space (s, c)) is non-monotone. As in the model of Dockner and Long [34], there is a continuum of steady states, each corresponding to a feedback Nash equilibrium strategy. To each steady-state stock s^*, the corresponding individual emission level is

$$c^* = \delta s^* - \frac{(s^*)^2}{(s^*)^2+1}.$$

[15] For details, see Kossioris et al. [62].

The steady state $s = 0.38$ is of particular interest, because it can be reached only from $s_0 > 0.38$. It is also found that if $s_0 \in (0.17, 0.38)$, then strategies that start just a little above the point $(s_0, \delta s_0 - s_0^2/(s_0^2 + 1))$ will result in a state trajectory that converges to a steady state to the right. Steady-state pollution stocks that are smaller than 0.17 are unstable. The important point is that pre-play communications allow the choice among feedback strategies, bringing the pollution stock closer to the social optimal one (0.38 is close enough to 0.353).

Note that the result of Dockner and Long [34], that the best feedback equilibrium steady state is arbitrarily close to the social optimal one if the discount rate tends to zero, carries over to the lake-pollution model. Nevertheless, we should not forget that there is a distinction between welfare at a steady state, and total welfare, which takes into account the welfare flows along the path to the steady state.

Given the feedback information structure, it is natural to consider the design of efficiency-inducing taxation where the tax rate on emissions is made dependent only on the state variable: $\tau = \tau(s)$. This issue was considered by Benchekroun and Long [11] in the context of a polluting oligopoly. They found that there exists a feedback tax scheme that ensures that the oligopolists replicate the socially optimal path. In the context of lake pollution, where the transition dynamics is more complicated, Kossioris et al. [63] focus on polynomial functional forms for the tax rate $\tau(s)$. They found that it is not possible to completely mimic the social optimal path when the polynomial is of low order.

1.6 Dynamic Games Involving Potential Regime Shifts and Skiba Point: R&D Races and Sabotage to Prevent Entry

In industrial organization theory, R&D races between firms have been a subject of intensive study. The winner of a race becomes a monopolist, so that there is a regime shift from, say, duopoly, to limit-pricing monopoly. Early models of R&D races assume that the time of a successful innovation is exponentially distributed: past investments in R&D have no strategic implications because the accumulated knowledge has no value [71, 89, 102]. This is because of the memorylessness of the exponential distribution, under which, if the event has not occurred, the future always looks the same, regardless of past levels of R&D. The resulting races cannot feature history dependence. The idea of a firm being ahead of another firm cannot be formulated under the assumption of exponential distribution. To capture the idea of history dependence, some authors propose multistage race models: to win a race, a firm must be the first to complete all stages of an R&D project. Thus, at any point of time, a firm may be ahead of another one. Several papers consider only deterministic multistage race models [40, 45, 46]. In such models, the equilibrium result is drastic: once a firm has a slight advantage, the other firm drops out immediately. This is called the ε pre-emption property. To avoid this unrealistic feature, one could add the assumption that the stage-to-stage transition is probabilistic [44, 47, 77]. However, these authors continue to assume that the time to completion for each stage is distributed exponentially. This implies that at each stage, firms' investment in R&D is independent of past investments. In these models, the laggard firm (that has completed fewer stages than its rival) will find it optimal to invest less than the industry leader, and consequently one observes that if a firm is behind, it tends to remain behind. This is not consistent with real-world observations: there are instances of laggards' catching-up behavior. To capture this catching-up feature, Doraszelski [37] formulates a model in which the hazard rate depends on both the state variable and the control variable.

Consider two firms. The stock of knowledge of firm i is denoted by $k_i(t)$, and its current R&D effort (a control variable) is denoted by $I_i(t)$. Doraszelski [37] assumes that

$$\dot{k}_i(t) = I_i(t) - \delta k_i(t),$$

where $\delta > 0$ is the rate of depreciation of knowledge. The conditional probability that firm 1 makes a breakthrough over the interval of time $(t, t + \Delta)$, given that it has not been successful prior to time t, is $h_1(I_1(t), k_1(t)) \times \Delta$, where the hazard rate function h_1 is specified as follows:

$$h_1(I_1, k_1) = \lambda I_1 + \gamma k_1^{\psi},$$

with $\lambda \geq 0$, $\gamma \geq 0$, and $\psi > 0$. Here, the hazard function $h_1(I_1, k_1)$ is additive and increasing in both the current effort, I_1, and the past efforts, as represented by k_1. (A multiplicative specification, e.g., $h_1 = I_1^{\alpha} k_1^{\beta}$, could be an interesting alternative, as Doraszelski [37, p. 40] points out.) In the special case where $\gamma = 0$, past efforts do not influence the hazard rate, and we would obtain the memorylessness of the models of Reinganum [101, 102]. The function h_1 is concave, linear, or convex in the state variable k_1 according to whether ψ is smaller than, equal to, or greater than 1. The cost of exerting effort I_1 is denoted by $c(I_1)$. This cost function is assumed to take the form

$$c(I_1) = \frac{1}{\eta} I_1^{\eta} \text{ where } \eta > 1.$$

Assume that as soon as one firm makes a breakthrough, the game ends, at which point the successful firm wins a big prize, $\overline{P} > 0$, and the other firm wins a small prize, $\underline{P} < \overline{P}$. The interpretation of these prizes is as follows. The high prize, $\overline{P}$, is the present value of future profits of the successful firm. The other firm can imitate the discovery after the patent has expired, and $\underline{P}$ is the present value of future profits of the imitating firm. The case where $\underline{P} = 0$ means that the innovating firm has perfect patent protection. In general, the ratio $\underline{P}/\overline{P}$ is a measure of patent protection. When this ratio is zero, the patent protection is perfect.

Denote the equilibrium strategies by $I_1 = \phi_1(k_1, k_2)$ and $I_2 = \phi_2(k_2, k_1)$. Then the HJB equation for firm 1 is

$$rW^1(k_1,k_2)=\max_{I_1}\Big\{\big[\overline{P}-W^1(k_1,k_2)\big]h_1(I_1,k_1)+\big[\underline{P}-W^1(k_1,k_2)\big]h_2(\phi_2,k_2)$$
$$-c_1(I_1)+\frac{\partial W^1}{\partial k_1}[I_1-\delta k_1]+\frac{\partial W^1}{\partial k_2}[\phi_2-\delta k_2]\Big\}.$$

The first-order condition for I_1 is

$$\big[\overline{P}-W^1(k_1,k_2)\big]\lambda+\frac{\partial W^1}{\partial k_1}=c_1'(I_1)=I_1^{\eta-1}.$$

It follows that firm 1's strategy is

$$\phi_1(k_1,k_2)=\left\{\big[\overline{P}-W^1(k_1,k_2)\big]\lambda+\frac{\partial W^1}{\partial k_1}\right\}^{\frac{1}{\eta-1}}.$$

Focusing on symmetric equilibrium, we can omit the subscript in the strategy functions ϕ_1 and ϕ_2 and the superscript in the value functions, W^1 and W^2, and thus we have

$$I_1^*=\phi(k_1,k_2)\text{ and }I_2=\phi(k_2,k_1).$$

It follows that

$$\phi(k_1,k_2)=\left\{\big[\overline{P}-W(k_1,k_2)\big]\lambda+\frac{\partial W(k_1,k_2)}{\partial k_1}\right\}^{\frac{1}{\eta-1}}$$

$$\phi(k_2,k_1)=\left\{\big[\overline{P}-W(k_2,k_1)\big]\lambda+\frac{\partial W(k_2,k_1)}{\partial k_2}\right\}^{\frac{1}{\eta-1}}$$

and the HJB equation can be written as the operator equation

$$\mathcal{N}(W)=0,$$

where

$$\mathcal{N}(W)(k_1,k_2)=\left(\lambda\phi(k_1,k_2)+\gamma k_1^{\psi}\right)\overline{P}+\left(\lambda\phi(k_2,k_1)+\gamma k_2^{\psi}\right)\underline{P}$$
$$-\frac{\phi(k_1,k_2)^{\eta}}{\eta}-\left(r+\lambda\phi(k_1,k_2)+\gamma k_1^{\psi}+\lambda\phi(k_2,k_1)+\gamma k_2^{\psi}\right)W(k_1,k_2)$$
$$+\frac{\partial W(k_1,k_2)}{\partial k_1}[\phi(k_1,k_2)-\delta k_1]+\frac{\partial W(k_1,k_2)}{\partial k_2}[\phi(k_2,k_1)-\delta k_2].$$

This is a non-linear first-order partial differential equation. Since a closed-form solution is not available, one must resort to numerical methods. Doraszelski [37] reports the following numerical results.

(i) For the case where $\gamma = 0$ (i.e., the hazard rate is a function of current R&D effort only), the equilibrium R&D efforts are constant, independent of the knowledge stocks k_1 and k_2. The value function W is then a constant (independent of k_1 and k_2). This corresponds to the memoryless R&D race models [101, 102].

(ii) When $\gamma > 0$, the accumulated knowledge stocks k_1 and k_2 matter. One can show that for any given finite k_2, $\lim_{k_1 \to \infty} W(k_1, k_2) = \overline{P}$, and for any given finite k_1, $\lim_{k_2 \to \infty} W(k_1, k_2) = \underline{P}$. With $\gamma > 0$, if $\lambda = 0$ (i.e., current R&D effort does not contribute directly to the hazard rate), the optimal R&D expenditure I_1 falls as the knowledge stock k_1 increases. That is, thanks to the "pure knowledge effect" on the hazard rate, the firm "can afford to scale back its investment in R&D as its knowledge stock increases" (p. 28).

(iii) When $\psi > 1$ so that h_1 is strictly convex and increasing in k_1, the increasing return to knowledge accumulation gives the firm a strong incentive to increase I_1 as k_1 rises from its low initial levels.[16] In particular, if firm 1 is a laggard (i.e., $k_1 < k_2$), it will try to catch up with firm 2 (i.e., investing more than firm 2) provided the gap between the two stocks is not too large. This catching-up feature is consistent with real-world experience. Doraszelski [37, p. 20] presented some evidence of catching up:

> Casual observation suggests that the laggard strives to catch up with the leader. When Transmeta unveiled its power-stingy Intel-compatible Crusoe chip in 2000, Intel pledged to introduce a version of its Pentium III processor that matched Crusoe's power consumption in the first half of 2001 and announced a new set of technologies for 2002 or 2003 that would give it the lead over Transmeta. Similarly, after Celera Genomics in 1998 challenged the Human Genome Project to be the first to sequence the human genome, the Human Genome Project announced that it would move up its target date from 2005 to 2003 and indeed dramatically stepped up its own pace during 1999. And yet, although Celera Genomics started the race as the underdog, it completed a draft of the human genome in 2000 and beat the Human Genome Project.

Doraszelski [37] relied on the (ex ante) symmetry between firms. Also, he did not attempt to explore the possibility of multiplicity of steady states and of Skiba points. As we have pointed out, the analysis of optimal control problems with multiple steady states involves the identification of a Skiba point. Skiba points can occur also in dynamic games. Dockner and Wagener [36] give an example of Skiba point in a differential game involving two symmetric players and a single capital stock. An interesting question is whether a Skiba point can exist when players are asymmetric. The paper by Dawid et al. [30] presents a dynamic game model which exhibits the Skiba point property with two asymmetric players and one capital stock.

Dawid et al. [30] pointed out that, in general, it is unlikely to have a Skiba point when players are asymmetric, because it would require the existence of a point at which two asymmetric players are indifferent between two courses of actions. Generically, it is impossible in an asymmetric game to have a single point where for each player, the two local value functions intersect. It follows that without very specific assumptions, it is unlikely that an MPE exhibiting Skiba points can exist.

[16] However, eventually when k_1 is large enough, I_1 begins to fall.

To illustrate this argument, consider an example discussed in Dawid et al. [30]. Suppose there are two firms, each investing in its own capital stock. Assume the firms produce goods that are perfect substitutes and they compete as Cournot rivals. Each firm would prefer that its rival invests less, because a low aggregate capital stock implies low aggregate output, which raises the price. *Given the strategy of firm 2*, suppose that a Skiba point, say k_1^S, exists for firm 1's optimal control problem. Then firm 2's value function would jump down as its rival's capital stock k_1 reaches k_1^S from below. Firm 2, therefore, has an incentive to prevent firm 1's capital stock to get close to k_1^S, and thus it would want to "overinvest" (to deviate from the given candidate strategy) in order to induce firm 1 to invest less. Such optimal behavior by firm 2 would then imply that firm 1 would invest less even for values of $k_1(0)$ that are slightly above k_1^S, i.e., k_1^S cannot be a Skiba point. While this argument is intuitively plausible, nevertheless a formal analysis of a two-state-variable differential game between two asymmetric players that would establish the existence, or impossibility of existence, of a Skiba point is unfortunately unavailable.

Dawid et al. [30] choose to work with a simpler model with two asymmetric players. They assume that there is only one stock of capital. There are two firms. The authors assume that firm 1, the incumbent firm, does not invest in R&D, and firm 2 is seeking to enter the market. Firm 2 can enter the market only if it is able to make a technological breakthrough. In order to make a breakthrough, firm 2 must invest in its stock of knowledge, k. If a breakthrough has not occurred at time t, the probability that it will occur during the time interval $(t, t + dt)$ is given by $h(k(t))dt$. The function $h(k)$ is called the hazard rate. Dawid et al. [30] assume that

$$h(k) = \alpha k^2, \alpha > 0.$$

This implies that there is increasing return to capital (in terms of probability of a breakthrough). They specify the following state dynamic equation:

$$\dot{k}(t) = I_2(t) - \lambda I_1(t) - \delta k(t),$$

where δ is the rate of depreciation, $I_2(t) \geq 0$ is firm 2's investment (R&D efforts), and $I_1(t) \geq 0$ is firm 1's sabotage effort. The positive parameter λ is a measure of the effectiveness of sabotage. The cost of I_i is $c_i(I_i) = \beta_i I_i + (\gamma_i/2)I_i^2$, with $\beta_i \geq 0$ and $\gamma_i \geq 0$.

A breakthrough by firm 2 implies a regime shift, from monopoly (under firm 1) to duopoly. Under duopoly, firm i earns a profit of π_i^d at each point of time. Under monopoly, $\pi_1 = \pi_1^m > 0$ and $\pi_2 = 0$. Assume that $\pi_1^m > \pi_1^d$, so that firm 1 has an incentive to sabotage firm 2's R&D efforts.

In order to establish the existence of a Skiba point, Dawid et al. [30] find it necessary to assume that there is an exogenous upper bound, denoted by $\overline{I}$, on I_i, $i = 1, 2$. This implies an upper bound on k: $k \leq \overline{k} = (1/\delta)\overline{I}$. The upper bound on investment is a crucial assumption, which results in a special property of the model: the value function of the incumbent is discontinuous at the Skiba point. The upper bound on sabotage makes it impossible for the incumbent to move the state variable

k from the lower branch of its value function to the upper branch. (If the upper bound on the control was removed, so that any player could move the state in both directions, then the value function of each player would be continuous under the equilibrium profile.)

Formally, the dynamic game considered by Dawid et al. [30] is a multi-mode game with two modes, m_1 (before entry) and m_2 (after entry), with $\pi_1(m_1) = \pi_1^m$, $\pi_2(m_1) = 0, \pi_1(m_2) = \pi_1^d$, and $\pi_2(m_2) = \pi_2^d$. (Dockner et al. [35] refer to such multi-mode games as piece-wise deterministic game.) Firm i's objective is to maximize

$$J_i = E\left[\int_0^{\infty} e^{-rt}\left[\pi_i(m(t)) - c_i(I_i(t))\right] dt\right]$$

subject to $\dot{k}(t) = I_2(t) - \lambda I_1(t) - \delta k(t)$ and subject to the mode process

$$\lim_{\Delta \to 0} \frac{\Pr\left\{m(t+\Delta) = m_2 |\, m(t) = m_1\right\}}{\Delta} = h(k(t)),$$

with $m(0) = m_1$ and $k(0) = k_0$. Both firms set $I_i = 0$ in mode 2, while in mode 1 they use feedback strategies $I_i = \phi_i(k)$.

Clearly, in mode 2, the value functions are independent of k:

$$V_i(m_2) = (1/r)\pi_i^d.$$

Denote firm i's value function in mode 1 by $W_i(k)$. Then, in mode 1, the HJB equation for firm 1 is

$$rW_1(k) = \alpha k^2\left[(1/r)\pi_1^d - W_1(k)\right] + \max_{I_1}\left[\pi_1^m - c_1(I_1) + W_1'(k)\left(\phi_2(k) - \lambda I_1 - \delta k\right)\right]$$

and the HJB equation for firm 2 is

$$rW_2(k) = \alpha k^2\left[(1/r)\pi_2^d - W_2(k)\right] + \max_{I_2}\left[-c_2(I_2) + W_2'(k)\left(I_2 - \lambda\phi_1(k) - \delta k\right)\right].$$

Then the first-order condition for firm 1 is

$$\beta_1 + \gamma_1 I_1 = -\lambda W_1'(k) \text{ if } I_1 \in \left(0, \overline{I}\right)$$

and, for firm 2,

$$\beta_2 + \gamma_2 I_2 = W_2'(k) \text{ if } I_2 \in \left(0, \overline{I}\right).$$

Assuming that the equilibrium strategies are almost everywhere continuous on $\left[0, \overline{k}\right]$, and writing

$$\phi_1(k) = -\frac{\lambda W_1'(k)}{\gamma_1} - \frac{\beta_1}{\gamma_1}$$

$$\phi_2(k) = \frac{W_2'(k)}{\gamma_2} - \frac{\beta_2}{\gamma_2}$$

one obtains a system of two first-order differential equations for $W_1(.)$ and $W_2(.)$. Unfortunately, no closed-form solution is available. The authors, therefore, resort to numerical methods. They use the homotopy method (see Vedenov and Miranda [123] for a discrete-time model and Dawid et al. [31] for a continuous time model).[17]

In the model, an increase in k has two qualitatively different and countervailing effects on the payoff of each player. First, since $h(k) = \alpha k^2$ is strictly convex and increasing, the effect of a marginal increase in k is more substantial at high levels of k. Therefore, a high k means a much greater chance of a regime switch. Second, a high k means the expected arrival time is closer to the present, which has the effect of reducing the impact of an increase in k on the expected future payoff stream of both players (bearing in mind that the size of k is irrelevant in mode 2, after entry). These opposing considerations suggest that equilibrium steady states with high and low investments for both players may co-exist.

Indeed, numerical calculations show that there are two locally stable steady states, one with high investment (or sabotage) by both players and one with low activities by both. The steady states are $k^* = 0$ and $k^{**} = 0.556$. There exists a Skiba point at $k^S < 0.556$.

1.7 Dynamic Games of Inducing Regime Shifts by a Big Push

Tornell [111] presented a model of economic growth that declines with endogenous switches in property-right regimes when rival fractions incur a lumpy cost to overthrow an existing regime. In his model, two groups of infinitely lived agents solve a dynamic game over the choice of property rights regime. He sought to find a possible equilibrium of the game involving multiple switching of regimes. Tornell allowed each group's share of aggregate capital to change after a switch takes place and introduced a once-off lump sum cost at switching time. Specifically, Tornell [111] specified three property rights regimes: common property, private property, and leader-follower. Under common property, both players have equal access to the aggregate capital stock. When one player incurs the once-off cost, it can convert the whole common property to its private property unless the other player is willing to incur the same cost. In the latter case, the result is the private property regime, where each player has access only to its own capital stock. In contrast, starting from the private property regime, if both players simultaneously incur each the once-off cost,

[17] This method yields polynomial approximations of value functions. One shortcoming is that such polynomial approximation gives continuous and smooth value functions, which may be incorrect. To deal with this issue, Dawid et al. [30] combine the homotopy method with another method that yields local value functions (each around a stable steady state).

the regime will revert back to common property. If one player incurs the once-off cost while the other does not, the former becomes the leader and has exclusive access to the economy's capital stock. Tornell [111] restricted the maximum number of regime switches to two. This simplifying assumption allows closed-form solutions. A key parameter in this game is σ, the elasticity of intertemporal substitution. The model generates a hump-shaped pattern of growth even though the underlying technology is linear and preferences exhibit a constant elasticity of intertemporal substitution. If $\sigma \leq 1$, the common property regime may last forever. (Alternatively, if the economy starts with the private property regime, this institution may also last for ever.) In contrast, if $\sigma > 1$, the economy exhibits a cycle: a switch from the common property regime to the private property regime, and later on, a re-switching back to common property. There is no equilibrium which involves a switch to the leader-follower regime.

While Tornell [111] assumed that the two players are symmetric, Long et al. [86] consider a model of regime-shift-inducing lumpy investments by asymmetric players. Each player can switch from one exploitation technology to another. They consider an economy that can operate under four possible regimes, denoted by I, N_1, N_2, and B. There are two players in this game, denoted by 1 and 2. Each player can make a big push only once during the game. Initially, the economy operates under regime I (where I stands for "initial"). Player 1 (he) can make a big push to switch the regime from I to N_1 which is to his advantage. However, player 2 (she) can pre-empt the rival's move by making a big push beforehand, thus switching the regime from I to N_2, to her advantage. In the case where both players make a big push at the same time, the economy's regime is switched from I to B (where B stands for "both"). Once the economy is in regime B, no further switch is possible. Regime B can also become operative after two consecutive big pushes, one by each player.

Let $\mathcal{S}$ denote the set of possible regimes, i.e.,

$$\mathcal{S} \equiv \{I, N_1, N_2, B\}.$$

Let $\mathcal{S}_i$ be the subset of regimes of $\mathcal{S}$ from which player i can make a big push. Then $\mathcal{S}_1 = \{I, N_2\}$ and $\mathcal{S}_2 = \{I, N_1\}$.

There is a continuous state variable, denoted by a vector $x \in \mathbb{R}^m_+$. For example, x is the economy's capital stock. To simplify the exposition, the authors set $m = 1$. In addition to a big push, each player also has a piece-wise continuous control variable c_i, with $c_i \in \mathbb{R}^n$. The instantaneous payoff $u_i(t)$ to player i at time t when the system is in regime $s \in \mathcal{S}$ is a differentiable function of the two control variables and the continuous state variable, and is, in general, different across regimes:

$$u_i(t) = U_i^s(c_i(t), c_{-i}(t), x(t)).$$

If player i, $i = 1, 2$, takes a regime change action at time $t_i \in \mathbb{R}_+$, he/she incurs a lumpy cost $K_i(x(t_i))$. If $0 < t_1 < t_2 < \infty$, the total payoff for player 1 is

$$\int_0^{t_1} U_1^I(c_1, c_2, x)e^{-rt}dt + \int_{t_1}^{t_2} U_1^{N_1}(c_1, c_2, x)e^{-rt}dt$$
$$+ \int_{t_2}^{\infty} U_1^B(c_1, c_2, x)e^{-rt}dt - K_1(x(t_1))e^{-rt_1}$$

with $r > 0$ is the discount rate.

The differential equation describing the evolution of the state variable x in regime s is

$$\dot{x} = G^s(c_1, c_2, x),$$

where, for each regime s, the function G^s is twice differentiable in the triplet (c_1, c_2, x).

For expositional purposes, Long at al. [86] focus on a specific sequence of regimes: I, N_1, and B. A natural way to proceed, for determining a MPE of this game, is to solve the problem recursively, starting from regime B, the last regime of the system. Recall that each player has two types of controls, a piece-wise continuous control variable c_i, and a big-push date, t_i. A Markovian strategy consists of a control policy and a big-push rule at every possible state of the system, $(x, s) \in \mathbb{R}_+ \times \mathcal{S}$. The *control policy* of player i is a mapping $\eta_i(.)$ from the state space $\mathbb{R}_+ \times \mathcal{S}$ to the set $\mathbb{R}^n$. To get an idea of a *Big-Push rule*, consider the following situation. Suppose player 1 thinks that if player 2 finds herself in regime N_1 at date t, she will make a big push at a date $t_2 \geq t$. Then player 1 conjectures that the interval of time between the current period and the switching date, $t_2 - t$, is a function of the state of the system. More generally, define player i's *time-to-go (before making a big push)*, given that $s \in \mathcal{S}_i$, as a mapping $\ell_2(.)$ from $\mathbb{R}_+ \times \mathcal{S}$ to $\mathbb{R}_+ \cup \{\infty\}$. For example, from the state (x, N_1), the real number $\ell_2(x, N_1)$ is the length of time that must elapse before player 2 makes her big push. If $\ell_2(x, N_1) = \infty$ for all x, this would mean that she does not want to make a big push if she finds herself under regime N_1.

Long et al. [86] introduce the concept of piece-wise feedback Nash equilibrium (PFBNE), defined as follows:

(i) A strategy vector of player i is a pair $\chi_i \equiv (\eta_i, \ell_i)$.
(ii) A strategy profile (χ_1, χ_2) is a piece-wise feedback Nash equilibrium (PFBNE) if starting at any time t and any state (x, s), the remaining lifetime payoff of player i is maximized by χ_i, given χ_{-i}.

As an application, Long et al. [86] consider a game of exploitation of exhaustible resources. There are two players. Each can choose a date at which she introduces a more efficient extraction technology. They find that the player with low investment cost is the first player to adopt a new harvesting technology. She faces two countervailing incentives: on the one hand, an early switch to a more efficient technology enables her to exploit the resources more cheaply; on the other hand, by inducing the regime change, which tends to result in a faster depletion, she might give her opponent an incentive to hasten the date of his technology adoption, if the opponent investment cost decreases as the stock decreases. As a consequence, in an equilibrium, the balance of these strategic considerations may make the low-cost player delay technology adoption even if her fixed cost of adoption is zero, contrary to what she would do (namely, immediate adoption) if she were the sole player.

Let us now contrast the big-push class of models (as considered in Tornell [111] and Long et al. [86]), with the other polar cases where a regime shift can occur only with gradual investments. For illustration, we review the model of Itaya and Tsoukis [54], who analyzed differential games involving symmetric agents who want to change their preferences away from envy-driven consumption. Itaya and Tsoukis [54] considered a community consisting of n infinitely lived agents who may contribute to the accumulation of a stock of "social capital", denoted by S. The higher is the stock, the lower is each individual's incentive to "out-do others" in terms of relative consumption. This incentive is captured by the term $(1-\theta_i(S)) \geq 0$, where $\theta_i(.)$ is an increasing function of S, with the property that $0 \leq \theta_i(S) \leq 1$ for all $S \geq 0$. The function $\theta_i(.)$ is the same for all i. Each individual i has 1 unit of time at each t. A fraction a_i of time is devoted to building up social capital (e.g., by spending time to socialize with other members of the community). The remaining fraction, $1-a_i$, is used to produce a consumption good, under the constant returns to scale technology $c_i = 1-a_i$. Production of c_i yields the utility of consumption, $\ln c_i$, from which the disutility of effort, βc_i, must be subtracted. The utility flow at time t is

$$\ln c_i(t) - \beta c_i(t) + (1-\theta_i(S(t))) \ln \left[\frac{c_i(t)}{C(t)/n} \right],$$

where C/n is the community's average consumption. While everyone knows that $\ln \left[\frac{c_i}{C/n} \right] = \ln(1) = 0$ in a symmetric equilibrium, it remains true that as long as $(1-\theta_i(S)) > 0$, each individual has an incentive to try to "out-do" others in terms of consumption, by spending a lot of time in production activities. This is the well-known "rat race" which reduces welfare. There is also an incentive to eliminate the rat race. If S is built up to the level $\overline{S}$ where $(1-\theta_i(\overline{S})) = 0$ and maintained at that level for ever, the rat race will be completely eliminated.

The authors assume that

$$\dot{S} = \left(\sum_{i=1}^{n} a_i \right) S - \delta S,$$

where δ is the rate of depreciation of S.

The authors describe the set of Markov-perfect equilibria (MPEs) of this game. They show that there are a continuum of MPEs, which can either involve a monotone decreasing path $S(t)$, ending up at $S = 0$, or a a monotone increasing path $S(t)$, ending up at $S = \overline{S}$. There is no stable equilibrium path that converges to an interior stock $S \in (0, \overline{S})$.

1.8 Directions for Future Research

The literature on regime shifts has contributed much to our understanding of the complexity of the problems that decision-makers face in a world where state dynamics are not immutable. We have learned from this literature that decision-makers should be very cautious when they face uncertainty about tipping points. Development planning should take account of threshold externalities, and foreign aids could be more useful if donor countries can coordinate on a big push. Analysis of political changes can benefit from models of how discontent might build up. While the literature on regime shifts is indeed very rich, there are a number of issues that deserve greater scrutiny.

The first issue concerns the analysis of changes in preferences. While the existing literature acknowledges that preferences may change, typically such changes are either assumed to be exogenous (e.g., Kemp and Long [57]), or triggered when an environmental threshold is crossed (e.g., Nkuiya and Costello [94]), or contemplated by infinitely lived agents, as in Itaya and Tsoukis [54]. However, a more important class of actions should be considered: how to influence the preferences of the future citizens so that environmental thresholds can be managed more efficiently. The literature on social investments that affect preferences of future generations is sparse. For models of intergenerational transmissions of preferences, see Bisin and Verdier [12–14] on the selection of traits, and Long [85] on the moral education to encourage pro-social behavior, switching players' preferences from Nashian to Kantian.[18] As Bowles [19] points out, a "moral economy" is more effective than an incentive-based economy in mitigating externalities and promoting investments in public goods.

The second issue concerns alternative paradigms for the analysis of regime shifts. Admittedly, the dominant paradigm in economic analysis is based on rational, forward-looking behavior. However, evolutionary game theory has been successfully used to explain many phenomena.[19] It would be interesting to model regime shifts in human societies from an evolutionary perspective.

Acknowledgements I would like to thank two anonymous reviewers for their careful reading of the paper and for their very helpful comments and suggestions.

References

1. Acemoglu, D., & Robinson, J. (2000). Inequality, growth and development: Democratization or repression? *European Economic Review*, *44*, 683–693.
2. Acemoglu, D., & Robinson, J. (2001). A theory of political transitions. *The American Economic Review*, *91*, 938–963.

[18]On Nashian behavior versus Kantian behavior, see, e.g., Grafton et al. [42].

[19]In a recent interesting paper, using evolutionary game theory, Wood et al. [128] outline a model that explains well a regime shift in the world oil market: the "Seven Sisters" were replaced by OPEC in the battle for oil market dominance.

3. Acemoglu, D., & Robinson, J. (2006). *Economic origins of dictatorship and democracy*. Cambridge University Press, New York: Cambridge Books.
4. Amit, R. (1986). Petroleum reservoir exploitation: Switching from primary to secondary recovery. *Operations Research, 34*, 534–549.
5. Anderson, E., Brock, W., Hansen, L., & Sanstad, A. H. (2014). *Robust analytical and computational explorations of coupled economic-climate models with carbon-climate response RDCEP* (Working Paper 13-05). Center for Robust Decision Making on Climate and Energy Policy.
6. Aguiar, M., & Amador, M. (2011). Growth in the shadow of expropriation. *Quarterly Journal of Economics, 126*(2), 651–697.
7. Azariadis, C., & Drazen, A. (1990). Thresholds externalities in economic development. *Quarterly Journal of Economics, 105*, 501–526.
8. AronssonT, Backlund K., & Löfgren, K. G. (1998). Nuclear power, externalities and non-standard pigouvian taxes. *Environmental and Resource Economics, 11*(2), 177–195.
9. Barro, R. (1999). Determinants of democracy. *Journal of Political Economy, 107*, 158–183.
10. Beltratti, A., Chichilnisky, G., & Heal, G. (1998). Uncertain future preferences and conservation. In G. Chichilnisky, G. Heal & A. Vercilli (Eds.). *Sustainability: Dynamics and uncertainty* (pp. 257–275). Heidelberg: Springer.
11. Benchekroun, H., & Long, N. V. (1998). Efficiency-inducing taxation for polluting oligopolists. *Journal of Public Economics, 70*, 325–342.
12. Bisin, A., & Verdier, T. (2001). The economics of cultural transmission and the dynamics of preferences. *Journal of Economic Theory, 97*(2), 298–319.
13. Bisin, A., Topa, G., & Verdier, T. (2009). Cultural transmissions, socialization and the population dynamics of multiple-trait distributions. *International Journal of Economic Theory, 5*(1), 139–154.
14. Bisin, A., & Verdier, T. (2017). *On the joint evolution of culture and institutions* (Working Paper 23375). Cambridge, MA: National Bureau of Economic Research.
15. Bohn, H., & Deacon, R. T. (2000). Ownership risk, investment, and the use of natural resources. *The American Economic Review, 90*(3), 526–549.
16. Boucekkine, R., Prieur, F., & Puzon, K. (2016). On the timing of political regime changes in resource-dependent economies. *European Economic Review, 85*, 188–207.
17. Boucekkine, R., Piacquadio, P. G., & Prieur, F. (2019). *A Lipsetian theory of voluntary power handover* (Working Paper). Aix-Marseille School of Economics, Aix-Marseille University.
18. Bourguignon, F., & Verdier, T. (2000). Oligarchy, democracy, inequality and growth. *Journal of Development Economics, 62*, 285–313.
19. Bowles, S. (2016). *The moral economy: Why good incentives are no substitute for good citizens*. New Haven: Yale University Press.
20. Brock, W. A., & Starrett, D. (2003). Managing systems with non-convex positive feedback. *Environmental and Resource Economics, 26*, 575–602.
21. Campante, F., & Chor, D. (2012). Why was the Arab world poised for revolution? Schooling, economic opportunities, and the Arab spring. *Journal of Economic Perspectives, 26*(2), 167–188.
22. Carpenter, S. R. (2003). *Regime shifts in lake ecosystems: Pattern and variation*. Oldendorf: International Ecology Institute.
23. Carpenter, S. R., Ludwig, D., & Brock, W. A. (1999). Management and eutrophication of lakes subject to potentially irreversible change. *Ecolo Applications, 9*(3), 751–771.
24. Clark, C. W. (1990). *Mathematical bioeconomics: The optimal management of renewable resources* (2nd ed.). New York: Wiley.
25. Clarke, H. R., & Reed, W. J. (1994). Consumption/pollution tradeoffs in an environment vulnerable to pollution-related catastrophic collapse. *Journal of Economic Dynamics and Control, 18*, 991–1010.
26. Couttenier, M., & Soubeyran, R. (2014). Drought and civil war in sub-Saharan Africa. *The Economic Journal, 124*(575), 201–244.
27. Couttenier, M., & Soubeyran, R. (2015). A survey of causes of civil conflicts: Natural factors and economic conditions. *Revue d'Economie Politique, 125*(6), 787–810.

28. Cropper, M. L. (1976). Regulating activities with catastrophic environmental effects. *Journal of Environmental Economics and Management, 5*(1), 1–15.
29. Dawid, H., Keoula, M., Kopel, M., & Kort, P. M. (2015). Product innovation incentives by an incumbent firm: A dynamic analysis. *Journal of Economic Behavior and Organization, 117*, 411–438.
30. Dawid, H., Keoula, M. Y., & Kort, P. M. (2016). Skiba phenomena in Markov perfect equilibria of asymmetric differential games. In H. Dawid, K. F. Doener, G. Feichtinger, P. M. Kort, & A. Seidl (Eds.), *Dynamic perspectives on managerial decision making: Essays in honor of Richard F* (pp. 63–76). Springer, Heidelberg: Hartl.
31. Dawid, H., Keoula, M. Y., & Kort, P. M. (2017). Numerical analysis of Markov perfect equilibria with multiple stable steady states: A duopoly application with innovative firms. *Dynamic Games and Applications, 7*(4), 555–577.
32. Dawkins, R. (1986). *The blind watchmaker: Why the evidence of evolution reveals a universe without design*. New York, W.W: Norton.
33. de Zeeuw, A., & Zemel, A. (2012). Regime shifts and uncertainty in pollution control. *Journal of Economic Dynamics and Control, 36*(7), 939–950.
34. Dockner, E., & Long, N. V. (1993). International pollution control: Cooperative versus non-cooperative strategies. *Journal of Environmental Economics and Management, 24*, 13–29.
35. Dockner, E., Jørgensen, S., Long, N. V., & Sorger, G. (2000). *Differential games in economics and management science*. Cambridge, U.K.: Cambridge University Press.
36. Dockner, E., & Wagener, F. (2014). Markov-perfect Nash equilibrium in models with a single capital stock. *Economic Theory, 56*, 585–625.
37. Doraszelski, U. (2003). An R&D race with knowledge accumulation. *The RAND Journal of Economics, 34*(1), 20–42.
38. Eldredge, N., & Gould, S. J. (1972). Puntuated equilibria: An alternative to phyletic gradualism. In T. J. M. Schopf (Ed.), *Models of paleobiology* (pp. 82–115). San Francisco: Freeman Cooper.
39. Feichtinger, G., & Wirl, F. (2000). Instabilities in concave, dynamic, economic optimization. *The Journal of Optimization Theory and Applications, 107*(2), 275–286.
40. Fudenberg, D., Gilbert, R., Stiglitz, J., & Tirole, J. (1983). Pre-emption, leapfrogging and competition in patent races. *European Economic Review, 22*, 3–31.
41. Gjerde, J., Grepperud, S., & Kverndokk, S. (1999). Optimal climate policy under the possibility of a catastrophe. *Resource and Energy Economics, 21*(3/4), 289–317.
42. Grafton, Q., Kompas, T., & Long, N. V. (2017). A brave new world? Interaction among Kantians and Nashians, and the dynamics of climate change mitigation. *European Economic Review, 99*, 31–42.
43. Granovetter, M. (1978). Threshold models of collective behavior. *The American Journal of Sociology, 83*(6), 1420–1443.
44. Grossman, G. M., & Shapiro, C. (1987). Dynamic R&D competition. *The Economic Journal, 97*, 372–387.
45. Harris, C. J., & Vickers, J. S. (1985a). Patent races and the persistence of monopoly. *The Journal of Industrial Economics, 33*, 461–481.
46. Harris, C. J., & Vickers, J. S. (1985b). Perfect equilibrium in a model of a race. *The Review of Economic Studies, 52*, 193–209.
47. Harris, C. J., & Vickers, J. S. (1987). Racing with uncertainty. *The Review of Economic Studies, 54*, 1–21.
48. Hartl, R., & Kort, P. M. (1996). Capital accumulation by a firm facing an emission tax. *Journal of Economics, 63*(1), 1–23.
49. Hartl, R., Kort, P. M., Feichtinger, G., & Wirl, F. (2004). Multiple equilibria and thresholds due to relative investment costs: Nonconcave–concave, focus—node, continuous–discontinuous. *The Journal of Optimization Theory and Applications, 123*, 49–82.
50. Haurie, A., Krawczyk, J. B., & Zaccour, G. (2012). *Games and dynamic games*. Singapore: World Scientific.
51. Heal, G. (1984). Interaction between economy and climate: A framework for policy design under uncertainty. In V. Kerry Smith & A. D. White (Eds.), *Advances in applied microeconomics* (Vol. 3, pp. 151–168). Greenwich: JIL Press.

52. Hillman, A. L., & Long, N. V. (1983). Pricing and depletion of an exhaustible resource when there is anticipation of trade disruption. *Quarterly Journal of Economics, 98*(2), 215–233.
53. Hung, N. M., Kemp, M. C., & Long, N. V. (1984). On the transition from an exhaustible resource stock to an inexhaustible substitute. In A. Ingham & A. M. Ulph (Eds.), *Demand, equilibrium and trade* (pp. 105–122). London: Macmillan.
54. Itaya, J., & Tsoukis, C. (2019). Social capital and status externalities. In *Paper presented at the Public Economic Theory Conference* (2017) Paris (Revised, 2019).
55. Kamien, M., & Schwartz, N. (1971). Optimal maintenance and sale age for a machine subject to failure. *Management Science, 17*(8), 495–504.
56. Keller, K., Bolker, B. M., & Bradford, D. F. (2004). Uncertain climate thresholds and optimal economic growth. *Journal of Environmental Economics and Management, 48*(1), 723–741.
57. Kemp, M. C., & Long, N. V. (1977). Optimal control problems with integrands discontinuous with respect to time. *Economic Record, 53*(142), 405–420.
58. Kemp, M. C., Long, N. V., & Shimomura, K. (1993). Cyclical and non-cyclical redistributive taxation. *International Economic Review, 34*, 415–430.
59. Klibanoff, P., Marinacci, M., & Mukerji, S. (2005). A smooth model of decision making under ambiguity. *Econometrica, 73*(6), 1849–1892.
60. Klibanoff, P., Marinacci, M., & Mukerji, S. (2009). Recursive smooth ambiguity preferences. *Journal of Economic Theory, 144*(3), 930–976.
61. Konrad, K., Olsen, T. E., & Schöb, R. (1994). Resource extraction and the threat of possible expropriation: The role of swiss bank accounts. *Journal of Environmental Economics and Management, 63*(2), 242–259.
62. Kossioris, G., Plexousakis, M., Xepapadeas, A., de Zeeuw, A., & Mäler, K. G. (2008). Feedback Nash equilibrium for non-linear differential games in pollution control. *Journal of Economic Dynamics and Control, 32*(4), 1312–1331.
63. Kossioris, G., Plexousakis, M., Xepapadeas, A., & de Zeeuw, A. (2011). On the optimal taxation of common-pool resources. *Journal of Economic Dynamics and Control, 35*(4), 1868–1879.
64. Kuran, T. (1989). Sparks and prairie fires: A theory of unanticipated political revolution. *Public Choice, 61*(1), 41–74.
65. Lang, J. C., & De Sterck, H. (2014). The Arab spring: A simple compartmental model for the dynamics of a revolution. *Mathematical Social Sciences, 69*, 12–21.
66. Lange, A., & Treich, N. (2008). Uncertainty, learning and ambiguity in economic models on climate policy: Some classical results and new directions. *Climate Change, 89*(1), 7–21.
67. Lasserre, P., & Soubeyran, A. (2003). A Ricardian model of the tragedy of the commons. *Journal of Economic Behavior & Organization, 50*, 29–45.
68. Laurent-Lucchetti, J., & Santugini, M. (2012). Ownership risk and the use of common-pool natural resources. *Journal of Environmental Economics and Management, 63*(2), 242–259.
69. Le Kama, A. A. (2001). Preservation and exogeneous uncertain future preferences. *Economic Theory, 18*(3), 745–752.
70. Le Kama, A. A., & Schubert, K. (2004). Growth, environment, and uncertain future preferences. *Environmental and Resource Economics, 28*(1), 31–53.
71. Lee, T., & Wilde, L. L. (1980). Market structure and innovation: A reformulation. *The Quarterly Journal of Economics, 94*, 429–436.
72. Lemoine, D., & Traeger, C. (2014). Watch your step: Optimal policy in a tipping climate. *American Economic Journal: Economic Policy, 6*(1), 137–166.
73. Lemoine, D., & Traeger, C. (2016). Ambiguous tipping points. *Journal of Economic Behavior & Organization, 132*, 5–18.
74. Leonard, D., & Long, N. V. (2012). Endogenous changes in property rights regime. *Economic Record, 88*(280), 79–88.
75. Leonard, D., & Long, N. V. (2014). Endogenous changes in tastes. *Economics Bulletin, 34*(2), 923–937.
76. Li, X., Narajabad, B., & Temzelides, T. (2014). *Robust dynamic optimal taxation and environmental externalities* (CESifo Working Paper Series, 4562).

77. Lippman, S. A., & McCardle, K. F. (1987). Dropout behavior in R&D races with learning. *The RAND Journal of Economics*, *18*, 275–287.
78. Lipset, S. (1959). Some social requisite of democracy: Economic development and political legitimacy. *American Political Science Review*, *53*, 69–105.
79. Lipset, S. (1960). *Political man: The social bases of modern politics*. New York: Doubleday.
80. Long, N. V. (1975). Resource extraction under the uncertainty about possible nationalization. *Journal of Economic Theory*, *10*(1), 42–53.
81. Long, N. V. (1977). Optimal exploitation and replenishment of a natural resource. In J. D. Pitchford & S. J. Turnovsky (Eds.), *Applications of control theory to economic analysis* (pp. 81–106). Amsterdam: North-Holland.
82. Long, N. V. (1992a). An addendum to the Feichtinger paper on limit cycles. In G. Feichtinger (Ed.), *Dynamic economic models and optimal control* (pp. 185–187). Amsterdam: North Holland.
83. Long, N. V. (1992b). Pollution control: A differential game approach. *Annals of Operations Research*, *37*, 283–296.
84. Long, N. V. (2013). The theory of contests: A unified model and review of the literature. *European Journal of Political Economy*, *32*, 161–181.
85. Long, N. V. (2019). Warm glow and the transmission of pro-socialness across generations. *Journal of Public Economic Theory*. https://doi.org/10.1111/jpet.12347.
86. Long, N. V., Prieur, F., Puzon, K., & Tidball, M. (2017). Piecewise closed loop equilibria in differential games with regime switching strategies. *Journal of Economic Dynamics and Control*, *76*(C), 31–42.
87. Long, N. V., & Sinn, H. W. (1985). Surprise price shift, tax changes and the supply behaviour of resource extracting firms. *Australian Economic Papers*, *24*(45), 278–289.
88. Lontzek, T. S., Cai, Y., Judd, K. L., & Lenton, T. M. (2015). Stochastic integrated assessment of climate tipping points indicates the need for climate policy. *Nature Climate Change*, *5*, 441–444.
89. Loury, G. C. (1979). Market structure and innovation. *Quarterly Journal of Economics*, *93*(3), 395–410.
90. Mäler, K. G., Xepapadeas, A., & de Zeeuw, A. (2003). The economics of shallow lakes. *Environmental and Resource Economics*, *26*(4), 603–624.
91. Mähler, K. G., & Li, C. Z. (2010). Measuring sustainability under regime shift uncertainty: A resilence-pricing approach. *Environment and Development Economics*, *15*, 707–719.
92. Michaeli, M., & Spiro, D. (2019). *The dynamics of revolutions*. Typescript: Department of Economics, Uppsala University.
93. Nkuiya, B., Marrouch, W., & Bahel, E. (2015). International environmental agreements under endogenous uncertainty. *Journal of Public Economic Theory*, *17*(5), 752–772.
94. Nkuiya, B., & Costello, C. (2016). Pollution control under possible future shift in environmental preferences. *Journal of Economic Behavior & Organization*, *132*, 193–205.
95. Nordhaus, W. (2008). *A question of balance: Weighing the options on global warming policies*. New Haven, CT: Yale University Press.
96. Olson, L. J., & Roy, D. S. (2002). The economics of controlling a stochastic biological invasion. *American Journal of Agricultural Economics*, *84*, 1311–1316.
97. Polasky, S., de Zeeuw, A., & Wagener, F. (2011). Optimal management with potential regime shifts. *Journal of Environmental Economics and Management*, *62*(2), 229–240.
98. Ren, B., & Polasky, S. (2014). The optimal management of renewable resources under the risk of potential regime shift. *Journal of Economic Dynamics and Control*, *40*(C), 195–212.
99. Reed, W. J. (1987). Protecting a forest against fire: Optimal protection patterns and harvest policies. *Natural Resource Modelling*, *2*, 23–54.
100. Reed, W. J. (1988). Optimal harvesting of a fishery subject to random catastrophic collapse. *IMA Journal Mathematics Applied in Medecine & Biology*, *5*(3), 215–235.
101. Reinganum, J. F. (1981). Dynamic games of innovation. *Journal of Economic Theory*, *25*, 395–410.

102. Reinganum, J. F. (1982). A dynamic game of R and D: Patent protection and competitive behavior. *Econometrica, 50*, 671–688.
103. Schelling, T. C. (1971). Dynamic models of segregation. *The Journal of Mathematical Sociology, 1*, 143–186.
104. Scheffer, M. (1997). *The ecology of shallow lakes*. London: Chapman and Hall.
105. Seabright, P. (2010). *The company of strangers: A natural history of economic life*. Princeton: Princeton University Press.
106. Sims, C., Finnoff, D., & O'Regan, S. M. (2016). Public control of rational and unpredictable epidemics. *Journal of Economic Behavior and Organization, 132*, 161–176.
107. Sinn, H. W. (2012). *The green paradox: A supply-side approach to global warming*. Cambrigde, MA: MIT Press.
108. Skiba, A. K. (1978). Optimal growth with a convex-concave production function. *Econometrica, 46*, 527–549.
109. Smith, A. (1776). An inquiry into the nature and causes of the wealth of nations. In E. Cannan (Ed.) (1937). New York: The Modern Library, Random House.
110. Tomiyama, K. (1985). Two-stage optimal control problems and optimality conditions. *Journal of Economic Dynamics and Control, 9*, 317–337.
111. Tornell, A. (1997). Economic growth and decline with endogenous property rights. *Journal of Economic Growth, 2*(3), 219–257.
112. Tornell, A., & Lane, P. (1999). The voracity effect. *The American Economic Review, 89*, 22–46.
113. Traeger, C. (2010). *Subjective risk, confidence, and ambiguity* (CUDARE Working Paper 1103). University of California, Berkeley.
114. Tsur, Y., & Zemel, A. (1996). Accounting for global warming risks: Resource management under event uncertainty. *Journal of Economic Dynamics and Control, 20*(6–7), 1289–1305.
115. Tsur, Y., & Zemel, A. (1998). Pollution control in an uncertain environment. *Journal of Economic Dynamics and Control, 22*, 961–975.
116. Tsur, Y., & Zemel, A. (2006). Welfare measurement under threat of environmental catastrophes. *Journal of Environmental Economics and Management, 52*(1), 421–429.
117. Tsur, Y., & Zemel, A. (2008). Regulating environmental threats. *Environmental & Resource Economics, 39*(3), 297–310.
118. Tsur, Y., & Zemel, A. (2016). Policy tradeoffs under risk of abrupt climate change. *Journal of Economic Behavior & Organization, 132*, 46–55.
119. Tsur, Y., & Zemel, A. (2017). Steady-state properties of multi-state economic models. *The Canadian Journal of Economics, 50*, 506–521.
120. van der Ploeg, F. (2010). *Rapacious resource depletion, excessive investment and insecure property rights* (CESifo Working Paper No 2981). University of Munich.
121. van der Ploeg, F. (2014). Abrupt positive feedback and the social cost of carbon. *European Economic Review, 67*, 28–41.
122. van der Ploeg, F. (2018). Political economy of dynamic resource wars. *Journal of Environmental Economics and Management, 92*(C), 765–782.
123. Vedenov, D. V., & Miranda, M. J. (2001). Numerical solution of dynamic oligopoly games with capital investment. *Economic Theory, 18*(1), 237–261.
124. Veliov, V. M. (2005). On the effect of population heterogeneity on dynamics of epidemic diseases. *Journal MathBiology, 51*(2), 123–143.
125. Wagener, F. O. O. (2003). Skiba points and heteroclinic bifurcations, with applications to the shallow lake system. *Journal of Economic Dynamics and Control, 27*, 1533–1561.
126. Wirl, F., & Feichtinger, G. (2005). History dependence in concave economies. *The Journal of Economic Behavior and Organization, 57*, 390–407.
127. Wirl, F. (2016). Indeterminacy and history dependence of strategically interacting players. *Economics Letters, 145*, 19–24.
128. Wood, A. D., Mason, C. F., & Finnoff, D. (2016). OPEC, the seven sisters, and oil market dominance: An evolutionary game theory and agent-based modeling approach. *Journal of Economic Behavior & Organization, 132*, 66–78.

129. Yanase, A., & Long, N. V. (2019). *Trade cost and strategic investment in infrastructure in a dynamic open economy*. Paper presented at the Australasia Trade Workshop: RMIT University, Melbourne.

Chapter 2
Institutional Change, Education, and Population Growth: Lessons from Dynamic Modelling

Gustav Feichtinger, Andreas Novak, and Franz Wirl

2.1 Introduction

This is one of the first papers that links population growth, education (one of the key factors affecting population growth, in particular, the education of girls), and institutional change (which is more specific than conflicts) within dynamic optimization models. It is motivated by political events like the Arab Spring and by papers (and presentations) of Raouf Boucekkine with different co-authors, in particular, Boucekkine et al. [7, 8]. Although the proposed models are far from trivial involving two (and more) states and in some cases two stages, we are able to characterize the inter-temporal policies. Therefore, our theoretical analysis is complementary to recent empirical papers like Acemoglu et al. [1] on population growth and conflict, and Boucekkine et al. [9] on education and illiberalism.

Of the different explanations of an uprising, we mention here two complementary ones. Kuran's theory of preference falsification, e.g., in Kuran [14], the decisions of individuals are addressed: Individuals hide their true preference for the opposition if this is individually opportune and any uprising or revolution requires that the support for the opposition surpasses a critical threshold. This theory can explain that

G. Feichtinger (✉)
Institute of Statistics and Mathematical Methods in Economics,
Operations Research and Control Systems, Vienna University of Technology, Wiedner Hauptstr. 8-10/E105-04, 1040 Wien, Austria
e-mail: gustav.feichtinger@tuwien.ac.at

A. Novak · F. Wirl
Faculty of Business, Economics and Statistics, University of Vienna,
Oskar Morgenstern Platz 1, 1090 Wien, Austria
e-mail: andreas.novak@univie.ac.at

F. Wirl
e-mail: franz.wirl@univie.ac.at

J. L. Haunschmied et al. (eds.), *Dynamic Economic Problems with Regime Switches*, Dynamic Modeling and Econometrics in Economics and Finance 25,
https://doi.org/10.1007/978-3-030-54576-5_2

revolutions and the subsequent changes of political leaders are often an event that surprises even the experts. Examples abound, from the French to the Russian and the Islamic (in Iran) revolution, the collapse of socialism in 1989 and more recently the Arab Spring in 2011 of which some (e.g., [11]) predict a second coming, e.g., with reference to the ongoing protests in Algeria. Although this is a very important and plausible explanation, we depart from a second and complementary approach that addresses the issue of revolutions from the perspective of a ruling elite and its possible voluntary power handover with theoretical reference to the American sociologist Lipset and a recent book of Albertus and Menaldo [5]. This second route has been proposed in Boucekkine et al. [7, 8]. They assume that an elite has access to a rent from selling abroad a resource and also at home by providing a vital input for domestic production in order to explain events like the Arab Spring that started in Tunisia in 2011 and are continuing until today in Algeria. Other potential applications are to past regime changes in South Africa, Chile (of Pinochet), maybe Russia, and the recent changes that Muhammad bin Salman, clearly a member of the ruling elite, initiated in Saudi Arabia.

The papers, Boucekkine et al. [7, 8], are our starting point. Both consider an economy with constant capital stock and population ruled by an elite. The elite has access to the revenues from selling a resource, at home and abroad. Gross domestic product is the output of two variable inputs (aside from constant capital and labour): the domestic resource use (a flow) and human capital (a stock build up by investment into education). Although the elite's concern is its own level of consumption, it invests into human capital, which expands output, raises wages, and also the population's demand for its share of the resource rent. A central assumption is that the elite can determine the transition and can retain some stakes under the future (democratic?) regime. Arithmetically, the elite solves an optimal control problem, possibly involving two stages before and after the handover. Although we follow this approach too, the role of elites is seen differently by different authors. It is not entirely different from Acemoglu and Robinson [3] who argue that "democracy consolidates when elites do not have strong incentive to overthrow it". Recently, Guriev and Treisman [13] argue that an informed elite is in opposition and constrains the actions of recently emerging autocrats within formal democracies. Similarly, Boucekkine et al. [9] observe that rising educational levels are incompatible with illiberalism and thus increase the probability of a regime change.

Our first extension is to address how population growth affects and constrains an elite's policies. This extension is crucial from an applied perspective, because population growth is a major force and even a threat to any elite. The reason is that many young people, in particular, too many 3rd and 4th sons without any economic and social perspective are prone to revolt. Acemoglu et al. (in press) is an empirical confirmation on how the drop in morbidity since 1940 (due to breakthroughs in medicine, hygiene, and fights against malaria, which are all exogenous for developing countries) led to population growth, which in turn had sizeable effect on civic conflicts. While population declines in the industrialized countries, it grows dramatically in most Sub-Saharan (doubling within a few decades) and Arab countries of which Jordan, Oman, Kuwait, Bahrain, United Arab Emirates, and Qatar have the world's highest population growth for 2010 over 2005 (of course, the huge annual growth rates of 12 and 13% per annum for the latter two includes migration). Indeed, the ignorance of population growth is, according to Lianos [15], 'the elephant in the living room' in economics (in his case, in environmental context, but we think much beyond that). The second is the modification of the optimistic assumption that the resource is used as a productive input for domestic production. Instead, its use is by and large in the form of cheap, actually absurdly cheap, petrol, which will lead to enormous political problems when prices must be raised eventually as the recent revolts in Ecuador (following increasing petrol prices to still moderate levels, $ 2.39 per gallon, see The Economist [21]) and in Iran (but from absurdly low levels below the price of bottled water) document. Even accounting for refining, petrochemicals, and the many airlines in the Gulf, we doubt that these industries deliver a profit based on a genuine comparative advantage because of the need for hiring foreign personnel (engineers from the West, workers from the Indian subcontinent, cabin crews and pilots from all over the world, etc.). Instead, they just benefit from the cheap input (natural gas, oil, and kerosene) which were presumably better sold abroad, compare Ghoddusi et al. [12].

The paper consists of a sequence of models, some of them only briefly sketched. The upshot of our analysis is that the elite's inter-temporal optimization problems do not have a long-run solution (i.e., the canonical equations implied by the, also sufficient, first-order optimality conditions do not lead to a saddle-point stable steady state) in most cases, including variants close to the ones suggested in Boucekkine et al. [7, 8]. The extension considered in this paper, population growth, renders (optimal) control by the elite in the long run even less possible unless the elite heavily penalizes population. Therefore, it is implicitly necessary to account for a second stage after the elite hands the power over to the 'people' unless one subscribes to the cynical description of the elite just in the above sentence. The elite can choose the time when to hand over to a future government and/or leaves the country. Given the impossibility that any kind of ruling, by elites or kings, can survive under the assumed constraints, it is crucial as to what are their stakes in a future government or in exile? Therefore, an elite facing the constraints addressed in this paper will only provide for a handover if it has an 'after life' after the handover. However, it is questionable nowadays whether an elite can accumulate resources outside the country for the exile. For example, *The Economist* (Africa's money launderers, 12 October 2019, p38) reports that anti-corruption campaigners make the stashing of illicit wealth, such as exercised by Sani Abacha of Nigeria and Bokassa of Central Africa, much harder nowadays. This can have unintended consequences as dictators and elites have only the option to 'take the money and run' as Adelman hypothesized about OPEC rulers. Or formulated positively, the elite chooses to behave well if that allows it to retain some benefits after a negotiated handover. According to Guriev and Treisman [13] describing how 'human' autocrats became in the recent decade (political killings and the number of political prisoners declined substantially), it is the elite (the informed ones, say in Russia) that opposes the autocrats and limits their actions. However, given the formal setup of an inter-temporal optimization problem, our models are, at least to some extent, also applicable to a ruler facing an informed elite.

The model departing from Boucekkine et al. [7, 8] and its extensions and variants are introduced and analysed in Sect. 2.2. The importance and even the necessity of a second stage arises due to the lack of (saddle-point) stable long-run solutions. Therefore, Sect. 2.3 analyses a two-stage problem: The elite rules during the first phase, uses a resource rent for consumption, investment (into education), and the build-up of foreign assets. In the second phase, the elite lives in exile from the assets transferred abroad during the period of ruling.

2.2 Models and Implications

2.2.1 *No Population Growth*

The papers Boucekkine et al. [7, 8] consider an elite having access to the revenues from resource sale. The elite uses the revenues from sales (export, domestic) for own consumption, for financing education, for subsidizing the domestic use of the resource, and directly for transfers to the population (e.g., including little or no taxes). Gross domestic product (Y) is the output of the inputs of physical and human capitals and of the domestic resource use. Our first modification is a simplification that ignores the contribution of domestic resource use for economic activity, because driving large SUVs through the desert provides little value added. Therefore, total output (Y) is given by the production function F with only the two inputs of physical (K) and human (H) capitals,

$$Y = F(K, H) = Pf(k, h).$$

Introducing per capita terms, P denoting the population (capital letters refer to aggregate and small letters to per capita values),

$$k := \frac{K}{P}, h := \frac{H}{P}; \tag{2.1}$$

and fixing per capita capital k (as implicitly in Boucekkine et al. [8] and [7]) yields:

$$y = \frac{Y}{P} = f(k, h) = a\varphi(h) = ah^{\alpha}, \ a := Ak^{1-\alpha}.$$

Assuming, $\alpha = 1$ (i.e., ah-technology in analogy to the AK-framework, see Rebelo [17]), the equilibrium wage is

$$w = y. \tag{2.2}$$

El-Mattrawy and Semmler [10] find a large Solow residual for Egypt and find that a substantial (but not the entire part) can be explained by human capital.

Elite and rulers own a resource that gives the elite access to the revenues R, e.g., oil revenues. However, also aid from governments and NGOs could be the source of the elite's rent, which broadens the applicability of the model. Indeed many rulers and their elite turned into kleptocrats pocketing aid money instead of using it for development, e.g., Acemoglu et al. [2] mention Democratic Republic of the Congo (Zaire) under Mobutu Sese Seko, the Dominican Republic under Rafael Trujillo, Haiti under the Duvaliers, Nicaragua under the Somozas, Uganda under Idi Amin, Liberia under Charles Taylor, and the Philippines under Ferdinand Marcos, not to forget Jean-Bedel Bokassa of Central Africa who "would slip his guest diamonds to thank him for France's support" according to The Economist [20].

The elite spends the rent R for own consumption C, the maximization of which is its objective (using the constant discount rate, $\rho > 0$, and for concreteness logarithmic utility),

$$\max \int_0^\infty \exp(-\rho t) \ln C(t)\, dt, \tag{2.3}$$

for transfers to the workers Θ and $\theta := \Theta/P$ per capita, and for education E , $e := E/P$ per capita,

$$R = C + \Theta + E. \tag{2.4}$$

Human capital (H) follows the standard capital accumulation rule (also used in Boucekkine et al. [7, 8]),

$$\dot{H} = \beta E - \delta H, \;\; H(0) = H_0 \text{ given}, \; H \geq 0 \tag{2.5}$$

because workers (=domestic population except for the elite) cannot invest in education. However, they are not entirely passive, because they will revolt if their income, consisting of their wage (w) plus the handout θ, is considered to be too low. Therefore, in order to deter a revolt, the following inequality,

$$w + \theta \geq z + \tau(h), \;\; \tau' > 0, \tag{2.6}$$

must be satisfied; z denotes the subsistence level and $\tau(h)$ a threshold, which is increasing with respect to education because it fosters awareness (as also Guriev and Treisman [13] and Boucekkine et al. [9] stress). Therefore, the inequality (2.6) must hold as long as the elite is in charge and the rent R must be sufficiently large, at least $R > zP$, to allow for positive consumption at all, $C > 0$. In the most simple version of a linear endogenous threshold level (again as in Boucekkine et al. [7, 8]),

$$\tau(h) = bh,$$

and assuming an ah-technology, the constraint (2.6) becomes in terms of the minimal per capita handout,

$$\theta \geq z + (b - a)\,h,\ \ b > a. \tag{2.7}$$

This above-assumed inequality states that the elite has to surrender some of its rent and that this amount per capita is increasing with human capital.

Remark 2.1 If $b < a$, the elite could extract taxes after investing into education in order to finance its consumption, hence the assumption, $b > a$. This possibility that the elite can tax its people, i.e., $\theta < 0$, arises also for a concave $\tau\,(h)$ if investments into education render human capital sufficiently productive and at the same time the population relatively docile, $h > \hat{h} : z + \tau\,(h) - ah = 0$, a situation that may apply to China's elite today. In fact, we wanted to rule this possibility out a priori, because we do not think that China's unique experience (e.g., decades of a one-child policy) applies to the countries we have in mind. Furthermore, if taxation became feasible, the elite would prefer a higher population.

Remark 2.2 As mentioned in the introduction and due to our point of departure from Boucekkine et al. [7, 8], we refer to the elite as the decision maker although it could be a much smaller set of an autocrat (to use the term of Guriev and Treisman [13]) and his inner circle.

Proposition 2.1 *Considering this slightly simplified version of Boucekkine et al. [7, 8], and still a constant population, P is constant, no long-run interior solution exists for the elite's inter-temporal optimization problem (2.3)–(2.7), i.e., the canonical equations implied by the (also sufficient) first-order conditions do not converge to an interior steady state.*

The economic intuition explaining this lack of a sustainable rule by the elite is that the long-run objective, i.e., substituting the steady-state value of human capital, $H = \beta E/\delta$, and accounting for the budget constraint, i.e., (2.7), yields a strictly concave objective,

$$\max_{E} U\,(E) := \ln\left(R - (b - a)\,\frac{\beta}{\delta}E - E - zP\right),$$

but one, which has no generic interior solution since the corresponding first-order condition (foc), $U' = 0$, cannot be met since $U' < 0$ at $E = 0$ and U' is declining.

2.2.2 *Population Growth*

The first extension is the account for exogenous population growth (at the constant rate g). Given a finite rent and a growing population, it is a no-brainer that this process cannot go on forever and that the elite can at best enjoy its rent for a limited time.

Indeed, *the elite's optimal policy is to not invest into education and to offer only the transfers (as long as feasible),* $R > \Theta = zP + (b - a) H$, *that just avoid and thereby delay the revolution.* There are examples, e.g., 'Tsarist Russia' and the elites in some developing countries enjoy a luxury life but spend little for education, e.g., Mobuto Sese Seko and some of the others mentioned in the introduction.

The second addition is that education lowers population growth (g denotes the population growth rate),

$$\dot{P} = g(h) P, \ P(0) = P_0, \ g' < 0, \ P \geq 0, \tag{2.8}$$

$$g(h) = \gamma - \pi h, \ \gamma > 0, \bar{h} := \frac{\gamma}{\pi}. \tag{2.9}$$

This link between population growth and education is not only an assumption but an empirical regularity, in particular, educating girls lowers fertility rates, even drastically in some countries. The assumption of a linear relation is chosen for reasons of simplicity. It implies a stationary population at a unique level of individual education denoted by $\bar{h}$ in (2.9). As a consequence, the elite has to pass between Scilla (an uneducated and growing population which cannot be fed from some future point onwards from the constant rent R) and Charybdis (educating the population sufficiently in order to achieve a stationary population but which will demand larger individual handouts).

Given the exogenous rent R, then $\bar{P}$ is the maximum level of population that can be educated, which is further diminished to $\hat{P}$ if accounting for the need for transfers according to (2.7),

$$\hat{P} := \frac{\beta\pi R}{\beta\pi z + \gamma(\beta(b-a) + \delta)} < \bar{P} := \frac{\beta\pi R}{\delta\gamma}.$$

Of course, only initial conditions below $\hat{P}$ make sense.

One possibility is to add a soft constraint: the elite values the subjective welfare of its people (their income minus their education-dependent demands), or respectively, accounts for the costs from demonstrations, uprisings, and the risk of a revolution of an unhappy population. However, we bypass this (anyway also not allowing for an interior long-run solution) and replace the constraint in (2.7) about avoiding uprisings or revolutions with equality. This yields the objective,

$$\max_{e(t)\geq 0} \int_0^\infty \exp(-\rho t) \ln(R - (b-a) H - eP - zP) dt, \tag{2.10}$$

with the single control, per capita expenditures for education (e), and the two states H and P. Even this problem does not allow for an optimal long-run rule by the elite:

Proposition 2.2 *No steady state exists for the optimal long-run interior solution, $e > 0$, for the optimal control problem with the objective (2.10) and the state differential equations (2.5) and (2.8).*

Proof Setting up

$$\mathcal{H} = \ln(R - (b-a)\,H - eP - zP) + \lambda\,(\beta eP - \delta H) + \mu\,(\gamma P - \pi H) \quad (2.11)$$

the first-order conditions are

$$\begin{aligned} \mathcal{H}_e &= \beta\lambda P - \frac{P}{R - (b-a)\,H - eP - zP} \Longrightarrow \\ e^* &= \max\left\{\frac{R - (b-a)\,H - zP}{P} - \frac{1}{\beta\lambda P}, 0\right\}, \end{aligned} \quad (2.12)$$

$$\dot{\lambda} = (\rho + \delta)\,\lambda + \pi\mu + \frac{b-a}{R - (b-a)\,H - eP - zP}, \quad (2.13)$$

$$\dot{\mu} = (\rho - \gamma)\,\mu + \frac{z+e}{R - (b-a)\,H - eP - zP} - e\beta\lambda. \quad (2.14)$$

Note that $\mathcal{H}_{ee} < 0$ for both solutions in (2.12). Assuming (indirectly) an interior solution of e^* and substituting it into the state and costate equations yields the canonical equation system,

$$\dot{H} = \beta\,(R - (b-a)\,H - zP) - \frac{1}{\lambda} - \delta H, \quad (2.15)$$

$$\dot{P} = \gamma P - \pi H, \quad (2.16)$$

$$\dot{\lambda} = (\rho + \delta + \beta\,(b-a))\,\lambda + \pi\mu, \quad (2.17)$$

$$\dot{\mu} = (\rho - \gamma)\,\mu + \beta z\lambda. \quad (2.18)$$

Solving first for steady states of the costates yields for the linear equation system (2.17) and (2.18),

$$\lambda = \mu = 0, \quad (2.19)$$

as the unique solution. This rules out the interior solution and thus implies $e^* = 0$ in the long run. ■

Remark 2.3 This finding extends to a concave threshold, e.g., $\tau\,(h) = \tilde{b}\sqrt{h}$, as well as to a convex specification of τ. The reasons, at least the arithmetical ones, are (i) τ appears only as a negative element in consumption and (ii) utility is logarithmic. Therefore, the specification of τ affects the above interior solution of e^* only indirectly (a different term is subtracted from consumption), but plays no role in the adjoint equations after substituting the optimal control. Thus, it allows again only for the trivial solution (2.19) for the steady states of the costate differential equation system and thus implies the boundary solution, $e^* = 0$ in the long run.

Why is it uneconomical (from the elite's perspective) to sustain education in the long run given this more or less straightforward optimal control model with a concave objective? The reason is that the implied static/stationary objective (replacing H by its steady-state value conditional on P, however, the same holds for carrying out the analysis with respect to H),

$$\max_P U(P) := \ln\left(R - (b-a)\frac{\gamma}{\pi}P - \frac{\gamma\delta}{\beta\pi}P - zP\right), \tag{2.20}$$

does not allow for an interior solution with the choice of population $P \geq 0$, as instrument, because

$$U' = -\frac{\beta\left((b-a)\gamma + z\pi\right) + \delta\gamma}{\beta\pi R - \beta\left((b-a)\gamma + z\pi\right)P - \gamma\delta P} < 0$$

since U' is declining ($U(P)$ is a concave objective) and $U'(0) < 0$. Now what is optimal in the long run? The boundary solution, $e^* = 0$, implies the state dynamics,

$$\dot{H} = -\delta H,$$
$$\dot{P} = \gamma P - \pi H,$$

which have the origin as the only steady state. This steady state is a saddle point (the eigenvalues are γ and $-\delta$), which is reached along the saddle, $H = (\gamma + \delta)P/\pi$. Therefore, the following policy results from joining the interior and the boundary ones.

Proposition 2.3 *Given a feasible solution, i.e., a sufficient rent and a not too large initial population, the objective of the interior policy $e^* > 0$ from (2.12) is to drive the states to the saddle, $H = (\gamma + \delta)P/\pi$, at which the decline of both states starts and continues, $H \to 0$ and $P \to 0$, until the elite can spend the entire rent on own consumption.*

This confirms the above finding from the analysis of the elite's stationary objective (2.20), namely, that the elite's objective is to drive the size of its population down to zero in order to keep the entire, then also uncontested, rent. Investment into education is only an instrument to achieve this. Of course this drastic policy stresses first, admittedly, the limits of our model. Secondly, it highlights again why the papers of Boucekkine et al. [7, 8] have to rely on a theory of a handover by the elite. Thirdly, our extension for population growth hardens this task, and maybe, also the hearts of the elite.

2.2.3 Optimal Timing of Abdication

Since there need not be either a feasible nor an optimal solution running up to infinity, the elite has to, and therefore will, exercise the option to quit in finite time (T) in particular if facing unstoppable population growth. From the above we know that quitting will be definitely optimal if population exceeds $\hat{P}$ introduced above since there is nothing then left for the elite for consumption. The following optimization problem captures this decision problem of an elite that considers quitting if the going gets rough and too little is left for consumption,

$$\max_{e(t)\geq 0,T} \int_0^T \exp(-\rho t) \ln(R - (b-a) H - eP - zP)dt, \tag{2.21}$$

$$\dot{H} = \beta e P - \delta H, \; H(0) = H_0, \; H(T) \text{ free}, \tag{2.22}$$

$$\dot{P} = g(h) P, \; P(0) = P_0, \; P(T) \text{ free}. \tag{2.23}$$

The change to finite and optional terminal time changes nothing in terms of the first-order condition (2.11)–(2.14) except for adding the boundary conditions to the costates,

$$\lambda(T) = 0,$$
$$\mu(T) = 0,$$

which imply immediately that the boundary policy must apply for $t \to T$, i.e., no investment into education, $e^*(t) = 0$ for $t \to T$. Assuming the boundary strategy and applying the transversality conditions, the condition for optimal stopping,

$$\mathcal{H} = \ln(R - (b-a) H - eP - zP) + \lambda(\beta e P - \delta H) + \mu(\gamma P - \pi H),$$

$$\mathcal{H}(T) = \ln(R - (b-a) H(T) - zP(T)) = 0,$$

i.e., terminal consumption utility is zero.

Proposition 2.4 *Even allowing for optimal stopping does not allow for an interior optimal policy at least close to the optimal termination date. Therefore, the optimal policy of the elite is no investment into education for $t \to T$ and to leave when utility from consumption turns zero due to the necessary transfers to the growing population.*

Apparently, this outcome seems to be the reason why a salvage value (in current value terms), $S = \sigma H(T)$, is introduced in the papers of Boucekkine with different co-authors. This means economically and politically that the elite must have some stakes in the future regime in order to invest in at least some development, here human capital. Accounting for population suggests that less remains for the elite if handing over to a larger population, e.g.,

$$S = \sigma \frac{H}{P}, \tag{2.24}$$

is the simplest version of such a salvage function. This extension changes only the objective in (2.21)–(2.23),

$$\max_{e(t)\geq 0,T} \int_0^T \exp(-\rho t) \ln(R - (b-a) H - eP - zP)dt + \exp(-\rho T)\, \sigma \frac{H(T)}{P(T)}. \tag{2.25}$$

However, we skip this analysis because we consider an explicit two-stage framework in the next section.

2.2.4 *Penalizing Larger Population*

In order to allow long-run ruling by the elite (and to apply standard techniques), we (have to) subtract a penalty, e.g., κP, from the elite's objective,

$$\max_{e(t)\geq 0} \int_0^\infty \exp(-\rho t) \left[\ln(R - (b-a) H - eP - zP) - \kappa P\right] dt. \tag{2.26}$$

This penalty is not only introduced for the above formal reason but also in order to account for conceivable economic and political constraints: a larger population is more difficult and also more costly to contain and to suppress by any elite.

Proposition 2.5 *Low discount rates, more precisely, discount rates below the maximal population growth rates, $\rho < \gamma$, ensure a stable long-run policy. Even higher discount rates, $\rho > \gamma$, allow for stable outcomes and positive population but only for large rents R. Comparative statics are as expected: a higher penalty (κ) lowers population (for $\rho < \gamma$) while higher rents (R) as well as a higher productivity (a, at least for $\gamma > \rho$) allow for a larger stationary population. The effect of higher discounting is negative (as expected) iff*

$$\beta(b-a) + \delta + 2\rho - \gamma > 0.$$

The qualitative implications on human capital are the same since $H_\infty = \gamma P_\infty / \pi$.

Proof Setting up the Hamiltonian (again using the symbol $\mathcal{H}$),

$$\mathcal{H} = \ln(R - (b-a)\,H - eP - zP) - \kappa P + \lambda\,(\beta e P - \delta H) + \mu g\,(H/P)\,P, \tag{2.27}$$

and deriving the first-order optimality conditions, the optimal control remains as in (2.12), and the costate equations are

$$\dot{\lambda} = (\rho + \delta)\,\lambda + \pi\mu + \frac{b-a}{R - (b-a)\,H - eP - zP},$$
$$\dot{\mu} = (\rho - \gamma)\,\mu + \kappa + \frac{z+e}{R - (b-a)\,H - eP - zP} - e\beta\lambda.$$

The following canonical equation system is derived for an interior solution, $e^* > 0$ from (2.12),

$$\dot{H} = \beta\,(R - (b-a)\,H - zP) - \frac{1}{\lambda} - \delta H, \tag{2.28}$$
$$\dot{P} = \gamma P - \pi H, \tag{2.29}$$
$$\dot{\lambda} = (\rho + \delta + \beta\,(b-a))\,\lambda + \pi\mu, \tag{2.30}$$
$$\dot{\mu} = \kappa + (\rho - \gamma)\,\mu + \beta z\lambda. \tag{2.31}$$

This system has a unique steady state,

$$H_\infty = \frac{(\gamma - \rho)\,(\rho + \delta + \beta\,(b-a)) + \beta\pi\,(\kappa R + z)}{(\gamma\,(\delta + \beta\,(b-a)) + \beta\pi z)\,\kappa}\frac{\gamma}{\pi}, \tag{2.32}$$
$$P_\infty = \frac{(\gamma - \rho)\,(\rho + \delta + \beta\,(b-a)) + \beta\pi\,(\kappa R + z)}{(\gamma\,(\delta + \beta\,(b-a)) + \beta\pi z)\,\kappa}, \tag{2.33}$$
$$\lambda_\infty = -\frac{\kappa\pi}{(\gamma - \rho)\,(\rho + \delta + \beta\,(b-a)) + \beta\pi z}, \tag{2.34}$$
$$\mu_\infty = \frac{\kappa\,(\rho + \delta + \beta\,(b-a))}{(\gamma - \rho)\,(\rho + \delta + \beta\,(b-a)) + \beta\pi z}. \tag{2.35}$$

Therefore, positive steady states, $P_\infty > 0$ and $H_\infty > 0$, result if either $\gamma > \rho$ such that the shadow price of human capital is negative ($\lambda_\infty < 0$), or also if $\rho > \gamma$ but then only if $\beta\kappa\pi R$ is sufficiently large.

The comparative static properties of the stationary population (and thus stationary human capital) follow from elementary partial differentiation of (2.33), for example,

$$\frac{\partial P_\infty}{\partial \kappa} = -\frac{(\rho+\delta)(\gamma-\rho)+\beta((b-a)(\gamma-\rho)+z\pi)}{\kappa^2(\delta\gamma+\beta(\gamma(b-a)+\pi z))} < 0 \text{ for } \gamma > \rho,$$
$$\frac{\partial P_\infty}{\partial R} = \frac{\beta\pi}{\gamma(\delta+\beta(b-a))+\beta\pi z} > 0,$$
$$\frac{\partial P_\infty}{\partial \rho} = -\frac{\beta(b-a)+(\delta+2\rho-\gamma)}{(\delta\gamma+\beta(\gamma(b-a)+\pi z))\kappa} < 0 \text{ unless } \gamma > \beta(b-a)+\delta+2\rho,$$
$$\frac{\partial P_\infty}{\partial a} = \frac{(\gamma\rho(\gamma-\rho)+\beta\pi(\rho z+\gamma\kappa R))\beta}{\kappa(\gamma\delta+\beta(\gamma(b-a)+\pi z))^2} > 0 \text{ at least for } \gamma > \rho.$$

The Jacobian,

$$J = \begin{pmatrix} -\beta(b-a)-\delta & -\beta z & 1/\lambda_\infty^2 & 0 \\ -\pi & \gamma & 0 & 0 \\ 0 & 0 & \beta(b-a)+\rho+\delta & \pi \\ 0 & 0 & \beta z & (\rho-\gamma) \end{pmatrix}, \tag{2.36}$$

has the (four) eigenvalues,

$$ev_{12} = \frac{1}{2}\left(-\xi \pm \sqrt{\xi^2+4\zeta}\right),$$
$$ev_{34} = \frac{1}{2}\left(2\rho+\xi \pm \sqrt{\xi^2+4\zeta}\right),$$

in which

$$\xi := \beta(b-a)+\delta-\gamma,\ \zeta := \delta\gamma+\beta((b-a)\gamma+z\pi) > 0.$$

Saddle-point stability results if two and only two of the four eigenvalues are negative. The first and the third eigenvalues must be positive. The second is for sure negative (even if $\xi < 0$), and the fourth iff,

$$\begin{aligned}(2\rho+\xi)^2 &< \xi^2+4\zeta \iff \rho^2+\rho\xi < \zeta \\ &\iff (\rho-\gamma)(\rho+\delta+\beta(b-a)) < \beta z\pi\end{aligned}$$

Assuming $\rho < \gamma$ then the unique steady state[1] must be a saddle point (the second and fourth eigenvalues are negative and the other two are positive). The case of $\rho > \gamma$ still allows for stability, but only if the coverage of the subsistence level is sufficiently large, more precisely,

$$z > \frac{(\rho-\gamma)(\beta(b-a)+\delta+\rho)}{\beta\pi}.$$

[1]Of course, in order to be meaningful, the steady states of population and human capital must be positive. The stability properties extend arithmetically to negative but meaningless solutions.

If this condition is not met, the unique steady state turns unstable that can be only reached along a one-dimensional manifold in the state space. ■

The stability finding in Proposition 2.5 is surprising in the light of Proposition 2.2: *the implicit penalty for population of* (zP) *subtracting from the elite's consumption in order to cover subsistence consumption of the population does not lead to a stationary outcome. However, the minor addition of an explicit and linear penalty renders a long-run stable outcome, and this for sure for discount rates below the maximal growth rate of population.* At least arithmetically, we can explain this. The penalty parameter κ appears now in the costate equation (2.31) which allows for steady states of the costates different from zero and therefore for a long-run interior policy.

Therefore, an elite, which is not too impatient but sufficiently cynical about its own population, is able to stir its course towards a stationary population and keep thereby its rule. Indeed, the assumption that the elite accounts for endogenous population growth as in (2.8) stipulates implicitly a far-sighted elite. However, this assumption may be not applicable to many resource-dependent countries as the ruling elites 'must take the money and run', as Adelman quipped about OPEC countries (see also Wirl [22] about positive objectives of OPEC politicians).

2.3 Adding a Second Stage: Going Abroad

Given the difficulties or even the infeasibility of keeping the power in the long run and the difficulties to obtain at least some stakes in the future (democratic) regime, dictators and a small part of the elite have the option, and almost always use it, to amass money abroad (A) and to leave if the going gets rough. For example, the Iranian revolution in 1978 led many people linked to the Shah regime to emigrate and not only the Shah but also his family left Iran. Idi Amin of Uganda did the same when leaving for exile in Saudi Arabia.

After leaving the country with no stakes left at 'home', the elite or the ruler exercises a standard Ramsey program:

$$V(A_0) := \max_{C} \int_{0}^{\infty} e^{-\rho t} \ln C dt, \tag{2.37}$$

subject to the budget constraint (A denotes the assets held in foreign accounts), the initial condition, and the no-Ponzi-game condition

$$\dot{A} = rA - C, \qquad A(0) = A_0, \qquad \lim_{t\to\infty} A(t) e^{-rt} \geq 0, \tag{2.38}$$

in which $r < \rho$ is the exogenously given interest rate earned in capital markets and A_0 the money accumulated outside the country during the periods of ruling. The solution is well known (see, e.g., Barro and Salai-Martin [6]), $A \to 0$ and $C \to 0$ (thus the corresponding costate $\nu \to \infty$) all asymptotically given the canonical equations of the second stage,

$$\dot{A} = rA - \frac{1}{\nu},$$
$$\dot{\nu} = (\rho - r)\,\nu.$$

Therefore,

$$A(t) = A_0 e^{(r-\rho)t},\;\; C(t) = \rho A_0 e^{(r-\rho)t},\;\; \nu(0) = \frac{1}{A_0 \rho}, \tag{2.39}$$

in which the initial levels of consumption, state, and costate are determined by the no-Ponzi game condition. This explicit solution allows to determine the value function by integrating over the net present value of future utility,

$$V(A_0) = \int_0^\infty e^{-\rho t} \ln(C(t))\, dt = \int_0^\infty e^{-\rho t} \ln\left(\rho A_0 e^{(r-\rho)t}\right) dt = \frac{\ln(\rho A_0)}{\rho} + \frac{r-\rho}{\rho^2}. \tag{2.40}$$

Accounting for the value obtainable from the second stage yields the optimal control problem,

$$\max_{e \geq 0, I, T} \int_0^T \exp(-\rho t) \ln(R - (b-a)H - eP - zP - I)dt$$
$$+ \exp(-\rho T)\, V(A(T)), \tag{2.41}$$
$$\dot{H} = \beta e P - \delta H,\;\; H(0) = H_0, \tag{2.42}$$
$$\dot{P} = g\left(\frac{H}{P}\right) P,\;\; P(0) = P_0, \tag{2.43}$$
$$\dot{A} = rA + I,\;\; A(0) = 0, \tag{2.44}$$

and all terminal values of the states are free. While expenditures for education must be non-negative, investment can be positive or negative. If $I > 0$, the rulers move money abroad if $I < 0$, then they draw on their foreign account or take foreign credits. That is, from the very beginning, the rulers must take into account their definite finite and often rather short time of ruling. Therefore, they start transferring money subject to the political constraint of avoiding an uprising. Given population growth, the elite will presumably transfer money to its foreign accounts from the beginning and may be even larger at the beginning given a growing population.

Proposition 2.6 *Assuming a feasible solution, the elite's optimal policy is*

(i) to stop investing into education (presumably long) before leaving ($e^* = 0$ for $t \to T$),

(ii) *to stay as long as the resource rent R contributes to pay for the elite's consumption (i.e., as long as* $R > \Theta$*) and leaves when this contribution vanishes* $(R - \Theta \to 0 \text{ for } t \to T)$*,*
(iii) *to enjoy continuity in consumption across the two stages, which means that the consumption towards the end is financed from revenues abroad, and,*
(iv) *but already an implication from (iii), to stop investing abroad (long) before leaving, i.e., negative investment, more precisely,* $I(T) = -C(T)$.

Proof Defining the Hamiltonian,

$$\begin{aligned} \mathcal{H} = & \ln(R - (b - a) H - eP - zP - I) + \lambda (\beta eP - \delta H) \\ & + \mu (\gamma P - \pi H) + \nu (rA + I), \end{aligned} \tag{2.45}$$

the first-order conditions for solutions are as follows: The Hamiltonian maximizing condition for the expenditures for education,

$$e^* = \max \left\{ \frac{R - (b - a) H - zP - I}{P} - \frac{1}{\beta \lambda P}, 0 \right\}, \tag{2.46}$$

is similar to (2.12). It is for the second control investment (I) with the optimal level,

$$\begin{aligned} \mathcal{H}_I & = -\frac{1}{R - (b - a) H - eP - zP - I} + \nu = 0 \\ & \Longrightarrow I^* = R - (b - a) H - eP - zP - \frac{1}{\nu}. \end{aligned} \tag{2.47}$$

The costates evolve according to,

$$\dot{\lambda} = (\rho + \delta) \lambda + \mu \pi + \frac{b - a}{R - (b - a) H - e^* P - zP - I}, \tag{2.48}$$

$$\dot{\mu} = (\rho - \gamma) \mu + \frac{e^* + z}{R - (b - a) H - e^* P - zP - I} - \beta \lambda e^*, \tag{2.49}$$

$$\dot{\nu} = (\rho - r) \nu, \tag{2.50}$$

(using e^* as shortcut for the optimal control from (2.46)), and finally for optimal stopping they require,

$$\mathcal{H} = \rho V = \ln(\rho A) + \frac{r - \rho}{\rho} \text{ at } t = T. \tag{2.51}$$

Defining with

$$MU = \frac{1}{C} = \frac{1}{R - (b - a) H - eP - zP - I}$$

the elite's marginal utility of consumption (i.e., the benefit from being able to spend an additional $ for own consumption given the logarithmic utility), then

$$MU = \beta\lambda \wedge MU = \nu.$$

Starting with the second equation, the elite is indifferent between spending or saving (foreign assets, of course) this incremental $. The first condition equates marginal utility to the marginal benefit from larger human capital (λ) multiplied by the efficiency of education investments (β). As a consequence, $\nu = \beta\lambda$, i.e., the two costates differ only by the efficiency of investment into education. However, this characterization holds only along the interior solution, i.e., $e^* > 0$.

Given our assumption that the elite has access to all its assets transferred abroad, we can apply the theory of two-stage dynamic optimization, Makris [16], see also Tomiyama [18], Tomiyama and Rosanna [19], in which the two stages are linked by continuity conditions, value matching, and smooth pasting. These conditions apply only to the state carried forward, namely the assets,

$$\nu (T) = V' (A (T)) = \frac{1}{\rho A (T)},$$

and simplifies the boundary conditions of the other costates in stage 1,

$$\lambda (T) = 0 \wedge \mu (T) = 0.$$

They imply immediately that the elite spends again nothing on education before leaving,

$$e^* (t) = 0, \ t \to T.$$

Therefore, reductions in educational investments are a strong signal that elite considers leaving. The continuity conditions are

$$\nu_1 (T) = \nu_2 (T), \tag{2.52}$$

$$\mathcal{H}_1 (T) = \mathcal{H}_2 (T), \tag{2.53}$$

in which the subscripts (which can be dropped for the costate ν due to the above) refer to the two stages:

$$\begin{aligned} \mathcal{H}_1 (T) &= \ln(C (T_-)) + \nu (T) (rA (T) + I (T)), \\ C (T_-) &= R - (b - a) H (T) - zP (T), \\ \mathcal{H}_2 (T) &= \ln(C (T_+)) + \nu (T) (rA (T) - C (T_+)). \end{aligned}$$

Applying the 'smooth pasting' condition (2.52) implies $C(T_+) = C(T_-)$ and thus

$$I(T) = -C(T), \tag{2.54}$$

due to 'value matching' (2.53). ■

The optimal stopping conditions imply that the elite stays as long as the rent R can contribute to its consumption. However, it has to rely increasingly on foreign assets to pay for its consumption (negative investments, $I < 0$ for $t \to T$). Furthermore, it will smooth its consumption level across the two stages, because otherwise an arbitrage opportunity would arise. Therefore not only lacking investment in education, but increasing reliance on money from abroad signal that an elite considers leaving.

The property of continuous consumption depends crucially on some of the assumptions. It is eliminated for switching costs, since the elite's exodus is for sure costly, so that the elite will face a discontinuous drop in consumption. Furthermore, with international banks not anymore protecting accounts fed from stolen money, the above-derived strategy becomes risky as mentioned in the introduction (e.g., the quoted article in *The Economist* reports that the auction revenues of \$ 27 millions from sports cars seized from Mr. Obiang, the son of the president of Equatorial Guinea, were returned). Indeed, this exit option is only possible for rulers retaining some goodwill when leaving, i.e., what is called a golden handshake in the case of managers. Let $V(A)$ again denote the value function of the above Ramsey program and $p(H, P, A)$ the probability (or share) that the former rulers expect to keep after the revolution and when leaving. Therefore define with

$$\begin{aligned} \Pi(H, P, A) &:= p(H, P, A)\, V(A), \\ &\Pi_H > 0,\ \Pi_P < 0,\ p_A < 0 \text{ yet } \Pi_A > 0 \end{aligned}$$

the expected net present value payoff of the ruler at the moment going into exile. The assumption $\Pi_A > 0$ is for sure violated for large values of A, but accumulating large amounts such that the expected benefit declines is clearly suboptimal. This change in assumptions, does not change the above state and costate equations in their dynamics but affects the boundary conditions,

$$\begin{aligned} \lambda(T) &= \Pi_H > 0, \\ \mu(T) &= \Pi_P < 0, \\ \nu(T) &= \Pi_A > 0, \end{aligned}$$

in which case the elite may continue investing in education until the very end but consumption need not remain continuous.

2.4 Concluding Remarks

Departing from the interesting papers of Boucekkine et al. [7, 8] on how an elite might run a resource exporting economy, this paper introduced a few modifications. Firstly, we question the productive use of the resource in domestic industry and secondly, we account for population growth, which seems crucial for many countries to which such socio-politico-economic scenarios are applicable. The major upshot of the different setups considered in this paper is that an elite will have a hard time in most cases to survive. Only a far-sighted and cynical elite with negative concerns about a larger population will be able to sustain its rule in the long run and only if it can stabilize its population.

Offering the elite some stakes in the future government could induce them to take domestic issues, here modelled as investing into education, into account. In this sense, the opportunities offered in the second stage, the after life of the elite, can cast an important and positive shadow over the first phase when it rules but that is not granted, or put the other way round. Only an elite that is sufficiently far-sighted, patient, and has some stakes in the country's future could lead to a peaceful handover. However, our analysis indicates that this mitigation policy of an elite is difficult and impossible in some cases in our framework.

Partially to our own surprise, we are able to characterize the elite's optimal intertemporal policies. The policies derived from our models reveal the 'bare bone' and thus drastic intentions of a greedy elite, which are presumably masked and mitigated by additional considerations or further constraints. Therefore, extensions and improvements are not only possible but necessary in order to understand such topical political events. Obvious candidates are: broader objectives that include social benefits (however, this is more likely for autocrats worrying about their remembrance in the future books of history than for a diffuse elite); a threshold function (i.e., a well-educated and also relatively rich population raises less demands) that could explain the transition, in particular, in oil exporting countries from subsidies to taxes; accounting for strategic issues applying (dynamic) game theory (Boucekkine et al. [7] include a cooperative handover which is ultimately reduced to endogenize the salvage value); for uncertainty, e.g., using Ito processes, possibly combined with jumps, instead of deterministic differential equations. The dramatic drop in oil prices in March 2020 to almost \$20 per barrel (for Brent) from above \$60 in December 2019 is a very recent reminder of the uncertainty associated with resource revenues and also for the need for taxation in oil exporting countries. In addition, an integration of the objectives and possibilities of individuals as addressed in Kuran [14] should be part of such an analysis.

Acknowledgements All authors acknowledge the ideas brought to us from Raouf Boucekkine and his co-authors not only in their published papers but also from a stimulating presentation of Raouf (in Rabat) and the subsequent discussions including the comments from the organizer Fouad El Ouardighi and ongoing (electronic) discussions with Raouf including helpful references (e.g., Albertus and Menaldo [5] and the very recent work of Acemoglu et al. [7]). We also thank two anonymous referees for their very helpful comments and detailed suggestions.

References

1. Acemoglu, D., Fergusson, L., & Johnson, S. Population and conflict. *Review of Economic Studies, 87*(4), 1565–1604.
2. Acemoglu, D., Robinson, J. A., & Verdier, T. (2004). Kleptocracy and Divide-and-Rule: A Model of Personal Rule, Afred Marshall Lecture. *Journal of the European Economic Association, 2–3*, 162–192.
3. Acemoglu, D., & Robinson, J. A. (2006). *Economic origins of dictatorship and democracy*. Cambridge University Press.
4. Adelman, M. A. (1990). The 1990 oil shock is like the others. *Energy Journal, 11*(4), 1–14.
5. Albertus, M., & Victor, M. (2018). *Authoritarianism and the elite origins of democracy*. Cambridge University Press.
6. Barro, R. J., & Sala-i-Martin, X. (1995). *Economic growth*. Mc Graw Hill.
7. Boucekkine, R., Piacquadioy, P. G., & Prieur, F. (2019a). A Lipsetian theory of voluntary power hand-over. *Journal of Economic Behavior and Organization, 168*(2019), 269–329.
8. Boucekkine, R., Prieur, F., & Puzon, K. (2016). On the timing of political regime changes in resource-dependent economies. *European Economic Review, 85*, 188–207.
9. Boucekkine, R., Rodolphe, D., & Melindi-Ghidi, P. (2019b). *The intolerance of education for political illiberalism*. Mimeo
10. El-Mattrawy, K., & Semmler, W. (2006). *The role of education and human capital for economic growth in middle income countries: Egypt's case*. Mimeo.
11. Fahmi, G. (2019). Wahlen können gefährlich sein, NZZ from Oct. 27th, p. 5.
12. Ghoddusi, H., Moghaddam, H., & Wirl, F. (2020). Going downstream: An economical option for oil and gas exporting countries? Mimeo, University of Vienna.
13. Guriev, S., & Treisman, D. (2019). Informational Autocrats. *Journal of Economic Perspectives, 23*(4), 100–127.
14. Kuran, T. (1989). Sparks and prairie fires: A theory of unanticipated political revolution. *Public Choice, 61*, 41–74.
15. Lianos, T. P. (2020). World population: The elephant in the living room. *Real-world economics review* 91. http://www.paecon.net/PAEReview/issue91/Lianos91.pdf
16. Makris, M. (2001). Necessary conditions for infinite-horizon discounted two-stage optimal control problems. *Journal of Economic Dynamics and Control, 25*(12), 1935–1950.
17. Rebelo, S. (1991). Long-Run Policy Analysis and Long-Run Growth. *Journal of Political Economy, 99*, 500–521.
18. Tomiyama, K. (1985). Two-stage optimal control problems and optimality conditions. *Journal of Economic Dynamics and Control, 9*, 317–337.
19. Tomiyama, K., & Rossana, R. (1989). Two-stage optimal control problems with an explicit switch point dependence: Optimality criteria and an example of delivery lags and investment. *Journal of Economic Dynamics and Control, 13*, 319–337.
20. The Economist. (2019a) Catch me if you can. African kleptocrats are finding it tougher to stash cash in the West, Oct. 10th.
21. The Economist. (2019b). Ecuador's state of emergency, October 12th
22. Wirl, F. (2009). OPEC as a Political and Economical Entity. *European Journal of Political Economy, 25*, 399–408.

Chapter 3
Poverty Traps and Disaster Insurance in a Bi-level Decision Framework

Raimund M. Kovacevic and Willi Semmler

JEL Classification: C 61 · C 63 · L 10 · L 11 · L 13

3.1 Introduction

There is a long tradition of economic literature where it is argued that economic agents can be driven into long-lasting poverty traps—or even into ruin—as a result of large negative shocks or disaster events. This often involves random catastrophic losses, leading into absorbing states from which an escape is not possible. Large negative shocks can result from large income or wealth contractions, such as economic and financial melt-downs, from disasters such as earthquakes and climate-related disasters. In each case, a significant percentage of GDP, public and private capital, essential infrastructure, as well as regional damages and life losses are occurring.

As to the economic strand of literature that has studied the likelihood of countries and regions to fall into poverty traps recently, the new growth theory[1] has redirected

[1] One approach of the new growth theory views persistent economic growth arising from learning by doing, externalities in investment and increasing returns to scale. This idea had been formalized by Arrow [1] and rediscovered by Romer [29], who argues that externalities—arising from learning by doing and knowledge spillover—positively affect the productivity of labor and thus the aggregate level of income of an economy. Lucas [22], whose model goes back to Uzawa [36], stresses education and the creation of human capital, Romer [29] focus on the creation of new technological knowledge as important sources of economic growth. And others emphasize productive public capital and investment in public infrastructure. For a more extensive survey, see Greiner et al. [10].

R. M. Kovacevic (✉)
Vienna University of Technology, Vienna, Austria
e-mail: raimund.kovacevic@tuwien.ac.at

W. Semmler
New School of Social Research, Newyork, USA

Bielefeld University, Bielefeld, Germany

IIASA, Laxenburg, Austria

J. L. Haunschmied et al. (eds.), *Dynamic Economic Problems with Regime Switches*, Dynamic Modeling and Econometrics in Economics and Finance 25,
https://doi.org/10.1007/978-3-030-54576-5_3

our attention to important mechanisms that can generate multiple equilibria, and among them an attractor which has been called a poverty trap. Those could arise from externalities and increasing returns to scale, constraints in the financial and credit markets, as well as from population movements after disaster shocks that, in the long run, can give rise to diverse per capita incomes across regions and countries. Such mechanisms may be able to explain the forces that bring about a twin-peak distribution of per capita income in the long run, namely the convergence of the size distribution to countries and regions of small per capita income and countries with large per capita income, predicting some twin-peak distribution.[2]

Financial studies have focused on large negative shocks that can result from large income or wealth shocks and contractions, with effects on credit markets, income flow, consumption and investment. In particular financial melt-downs and the destruction of capital as well as jumps in risk premia after rare large economic and financial crises are investigated in great detail. For example, Rietz [28] studies rare market crashes and their effect on equity risk premia. Barro [5] uses as disaster measure the decline of GDP growth, while Barro and Ursua [6] and Gabaix [9] investigate the decline of consumption spending due to large financial and economic disasters.

Climate change and weather extremes studies explore disaster effects from large scale floods, storms, landslides, heat waves and droughts, and forest fires. This literature also stresses nonlinearitites and tipping points, leading to the phase shift, and long period lock-ins. This work goes back to extreme event studies initiated by Gumbel [13, 14]. Recent important contributions are the one by Burke et al. [7, 8], Hochrainer et al. [15], Hochrainer-Stigler et al. [16], Independent Evaluation of the Asian Development Bank [17], Yumashev et al. [37], and Mittnik et al. [20]. Much research work published in the IPCC assessment reports since 1988 have elaborated on those issues.

A further strand of literature is the insurance work on this topic. Here the question is pursued whether and for which type of shocks, insurance against large random shocks can aid to reduce the risk of large capital and income losses leading to dynamics to fall below the poverty trap. An overview of those models is given in and in Kovacevic and Pflug [18]. In the latter work a critical level of capital is introduced that can dampen the losses. Above the critical level, which depends on the fraction covered by insurance, the expansion of capital is feasible after a disaster shock. On the other hand, close to a critical point—at some cliff—insurance might be too expensive, because it disturbs the deterministic growth.

Our paper is related to the above literature. Our deterministic dynamics also has three equilibria: the outer two are stable while the middle one is unstable. The deterministic dynamics is driven by optimizing behavior, similar as in dynamic growth models, as in Semmler and Ofori [32]. As many recent growth models do, we start with a capital accumulation model with a mechanism that gives rise to multiple

[2] An early theoretical study of this problem can be found in Skiba [34]. Further theoretical modeling is in Azariadis and Drazen [3], Azariadis [2] and Aziaridis and Stachurski [4]. For recent empirical studies see, for example, Quah [27] and Kremer et al. [19]. A more recent study on this issue is Semmler and Ofori [32] where also empirical evidence is provided.

equilibria. It represents a basic model of the dynamic decision problem of countries where the capital stock is the state variable and investment is the decision variable. We explore mechanisms that may lead to thresholds and the separation of domain of attractions, predicting a twin-peak distribution of per capita income in the long run. We show that only countries that have passed certain thresholds may enjoy a rise of per capita income.[3]

The deterministic dynamics is overlayed by random dynamics, modeling the waiting times between catastrophic events. In this setup, the possibility of insurance is analyzed. While Kovacevic and Pflug [18] consider only the possibility of a fixed retention rate (respectively, a fixed proportion of insured capital), we allow a change of the retention rate after each catastrophic event. The search for an optimal process of retention rates gives rise to a bi-level decision problem. In our context of finite upper equilibria of the deterministic dynamics it turns out that it is not meaningful to concentrate on the trapping probability of falling below the cliff (which was the approach in Kovacevic and Pflug [18], where the upper equilibrium was infinite). It turns out that this probability is always one—that is in the long run ruin happens for sure. Therefore we aim at maximizing the expected discounted capital after the next jump and develop a numerical algorithm in order to analyze the optimal retention rate depending on the starting capital.

The remainder of the paper is organized as follows. Section 3.2 presents a deterministic model with stable outer equilibria and an unstable middle equilibrium. Moreover, the economic mechanisms that make such thresholds plausible are discussed. Section 3.3 introduces the stochastic process of catastrophic events, stylized by a Poisson process for the number of events and a beta distribution for the proportion of destroyed capital. Moreover, insurance is introduced in this part. In Sect. 3.4, we analyze the long-term trapping probability. Section 3.5 describes the numerical procedure used to calculate optimal decisions in the expected capital framework. Moreover, we present numerical results and implications for a stylized example. Section 3.6 concludes the paper. In the appendix, we sketch the numerical solution procedure.

3.2 The Deterministic Dynamic Model

The basic economic mechanisms to explain poverty traps frequently refer to technological traps. The idea of a technological trap is based on the work by Rosenstein-Rodan [30, 31], Singer [33], Nurske [24] and others. The starting point is a modified production function that has both increasing and decreasing returns to scale. Increasing returns can only be realized if a country is capable to build up a capital stock that is above a certain threshold. If this threshold is passed, and sufficient externalities are

[3]The working of the above mechanisms are then empirically explored by applying a kernel estimator and Markov transition matrices to an empirical data set of per capita income across countries, see Semmler and Ofori [32].

generated, the production function exhibits increasing returns. Countries converge to a higher steady state as compared to countries that have fallen short of the threshold. With reference to the technological trap the so-called "Big Push Theory" proceeds from the idea that industrial countries had in their past a massive capital inflow and, therefore, can converge to a steady state with a high income level. In contrast, less developed countries have a shortage of such massive capital inflow and accordingly stagnate at a low income level.

A related explanation is given by Myrdal [21] who points out that a tendency toward automatic stabilization in social systems does not exist and that any process which causes an increase or decrease of interdependent economic factors including income, demand, investment, and production will lead to a circular interdependence. Thus this would lead to a cumulative dynamic development that strengthens the effects of up—or downward movement. On this ground poor countries are in a *vicious circle*, becoming poorer. This is in contrast to rich countries who will profit by a positive feedback effect, the so-called "Backwash Effects" arising from capital movement and migration to get richer.[4]

3.2.1 The Deterministic Model

As previously mentioned, the idea of externalities and increasing returns to scale has been extensively employed in growth theory recently. It is shown that a variety of positive externalities arising from scale economies, learning by using, increasing returns to information and skills are set in motion if a country enjoys, for example, by historical accident, a "big push" and take-off advantages.

Our proposed variant of a model of dynamic investment decisions of countries builds on locally increasing returns to scale arising from externalities. Locally increasing returns due to positive externalities may be approximated by a convex–concave production function as proposed by Skiba [34] to illustrate those effects.

To present this idea of a convex–concave production function resulting from externalities and locally increasing returns to scale we use a model similar to Azariadis and Drazen [3].[5] With capital stock denoted by k, we can write a production function such as

$$y(k(t)) = ak(t)^{\alpha_k(t)} \tag{3.1}$$

$$\alpha_k(t) = \begin{cases} \overline{\alpha}_k & \text{if } k(t) > \overline{k}(t) \\ \underline{\alpha}_k & \text{otherwise,} \end{cases}$$

[4]Scitovsky's work in the 1950s is another example predicting poverty traps, thresholds and take-offs, see Scitovsky [40].

[5]See furthermore, Azariadis [2] and Azariadis and Stachurski [4].

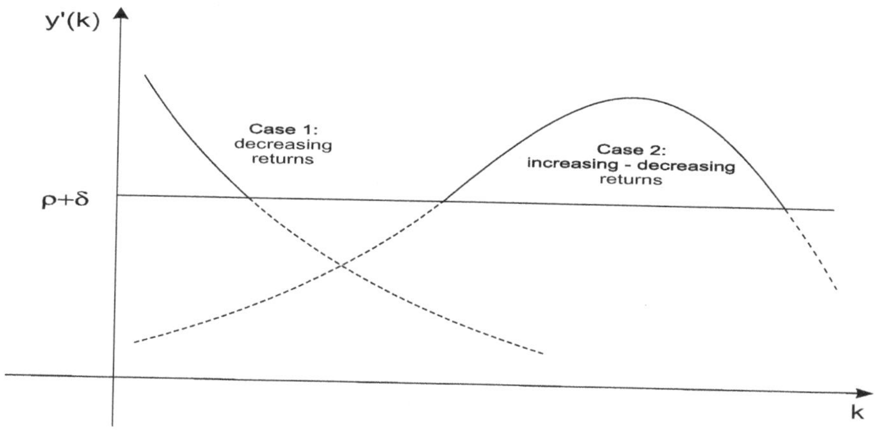

Fig. 3.1 Increasing and decreasing returns

where the coefficients $\alpha_k(t)$, vary with the underlying state (k) and the quantity $\overline{k}(t)$ is the threshold for the capital.

We consider an optimal control problem

$$V(k) = \underset{\{u(t)\}}{Max} \int_0^T e^{-\rho t}((y(t)u(t))^{(1-\sigma)}/(1-\sigma))dt \tag{3.2}$$

$$\dot{k}(t) = y(t)(1-u(t)) - \delta k(t), \quad k(0) = k, \tag{3.3}$$

which describes the economy based on the optimal allocation of income between consumption and investment. The control is the consumed fraction of income $u(t)$.

Equation (3.2) represents the related value function and Eq. (3.3) the evolution of capital stock, whereby the first term $y(t)(1-u(t))$ is gross investment and the second term $\delta k(t)$ is the depreciation of capital which will be augmented by the loss of capital due to the insurance premium. Finally, σ is the parameter of risk aversion of the economic agents which is in the literature assumed to be between 0.5 and 4 (We want to assume the risk aversion on the low side, so the dynamics are not much impacted by this parameter itself.[6]).

One can show, using Dechert and Nishimura [39] that if $\alpha_k < 1$ in Eq. (3.1) holds forever, the marginal product of capital, $y'(k)$ would approach the line given by the discount rate ρ plus capital depreciation, δ, if depreciation is allowed, from above or below, see case (1) in Fig. 3.1.

[6]Population growth could also be included, requiring a slight modification of the model.

On the other hand, presuming that the parameter α_k is state dependent and approximating the convex–concave production function by a smooth function one obtains the case 2 in Fig. 3.1.

For locally increasing returns to scale, case 2, the return on capital $y'(k)$ will first approach $\rho + \delta$ from below, then move above this line and eventually decrease again. In the first case, the return on capital below $\rho + \delta$, because of externalities, too small a capital stock will generate a too low return for the economy so that the capital stock will shrink.

Thus the case 2 has three equilibria, one unstable equilibrium where the horizontal axis $\rho + \delta$ intersects with the case 2 curve, and the other two equilibria are somewhere above and below the $\rho + \delta$ line.

As Fig. 3.1 demonstrates, increasing returns can be assumed to hold, as Greiner et al. [10, Chap. 3] show only up to a certain level of the capital stock. A region of a concave production function may be dominant thereafter where $y'(k)$ might start falling again.

If we compute the investment strategy for a model variant with a convex–concave production function as suggested above, the convex–concave production function is for our numerical purpose specified as a logistic function of k

$$y(k) = \frac{a_0 \, exp(a_1 k)}{exp(a_1 k) + a_2} - \frac{a_0}{1 + a_2}. \tag{3.4}$$

This convex–concave production function specifies the production function $y(k)$ in Eq. (3.1). We refer to the model (3.2)–(3.4) as the “deterministic dynamics $DP(\delta)$”

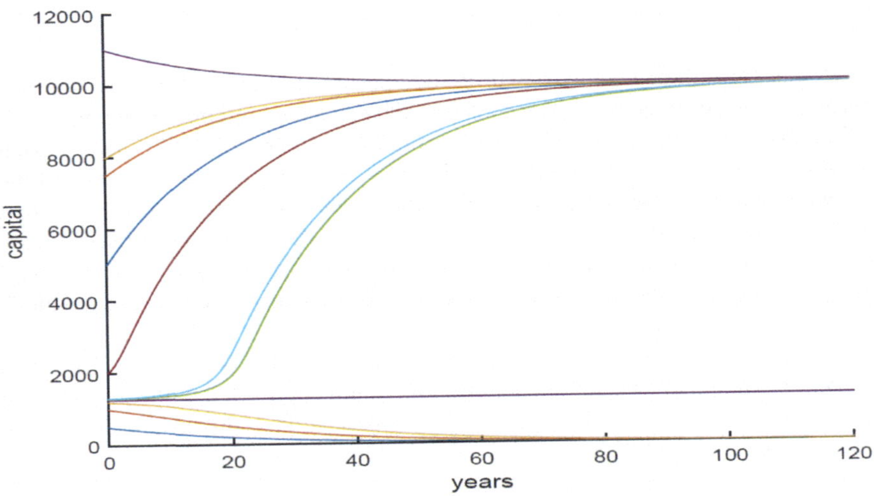

Fig. 3.2 Deterministic dynamics for several starting points

in order to emphasize the dependence on the depreciation rate, which later will be augmented by the insurance premium per unit of capital.

For the numerical computation of the solution of model (3.1)–(3.4) we employed the NMPC procedure of Gruene et al. [12], which is sketched in the appendix of this paper.

In a stylized numerical example, we use parameter values $a_0 = 2500$, $a_1 = 0.0034$, $a_2 = 500$. for the production function and set $\delta = 0.05$ and $\sigma = 0.5$. Figure 3.2 shows that there are indeed three equilibria, the middle one is unstable and the lower and upper equilibria are stable ones.

3.3 Stochastic Shocks and Insurance

We assume now that the capital normally evolves according to the dynamics described in the previous section, but is reduced at random points in time by shocks of random size. Such catastrophic events may imply substantial damages to the economy, but can be considered as rare events. Because of this rareness, we assume that all economic decisions by economic agents are made without accounting for the possibility of catastrophic events by the economic decision-makers.

Figure 3.3 provides an example of such dynamics, based on the numerical setup. A detailed description will be given below. Yet, we can already observe that even shocks far away from the poverty trap (the middle unstable equilibrium) can lead to dynamics ending up in the poverty trap.

Next, a second layer of decisions is added at this point: Governments, e.g., might be well aware of the possibility of catastrophic losses. One way to deal with such events

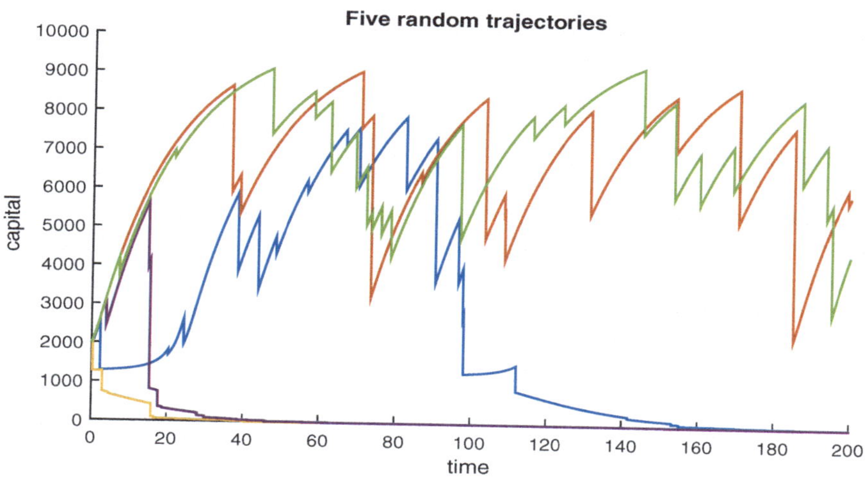

Fig. 3.3 Stochastic dynamics without insurance ($\eta = 1$)

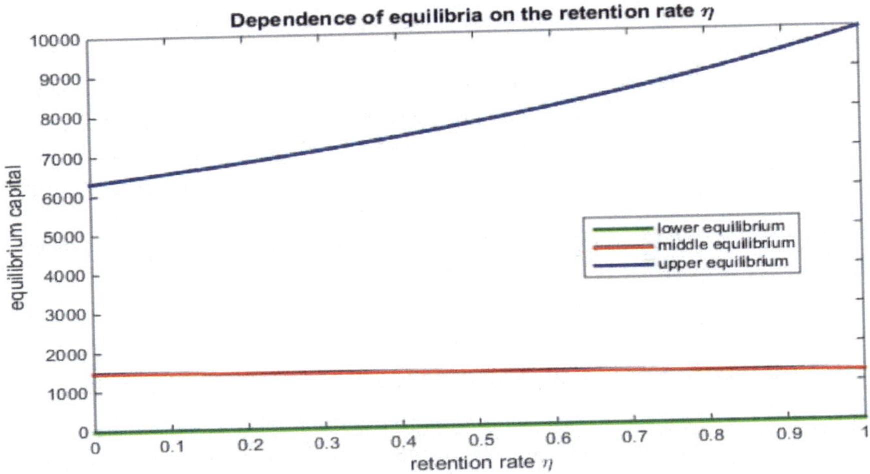

Fig. 3.4 Dependence of equilibrium values on the retention rate η

is the introduction of a mandatory insurance scheme. Hereby the economic agents would pay insurance premium (maybe as an additional tax) and receive insurance benefits in the form of a reduction of the capital loss, in case of a catastrophic event. We use the terminus "insurance" here, although the "insurance premium" might be also implemented as some kind of earmarked tax and the "insurance benefits" might be just payments of the government in case of catastrophic events.

The insurance premium per unit of capital depends on the retention rate η, i.e., the proportion of capital that is not insured[7] and is added to the deterministic model in form of additional capital depreciation. This slows down capital growth and shifts the equilibrium points of the deterministic dynamics, see Fig. 3.4. On the other hand, some capital is recovered after each event. In view of this tradeoff we ask the question, how much insurance—or which retention rate—is "optimal" depending on the start capital.

Dependence of η on capital size is critical, because insensitive introduction of insurance premium ($\eta > 0$ fixed) has the unwanted effect that small start capital just above the unstable middle equilibrium (Skiba point) without insurance would be below the Skiba point after introduction of insurance. This leads to a deterministically shrinking capital exposed to random shocks, hence accelerated extinction. This effect was observed for simpler dynamics in Kovacevic and Pflug [18].

The overall decision problem has the form of a bi-level optimization problem. The upper-level decision (insured fraction of capital) is taken by the government, whereas the lower level decisions (the solution of the control problem) are taken by the economic agents in view of the prescribed amount of premium payments. While

[7] If the capital is not insured at all, the retention rate equals one, if the whole capital is insured, the retention rate equals zero.

bi-level problems are generally difficult to solve, in the present context the task is facilitated by the fact that the control problem has unique solutions.

In the present paper, we take the size of the insurance premium (per insured unit of capital) as dependent on the loss distribution but exogenously given for the agents. Questions of exact financing of the insurance system and of adequate premium amounts are left to future research. We here concentrate on proportional insurance, which is the simplest insurance scheme.

3.3.1 Catastrophic Events, Insurance and the Modified Deterministic Dynamics

The catastrophic events happen at random points in time T_i. We assume that the related waiting times $\tau_i = T_i - T_{i-1}$ are i.i.d. according to an Exponential distribution with (constant) parameter $\lambda > 0$. This means that the probability density of each τ_i is $g_\tau(t) = \lambda e^{-\lambda t}$ and the expected waiting time is given by $\frac{1}{\lambda}$. Therefore, the number of events up to time t follows a homogeneous Poisson process.

In the stochastic model, we denote capital by $K(\cdot)$. When a catastrophic event i happens, the instantaneous capital $K(T_i^-)$ before the jump is reduced by a random fraction Z_i such that the Z_i are i.i.d. distributed according to a cumulative distribution function $G_Z(\cdot)$ and probability density $g_Z(\cdot)$, hence

$$K(T_i) = (1 - Z_i)K(T_i^-), \tag{3.5}$$

where $K(T_i^-)$ denotes the capital immediately before the event.

If $G_Z(\cdot)$ is differentiable, we denote the related probability density by $g_Z(\cdot)$. The support of the loss distribution is the interval $[0, 1]$. In addition, it is assumed that the Z_i are jointly independent of the waiting times τ_i.

With insurance, (3.5) is adjusted in the following way

$$K(T_i) = (1 - \eta Z_i)K(T_i^-), \tag{3.6}$$

where the retention rate $0 \leq \eta \leq 1$ is the proportion of damage beared by the insured entity.

Using the expectation premium calculation principle (see, e.g., Mikosch [23]), the insurance premium per capital unit, $c(\eta)$, then can be expressed as

$$c(\eta) = \lambda(1 + \gamma)(1 - \eta)\mathbb{E}[Z_1]. \tag{3.7}$$

Here $\gamma > 0$ is some risk loading parameter. Recall that the cession rate $1 - \eta$ is the proportion of damage beared by the insurer.

The distribution parameter λ, and the expectation $\mathbb{E}[Z_1]$ are assumed to be known throughout the paper. In real-world applications they have to be estimated from data.

The risk adjustment parameter γ is also given exogenously in the present paper. Basically, it has to be chosen, e.g., by an insurance company taking into account the riskiness of the loss distribution. Finally the retention rate η has to be decided by the government.

We assume that a new value of η can be chosen after each catastrophic event, dependent on the remaining capital $K(T_i)$. This justifies the notation $\eta_i = \eta(K(T_i))$.

After each jump, the capital starts with a value of $K(T_i)$. We assume $T_0 = 0$ and so

$$K(0) = k \tag{3.8}$$

is the start value of the whole process. Between two events $i-1$ and i, capital $k(t)$ develops according to the solution of the deterministic control problem (3.2)–(3.3), but the original depreciation rate δ is replaced by $\delta + c(\eta)$—which we denote by $DP(\delta + c(\eta))$. In consequence, we may write

$$K(t) = F(t - T_{i-1}, K(T_{i-1}); \eta) \text{ if } T_{i-1} < t < T_i, \tag{3.9}$$

where F denotes the optimal state $k(\cdot)$ for the deterministic problem $DP(\delta + c(\eta))$ with $k(0) = K(T_{i-1})$ and call the resulting control problem the modified deterministic problem $DP(\delta + c(\eta))$. In order to shorten notation, we may write $F(t - T_i, X_{T_i})$, when the dependence on η is not in the focus.

Altogether, $K(t)$ is a piecewise deterministic Markov process. The stochastic evolution of capital is described by the distribution assumptions on waiting times and proportional jump size, the premium principle (3.7) together with the resulting modification of the deterministic dynamics (3.2)–(3.3) and the defining equations (3.8), (3.6), and (3.9).

In the modified problem all the equilibria depend on the retention rate. With $\widetilde{k}^\eta$ we denote the unstable equilibrium of the dynamics as described above and $\hat{k}^\eta$ denotes the stable upper stationary equilibria of the modified deterministic system. It turns out that the upper equilibrium increases with the retention rate, whereas the Skiba point decreases with increasing retention rate, see Fig. 3.4.

The lowest Skiba point $\tilde{k} = \tilde{k}^1$ is the boundary between amounts of capital for which growth is possible with appropriately chosen η, and the region where capital shrinks for any η without a chance for recovery. We also will consider the largest upper equilibrium $\hat{k} = \hat{k}^1$. In principle the upper equilibrium of some dynamics might not exist, i.e., $\hat{k}^\eta = +\infty$. This was, e.g., the case for the dynamics used in Kovacevic and Pflug [18]. However, the deterministic problem $DP(\delta + c(\eta))$ definitely leads to finite upper equilibria.

For the deterministic part of our stylized example, we set the parameter of the exponential distribution (waiting times) to $\lambda = 0.1$. The parameters of the Beta distribution (proportional loss given a catastrophic event) are $\alpha = 1.92$ and $\beta = 3.39$. Moreover, we use $\gamma = 0.05$ for the risk adjustment parameter. These values were already used for producing Fig. 3.3.

3.3.2 The Remaining Capital and Its Transition Distribution

In order to simplify notation we introduce the random variables $V_i = 1 - \eta Z_i$ (which are also i.i.d.), taking values in $[0, 1]$. They have cumulative distribution function

$$H(v; \eta) = 1 - G_Z\left(\frac{1-v}{\eta}\right),$$

and probability density

$$h(v; \eta) = \frac{1}{\eta} g_Z\left(\frac{1-v}{\eta}\right).$$

V_i models the remaining fraction of capital after event i occurred and the insurer already repaid the insured sum. Equation (3.5) can be rewritten as

$$K(T_i) = V_i K(T_i^-).$$

In order to analyze the remaining capital after the jumps, we may consider the discrete-time process

$$K_i = K(T_i),$$

by sampling immediately after the occurrence times of catastrophic events. This is also a Markov process and given the above specification it is possible to characterize the related transition density $p(k_1, k_0; \eta)$, i.e., the conditional probability density for reaching capital level k_1 after the next catastrophic event, when the process starts with capital level k_0 after the last event. If a remaining capital $K_i = k_0$ is observed, it is possible to neglect all previous observations when calculating the density of capital $K_{i+1} = k_1$ (after the next jump). Because η_i is reconsidered after each jump, it will never be chosen such that the new Skiba point would be below the start capital K_i, because this would lead to a decreasing deterministic dynamics. Therefore, it suffices to consider the case $k_0 > \widetilde{k}^\eta$.

When $V_i = v$ is given, then the capital after the next loss fulfills

$$K_{i+1} = k_1 = vF(\tau, k_0; \eta),$$

which means that the waiting time τ until the next event can be calculated as a function of k_0, k_1, v by

$$\tau = F^{-1}\left(\frac{k_1}{v}, k_0\right),$$

where $F^{-1}(\cdot, k_0; \eta)$ is the inverse function of $F(\cdot, y_0 \eta)$ with respect to the first argument. Inversion is possible because $F(\cdot, y_0)$ is strictly increasing (recall the assumption $k_0 > \widetilde{k}^\eta$). Note also that we have $F : [0, \infty) \to [k_0, \hat{k}^\eta)$ and $F^{-1} : [k_0, \hat{k}^\eta) \to [0, \infty)$. Calculating the derivative leads to

$$\frac{\partial \tau}{\partial y_1} = \frac{1}{vF_1(\tau, k_0)} > 0.$$

3.3.3 Transition Densities

Knowing the density of τ, it is possible to calculate the conditional density of K_{i+1} given $K_i = k_0$ and $V = v$ by using density transformation. Taking expectation with respect to the c.d.f. $H(\cdot; \eta)$ of the random variable V leads to the transition density.

$$p(k_1, k_0; \eta) = \lambda \int_{\frac{k_1}{\hat{k}\eta}}^{1} \frac{e^{-\lambda F^{-1}\left(\frac{k_1}{v}, k_0 . \eta\right)}}{v F_1\left(F^{-1}\left(\frac{k_1}{v}, k_0; \eta\right), k_0; \eta\right)} \frac{1}{\eta} gz \left(\frac{1-v}{\eta}\right) dv. \qquad (3.10)$$

The integration boundary follows from the domain of F^{-1} together with the fact that $0 \leq V \leq 1$.

From this result one can also derive the related conditional distribution function

$$\begin{aligned} P(k_1, k_0; \eta) &= \mathbb{P}\left(K_i = k_1 | K_{i-1} = k_0\right) \qquad (3.11) \\ &= \int_{\frac{k_1}{\hat{x}\eta}}^{1} \left[1 - e^{-\lambda F^{-1}\left(\frac{k_1}{v}, k_0 . \eta\right)}\right] \frac{1}{\eta} gz \left(\frac{1-v}{\eta}\right) dv, \end{aligned}$$

which may be used in order to simulate realizations of the process K_i.

3.4 Aiming at the Trapping Probability

Based on the optimal dynamics of $DP(\delta + c(\eta))$ and using waiting times and proportional losses as described above, we aim at "optimal" retention rates η. One possible objective consists in minimizing the probability that the capital reaches or falls below the trapping point $\tilde{x}$—the smallest possible Skiba point (which results from setting $\eta = 1$). If this happens, then there is no chance for escaping from the lower stable equilibrium in the long run, because already the deterministic dynamics leads to decreasing capital, and all jumps decrease the capital further. We refer to this probability as the trapping probability in the following. Such an approach was suggested in Kovacevic and Pflug [18] for a simpler dynamics with a trapping point, without using an underlying optimal control problem and without the possibility to change the retention rate η after a catastrophic event.

In such a setup, one searches for a (point-wise) minimal trapping probability $Q(k)$, defined on $(\tilde{k}, \hat{k}]$, where $\tilde{k} = \tilde{k}^1$ is the smallest Skiba point and $\hat{k} = \hat{k^1}$—the largest upper equilibrium. This is the probability of eventually reaching $\tilde{k}$ or any point below at some point in time, after starting with a capital of $k > \tilde{x}$. Again it is assumed that fraction η of noninsured capital can be readjusted after each catastrophic event, hence $\eta = \eta(k)$ can be considered as a function of the starting capital. The function Q fulfills the functional equation

$$Q(k) = \min_{\eta} \left[\int_0^{\tilde{k}} p(y, k; \eta)\, dy + \int_{\tilde{k}}^{\hat{k}^{\eta}} Q(y) p(y, k; \eta)\, dy \right], \tag{3.12}$$

and the function defined by the argmin, $\eta(k)$, describes the optimal fraction of uninsured capital for each starting capital x. Basically, the trapping probability equals the probability of falling below the poverty line immediately after the next jump plus the expectation (with respect to the transition densities) of the trapping probabilities for capital values above the poverty line. We assume that $Q(x)$ is a bounded function $0 \leq Q(\cdot) \leq 1$ and denote the set of such functions by $\mathcal{B}$. Because of these bounds $Q(y)$ is integrable w.r.t. any probability density $p(y, k; \eta)$, hence the right-hand side of (3.12) is well defined. This integrability property also ensures a bounded minimum. With T we denote the operator, defined by the right-hand side, which is a mapping $\mathcal{B} \to \mathcal{B}$, because p is a probability density for all possible values of η. Any solution of (3.12) is a fixed point of T.

It should be noted that $Q(k) \equiv 1$ is always a (trivial) solution of (3.12), which shows the existence of a fixed point. However, basically we seek for a nontrivial solution that is smaller than one for at least some amounts of capital. Unfortunately, classical contraction arguments (e.g., applications of the Banach fixed point theorems) can not be applied, because they lead to unique fixed points.

However, let $(\mathcal{B}, \leq)$ denote the vector space $\mathcal{B}$ together with the point-wise partial order, i.e., $Q_1 \leq Q_2$ when $Q_1(k) \leq Q_2(k)$ for all $x \in (\tilde{k}, \hat{k}]$. Then we can show the following:

Proposition 1 *The operator T has a smallest and a largest fixed point, Q_* and Q^* in $(\mathcal{B}, \leq)$, which can be obtained by*

$$Q_* = \sup \{Q \in \mathcal{B} : TQ \geq Q\} \tag{3.13}$$

$$Q^* = \inf \{Q \in \mathcal{B} : TQ \leq Q\}. \tag{3.14}$$

Here the infimum has to be understood in the point-wise sense, induced by the point-wise order $\leq$.

Proof The operator T is monotone: assume $Q_1 \leq Q_2$ and define

$$\eta_2(k) = \arg\min_{\eta} \left[\int_0^{\tilde{k}} p(y, k; \eta)\, dy + \int_{\tilde{k}}^{\hat{k}^{\eta}} Q_2(y) p(x, k; \eta)\, dy \right],$$

then we have for any x.

$$
\begin{aligned}
(TQ_1)(k) &= \min_{\eta}\left[\int_0^{\tilde{k}} p(y,k;\eta)\,dy + \int_{\tilde{k}}^{\hat{k}^{\eta}} Q_1(y)p(y,k;\eta)\,dy\right]\\
&\leq \int_0^{\tilde{k}} p(y,k;\eta_2(x))\,dy + \int_{\tilde{k}}^{\hat{k}^{\eta_2(x)}} Q_1(y)p(y,k;\eta_2(x))\,dy\\
&\leq \int_0^{\tilde{k}} p(y,k;\eta_2(x))\,dy + \int_{\tilde{k}}^{\hat{k}^{\eta_2(x)}} Q_2(y)p(x,y;\eta_2(x))\,dy\\
&= \min_{\eta}\left[\int_0^{\tilde{k}} p(y,k;\eta)\,dy + \int_{\tilde{k}}^{\hat{k}^{\eta}} Q_2(y)p(x,k;\eta)\,dy\right] = (TQ_2)(k).
\end{aligned}
$$

Now $(\mathcal{B}, \leq)$ is a complete lattice with smallest element $Q(x) \equiv 0$ and largest element $Q(x) \equiv 1$ and we can apply the Knaster–Tarski theorem, see Tarski [35], in order to show the existence of a largest and a smallest fixed point together with properties (3.13)–(3.14). The smallest fixed point is found by starting with $Q^0(x) \equiv 0$ and applying the operator T until some stopping criterion is fulfilled. □

Based on the proposition, it is clear that the smallest trapping function Q is given by the smallest fixed point of (3.13), and can be obtained by (3.13). It is also a simple fact that the constant function $Q(x) = 1$ is in $\mathcal{B}$ and fulfills the functional equation. So this constant function is the largest element of $\mathcal{B}$ and hence also the largest fixed point of the operator T.

While Proposition 1 may lead to useful algorithms in case of infinite upper equilibria $\hat{k} = +\infty$, unfortunately it is not applicable when $\hat{x}$ is finite as for our optimal dynamics F. It can even be shown that in this case $Q(k) \equiv 1$ is the only solution (and fixed point) of (3.12).

Proposition 2 *If $\hat{k} < +\infty$ and the support of the random variable $\hat{k}V$ contains the value $\tilde{k}$ then $Q(k) = 1$ for any $k \in (\tilde{k}, \hat{k}]$.*

Proof Higher starting capital leads to a smaller ruin probability. Therefore, we have

$$
\begin{aligned}
Q(k) &\geq Q(\hat{k})\\
&= \min_{\eta}\left[\int_0^{\tilde{k}} p(y,\hat{k};\eta)\,dy + \int_{\tilde{k}}^{\hat{k}^{\eta}} Q(y)p(y,\hat{k};\eta)\,dy\right]
\end{aligned}
$$

$$= \int_0^{\tilde{k}} p(y, \hat{k}; \eta^*)\, dy + \int_{\tilde{k}}^{\hat{k}^*} Q(y) p(y, \hat{k}; \eta^*)\, dy$$

$$\geq \int_0^{\tilde{k}} p(y, \hat{k}; \eta^*)\, dy + Q(\hat{k}) \int_{\tilde{k}}^{\hat{k}^*} p(y, \hat{k}; \eta^*)\, dy.$$

Here η^* is the optimal solution of the second line.

This leads to

$$Q(\hat{k}) \geq P + (1 - P) Q(\hat{k})$$

or—after reordering and dividing by P—to

$$Q(\hat{k}) \geq 1,$$

and hence

$$Q(k) \geq 1.$$

Here $P > 0$ is ensured by the condition on the support of $\hat{k}V$, which ensures that regions below or at $\tilde{k}$ can be reached with positive probability after starting at $\hat{k}$. Because Q is a probability, it is possible to conclude

$$Q(k) = 1.$$

□

With a sure transition below the Skiba point in the long run, it is not meaningful to use the trapping probability as a measure of success in a meaningful way. An alternative would be to aim at expected first passage times (trapping times), i.e., the expectation of the random variable $\inf \left\{ t : K(t) \leq \tilde{k} \right\}$. Unfortunately first passage times are difficult to treat and in particular to optimize. For arbitrary processes it is already difficult to calculate first passage times using, e.g., recursive algorithms of Laplace transforms (see Nyberg et al. [25])). Applying such results in a situation where already the transition densities can be computed only in a numerically costly way and an important parameter of the process should change over time in an optimal way is not tractable with reasonable computational effort.

3.5 Optimizing the Expected Capital After Jumps

Observe now that starting with larger capital always must be better than starting with lower capital if decisions on the retention rate can be taken after catastrophic jumps. This is true whichever objective should be optimized (as long as larger capital is counted as better than smaller capital).

Fact 3 *Consider two processes K_i and K_i' as described in Sect. 3.3, and controlled by processes with retention rates η_i, η_i' chosen by a decision-maker. If $K_0 \geq K_0'$ (i.e., process K starts at the higher capital level), then setting $\eta_i = \eta_i'$ for all $i \in \mathbb{N}_0$ implies*

$$K_i \geq K_i' \text{ for all } i \in \mathbb{N}_0.$$

Proof Assume that after jump j the capital of the first process is not smaller than the capital of the second process, i.e., $K_j \geq K_j'$. Then

$$F(t - T_j, K_j; \eta') \geq F(t - T_j, K_j'; \eta')$$

until the next jump, because of continuity: $F(t - T_j, K_j; \eta') < F(t - T_j, K_j'; \eta')$ can only happen if

$$F(t_1 - T_j, K_j; \eta') = F(t_1 - T_j, K_j'; \eta')$$

at some t_1, but this would imply

$$F(t - T_j, K_j; \eta') = F(t - t_1, F(t_1 - T_j, K_j; \eta'); \eta') = F(t - T_j, K_j'; \eta')$$

for $t \geq t_1$until the next jump. So

$$K(T_{j+1}^-) \geq K'(T_{j+1}^-)$$

and hence

$$K(T_{j+1}) = V_{j+1}K(T_{t+1}^-) \geq V_{j+1}K'(T_{j+1}^-) = K'(T_{j+1}).$$

Therefore, because $K_0 \geq K_0'$ by assumption, the assertion $K_i \geq K_i'$ follows by induction. □

In consequence, a choice of $\eta(K_i)$ maximizing K_{i+1} given K_j in each step would maximize the trapping time. This is not possible, because of the random jumps and event times. However, in the following we will analyze a strategy that maximizes the discounted expectation $\mathbb{E}\left[e^{-\rho\tau}K_{j+1}\right]$ when K_j is known, which is motivated by discussed fact. The expectation can be replaced by other relevant acceptability measures in order to take into account the risk dimension of the problem in a better way. Moreover, it is possible to extend this approach to a fully dynamic decision problem. We leave such extensions for future research and stick to the myopic formulation in the present work.

The decision problem can be formulated as the bi-level problem

$$\begin{aligned} \max_{\eta} \quad & \mathbb{E}_{\tau,V}\left[e^{-\rho\tau}V(\eta)F(\tau, K; \eta)\right] \\ & s.t. F(\cdot, \cdot; \eta) \text{ is the optimal dynamics of } DP(\delta + c(\eta)) \\ & \quad \text{for retention rate } \eta, \text{ see (3.5).} \end{aligned}$$

Because already $DP(\delta + c(\eta))$ can be solved only numerically, the same is true for the overall bi-level problem. Because solutions of $DP(\delta + c(\eta))$ are unique for given parametrization, in a numerical setup the task simplifies considerably: it is possible to calculate approximations of the function F already in advance, such that the bi-level problem reduces to an optimization problem without constraints.

3.5.1 Interpolating the Function F

If the function values of F are calculated on a grid $\mathcal{T} \times \mathcal{K} \times \mathcal{E}$, where $\mathcal{T}$ contains (finitely many) points in time, $\mathcal{K}$ contains values for the start capital and $\mathcal{E}$ contains possible values for η, this information can be used to interpolate the function F over the relevant domain—at least if $\mathcal{T}$, $\mathcal{K}$, and $\mathcal{E}$ are chosen sufficiently fine. It should be ensured that $\mathcal{E}$ contains values in the interval $[0, 1]$. Moreover, K should include the value zero and sufficiently many possible capital values up to a level that contains the largest upper equilibrium $\hat{x}$.

It would be a very slow approach to calculate $F(t, K; \eta)$ for such a large grid by fully solving $DP(\delta + c(\eta))$ for all start values K and retention rates η over a large time range. Therefore we start with the derivative

$$f(K, \eta) = F_1(t, K; \eta). \tag{3.15}$$

Note that it does depend on t only via K (as the closed loop control u also depends on K). We calculate an estimate $\hat{f}$ by applying the NMPC procedure (see the appendix) with given η and start value K over a small time horizon $[0, \Delta]$ with step size δt and plug the resulting optimal control value u_0 (i.e., the optimal control at time $t = 0$, calculated by the NMPC procedure, see the appendix) into the right-hand side of equation (3.3) for $t = 0$. This leads to an estimate $\hat{f}(K, \eta)$ for all pairs $(K, \eta) \in \mathcal{K} \times \mathcal{E}$. In further calculations we use the function $\hat{f}_I(K, \eta)$, which takes values $\hat{f}(K, \eta)$ for $(K, \eta) \in \mathcal{K} \times \mathcal{E}$ and else interpolates by cubic splines (if (K, η) is at least in the range of $\mathcal{K} \times \mathcal{E}$). For our standard parametrization, the interpolating function $\hat{f}_I(K, \eta)$ is shown in Fig. 3.5. Using the estimated derivatives then gives a very convenient way to calculate the equilibrium points of the deterministic dynamics for any relevant start value K. These can be found by searching for stationary points K^s, i.e., by finding the solutions of the equations

$$\hat{f}_I(K^s, \eta) = 0,$$

separately for all $\eta \in \mathcal{E}$. By interpolation, we find estimates $\tilde{k}_I^{\eta}$, $\hat{k}_I^{\eta}$ for the unstable middle and the stable upper equilibrium given the retention rate for any $\eta \in [0, 1]$. The results for our standard parametrization were already shown in Fig. 3.4.

In the next step, the interpolated time derivatives $\hat{f}_i$ are used to reconstruct the paths of the deterministic dynamics for any $\eta \in [0, 1]$, i.e., the function F. Let now $\varepsilon > 0$ be a small real number. We define a function

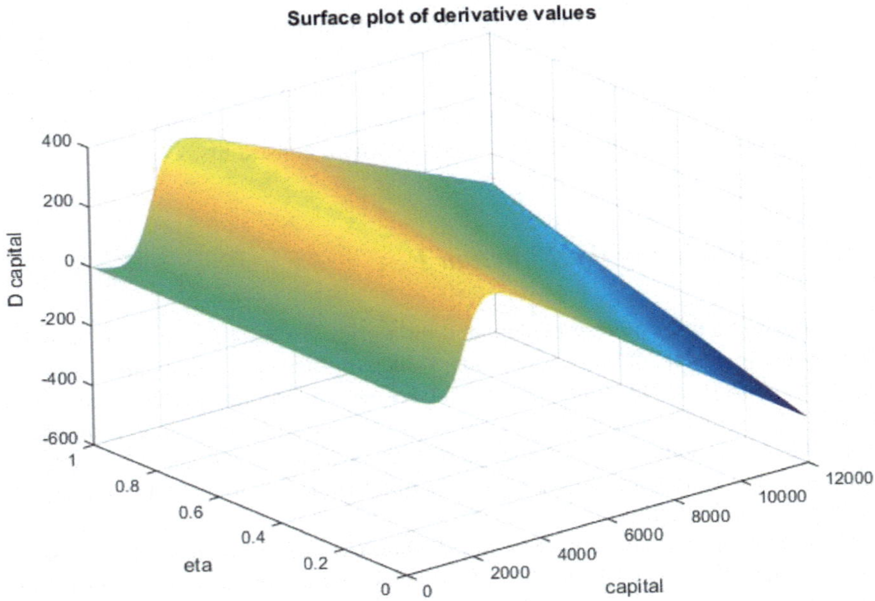

Fig. 3.5 The interpolating function $\hat{f}_I(K, \eta)$ of derivatives for the standard parametrization

$$\tilde{F}(t, \eta) = \begin{cases} \tilde{k}_I^\eta + \varepsilon & t = 0 \\ \tilde{F}(t_-, \eta) + \hat{f}_I(\tilde{F}(t_-, \eta), \eta)\,(t - t_-) & \text{for any other } t \in T \\ & \text{and } t_- = \max\{d \in \mathcal{T} : d < t\} \end{cases}$$

on $\mathcal{T} \times \mathcal{E}$ and its interpolated version $\tilde{F}_I(t, \eta)$. Figure 3.6 shows the interpolated function $\tilde{F}_I(t, \eta)$ for our standard parametrization.

Finally, we get an estimate $\hat{F}(t, K; \eta)$ for the function $F(t, K; \eta)$ which describes the optimal dynamics of the modified deterministic problem with start capital K and retention rate η. Because of the autonomous nature of the control problem, the trajectory for any start value $K \geq \tilde{k}_I^\eta + \varepsilon$ can be reconstructed from the trajectory with start value $\tilde{k}_I^\eta + \varepsilon$ using the relation

$$\hat{F}(t, K, \eta) = \hat{F}_I(\hat{F}_I^{-1}(K, \eta) + t, \eta),$$

where $\hat{F}_I^{-1}(\cdot, \eta)$ denotes the inverse function of F_I with respect to the first argument and with η fixed.

In this way it can be avoided to apply the full NMPC procedure for each possible starting capital over the full planning horizon, which is considerably faster. Due to some random test instances, the loss in accuracy compared to the full NMPC procedure turned out to be very minor. On the other hand, the substantial gain in

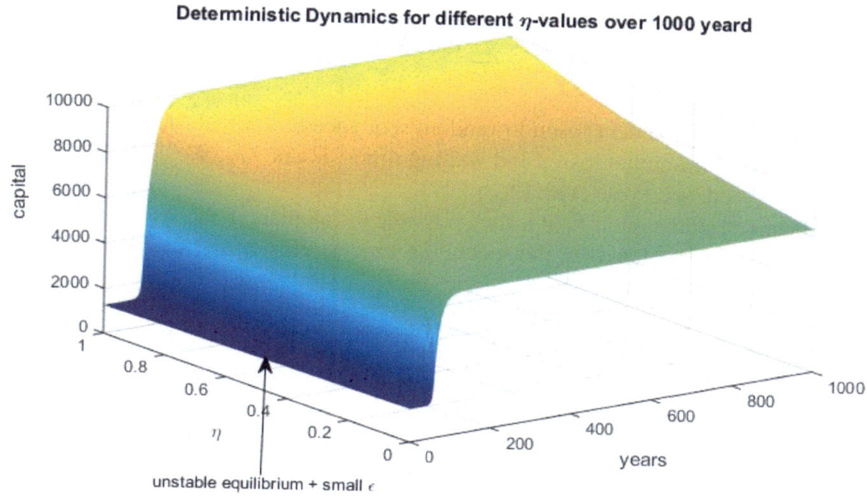

Fig. 3.6 The interpolating function $\hat{F}_I(t, \eta)$ of deterministic trajectories with retention rate η and start capital $\tilde{k}_I^{\eta}$

speed makes it possible to use the deterministic control problem for simulation and optimization of expected capital values after the jumps, as described below.

3.5.2 *Simulation of Capital Values and Optimization of the Remaining Capital*

If the aim is to maximize the expected value of remaining capital after the next catastrophic event, different approaches are possible to calculate the expectation. One may plug the interpolating functions $\hat{f}_I$, $\hat{F}_I$ into Eq. (3.10) for F_1, F and use the resulting interpolating estimate of the transition density p in order to calculate the relevant expectations dependent on η as integrals. However, already calculation of the estimated densities affords integration, so the calculations become numerically very involved, especially when the final aim is to optimize over the resulting expectations.

Therefore, in this study the expectations for any relevant retention rate η are calculated by simple Monte Carlo simulation: the expectation for given start capital K_0 and retention rate η, i.e.,

$$\mu(K_0, \eta) = \int_0^{\infty} \int_0^{\hat{x}^{\eta}} \lambda e^{-(\rho+\lambda)t} y p(y, K_0; \eta) \, dy \, dt,$$

is estimated by the mean

$$\hat{\mu}(K_0, \eta) = \frac{1}{n} \sum_{i=1}^{n} e^{-\rho\tau_i} V_i \hat{F}_I(\tau_i, K_0, \eta).$$

Here V_i, τ_i are (independent) pseudo-random sequences, obtained from the distribution functions of fractional losses and waiting times, H and G_τ. We used the simple inversion method, see Press et al. [26, p. 27].

For the given start capital K_0 it is then possible to calculate $\hat{\mu}(K_0, \eta)$ for $\eta \in E$ and to use a spline-interpolated (in the second argument) version $\hat{\mu}_I(K_0, \eta)$ for finding the optimal retention rate

$$\hat{\eta}^*(K_0) = \arg\min_{\eta} \left\{ \hat{\mu}_I(K_0, \eta) : 0 \leq \eta \leq 1 \right\} \text{ for } K_0 \in \mathcal{K}.$$

Assuming a singular optimizer in the interval [0, 1], golden section search (see, e.g., Press et al. [26], Sect. 10.2) was used to find $\eta^*(K_0)$ in an efficient way.

Figure 3.7 demonstrates the approach and the resulting optimal strategy for our standard example. The upper part gives an overview for the whole range of start capital values, while the lower part analyzes the situation for smaller start capital, in particular start capital near the trapping point $\tilde{k}$. In both pictures, the colors indicate the approximate size of the expected present value $\hat{\mu}(K_0, \eta)$ of capital after the next catastrophic event. Lighter colors are related to higher values according to the color codes shown at the right margin. In addition the contour lines show points (K_0, η) with constant values $\hat{\mu}(K_0, \eta) = c$.

The violet regions have special interpretations. The violet region at the right side of the upper picture contains combinations of start capital K_0 and retention rate η where the stable upper equilibrium $\hat{k}^\eta$ lies below the start capital. This leads to a decreasing deterministic dynamics and, therefore, can not be optimal when compared with η chosen such that the upper equilibrium equals the starting capital. The violet region to the left (of both pictures) is related to points (K_0, η) where $K_0 \leq \tilde{k}^\eta$, i.e., the start capital is below the trapping point of the deterministic dynamics with retention rate η. Here again already the deterministic dynamics leads to a decrease in capital, and η cannot be optimal (e.g., compared with η chosen such that the start capital lies already below the resulting trapping point).

The red line shows (in both pictures) the optimal retention rate $\eta^*(K_0)$ for each relevant start capital. One can see (lower picture) that near the trapping point $\tilde{k}$, in a region where the contour lines are strict monotone decreasing, it is optimal to require no insurance at all ($\eta^* = 1$). The costs of insurance lead to a braked deterministic dynamics, which is especially dangerous near the unstable equilibria. Stated in a different way, near the lower boundary any insurance premium costs more in terms of growth than what the related claims payments could replace in average.

There is also a region, where the contour lines are not unequivocally curved in one direction. Here the optimal $\eta^*(K_0)$ decreases almost linearly from $\eta = 1$ to $\eta = 0$, i.e., from no insurance to full insurance. In the (largest) third regions (see the upper picture) the contour lines are curved to the right, hence here it is optimal to choose

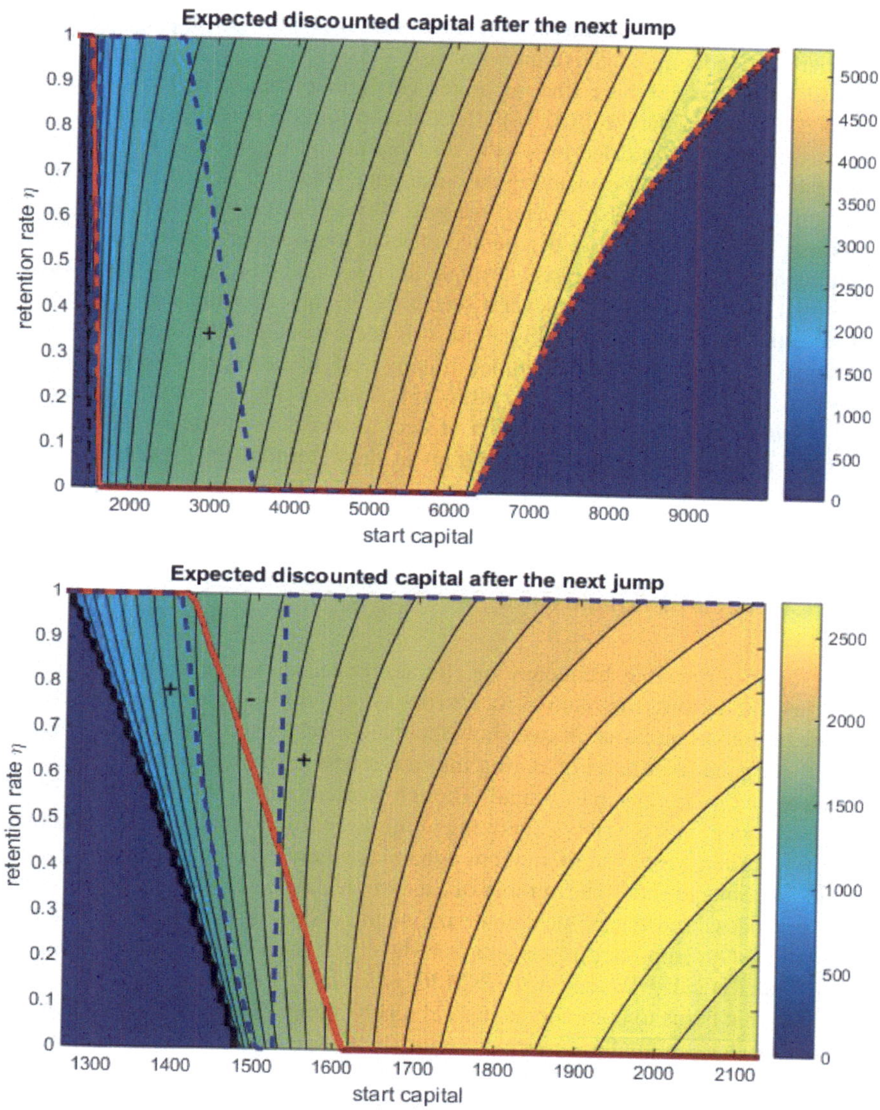

Fig. 3.7 Contour plots of expected discounted capital (EPV), i.e., $\hat{\mu}(K_0, \eta)$ after the next catastrophic event dependent on start capital K_0 and retention rate η. Lighter regions are related to higher values of EPV. The red line indicates the optimal retention rate $\eta^*(K_0)$. The dashed blue line shows combinations of K_0 and η where the EPV equals K_0 and separates regions with $EPV > K_0$ (indicated by +) from regions with $EPV < K_0$ (indicated by −). **Upper picture**: full range of possible start capital, **lower picture**: start capital in the range [1264.7, 2127.5]

the retention rate as low as possible without starting below the upper equilibrium (and therefore inducing negative growth).

Finally, the dashed blue line separates (in both pictures) regions where the expected discounted capital $\hat{\mu}(K_0, \eta)$ after the next jump is below (indicated by +), respectively, above (indicated by +) the start capital K_0. It can be seen that near the lower bound the strategy of denying any insurance leads to a positive effect. When the optimal retention rate decreases, we pass through a region where the optimal amount of insurance leads to a discounted expectation smaller than the start capital, which means that it is hard to pass through this region when repeating the selection of optimal retention rates after several successive events. In some sense this region separates the poor from the wealthy: With still decreasing optimal retention rate the blue line is passed again and the optimal strategy leads to the expectation of increased discounted capital. Finally, for very high capital values (upper picture) the optimal strategy again leads to the expectation of decreased discounted capital due to the effects of a finite stable upper equilibrium in the deterministic dynamics and the negative dynamics above this boundary.

3.6 Conclusions/Discussion

Given the recent rise in frequency of climate-related disasters, severely affecting countries and regions, the issue of recovering lost asset by an insurance scheme has become an important issue. As we show there are mechanisms after disaster shocks that enhance the likelihood of falling into a poverty trap, even with insurance—if the retention rate is chosen too small. Though we start with a stylized deterministic dynamic model, with possibly generating multiple equilibria paths, the deterministic dynamics is then overlayed by random dynamics where catastrophic events happen at random points of time. The number of catastrophic events follows a homogeneous Poisson process and the proportional size of the disasters are modeled by a beta distribution. Our approach represents a bi-level decision model which is hard to compute analytically. Based on the NMPC procedure, we, therefore, apply a new algorithm that helps to compute numerical results. Even if a fraction of capital loss is insured and an optimal insurance premium, including possibly an appropriate risk loading, can be computed, falling into a poverty trap is still feasible. The expected discounted capital after the next catastrophic event, if a certain fraction of capital is insured, is computed in dependence of the (changing) initial size of capital. As also shown insurance against disaster shocks close to the cliff might not pay-off, thus other policies are needed in this case, see Mittnik et al. [20]. However, for larger start capital insurance is a valuable strategy for reducing the probability of falling off the cliff. This feature that the optimal insurance premium of insuring a certain fraction of assets may be also a helpful device to compute state-dependent credit cost and to assess risk premia and creditworthiness of borrowers when a sequence of shocks at uncertain times and of uncertain size is expected.

Yet, further research is needed. In particular, the underlying economic control problem can be enhanced. So far, only the capital is taken into consideration. However, certain types of severe disasters may also have an impact on the workforce, which per se cannot be insured. Insurance of the capital stock might be even more attractive in such a setup because it can partially compensate for the decreasing workforce and may mitigate the effects on production.[8] In order to develop the general algorithmic approach, in the present Paper, we have restricted the analysis to expected values, which neglects risk. Therefore, the effect of optimizing risk-sensitive functionals instead of the expected value has to be analyzed next. An important step forward will be the introduction of the funds generated by the premium payments into the deterministic model as an additional state variable. This allows to optimize the risk loading (which was assumed as given in the present paper) in addition to the insured fraction of capital. In such a manner, it will be possible to explore risk loading drivers in its interaction with macroeconomic effects. Finally, despite the fact that one falls off the cliff for sure as shown, it is possible to analyze the time until falling below the cliff. Optimizing the acceptability of this first passage time will lead back from the myopic optimization to a dynamic approach for the upper-level problem.

Acknowledgements We want to thank Stefan Mittnik, Elena Rovenskaya and Georg Pflug for valuable discussions. Willi Semmler would also like to thank IMF researchers for stimulating discussions while he was a visiting scholar at the IMF.

Appendix: Numerical Solution Procedure (NMPC)

For the numerical solution of the deterministic model presented in Sect. 3.2, and used further in the next sections, we do not apply here the dynamic programming (DP) approach as presented in Grüne and Semmler [11] and as used in the original paper of Semmler and Ofori [32]. Though DP method also can find the global solution to an optimal growth model with multiple equilibria by using a fine grid for the control as well state variables but its numerical effort typically grows exponentially with the dimension of the state variable. Thus, even for moderate state dimensions it may be impossible to compute a solution with reasonable accuracy.

Instead computing the solution at each grid point as DP do we here use a procedure that is easier to implement. We are using what is called nonlinear model predictive control (NMPC) as proposed in Gruene and Pannek [38] and Gruene et al. [12]. Instead of computing the optimal solution and value function for all possible initial states, NMPC only computes single (approximate) optimal trajectories at a time. To describe the NMPC procedure we can write the optimal decision problem as

$$\text{maximize} \quad \int_0^\infty e^{-\rho t} \ell(x(t), u(t))dt, \tag{3.16}$$

[8]The authors thank an anonymous referee for pointing out this possibility.

where $x(t)$ satisfies

$$\dot{x}(t) = g(x(t), u(t)), \quad x(0) = x_0. \tag{3.17}$$

By discretizing this problem in time, we obtain an approximate discrete-time problem of the form

$$\text{maximize} \sum_{i=0}^{\infty} \beta^i \ell(x_i, u_i), \tag{3.18}$$

where the maximization is now performed over a sequence u_i of control values and the sequence x_i that satisfies $x_{i+1} = \Phi(h, x_i, u_i)$. Hereby $h > 0$ is the discretization time step. For details and references where the error of this discretization is analyzed we refer to Grüne et al. [12].

The procedure of NMPC consists in replacing the maximization of the infinite horizon functional (3) by the iterative maximization of finite horizon functionals

$$\text{maximize} \sum_{k=0}^{N} \beta^i \ell(x_{k,i}, u_{k,i}), \tag{3.19}$$

for a truncated finite horizon $N \in \mathbb{N}$ with $x_{k+1,i} = \Phi(h, x_{k,i}, u_{k,i})$. Hereby the index i indicates the number of iterations. Note that neither β nor ℓ nor Φ changes when passing from (3.18) to (3.19). The procedure works by moving ahead with a receding horizon.

The decision problem (3.19) is solved numerically by converting it into a static nonlinear program and solving it by efficient NLP solvers, see Gruene and Pannek [38]. In our simulations, we have used a modification of NMPC, as developed by Gruene and Pannek [38], in their routine `nmpc.m`, available from `www.nmpc-book.com`, which uses MATLAB's `fmincon` NLP solver in order to solve the static optimization problem. Our modification employs a discounted variant of the NMPC MATLAB version, see [12].

Given an initial value x_0, an approximate solution of the system (3.16)–(3.17) can be obtained by iteratively solving (3.19) such that for i = 1, 2, 3, that solves for the initial value initial value $x_{0,i} := x_i$ the resulting optimal control sequence by $u^*_{k,i}$, but uses only the first control $u_i := u^*_{0,i}$ and iterates forward the dynamics $x_{i+1} := \Phi(h, x_i, u_i)$ by employing only the first control. Thus, the algorithm yields a trajectory x_i, $i = 1, 2, 3, \ldots$ whose control sequence u_i consists of all the first elements $u^*_{0,i}$ of the optimal control sequences of the finite horizon problem (3.19). Under appropriate assumptions on the problem, it can be shown that the solution (x_i, u_i), which depends on the choice of N in (3.19), converges to the optimal solution of (3.16) as $N \to \infty$, see [12].

Figure 3.8 illustrates the working of the algorithm. The upper black line represents the solution at the step i = 1 with the decision horizon N = 4. This is iterated forward 6 times, thus we have i = 1 ... 6. The lower red line is the outer envelop of the piecewise

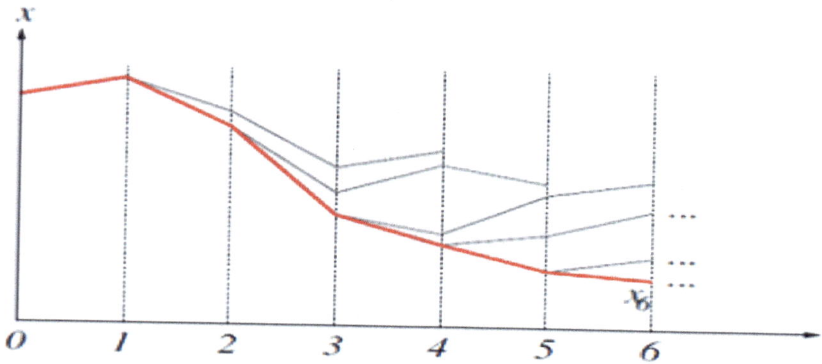

Fig. 3.8 Receding horizon solution

solutions using the horizon N = 4 multiple times, in our case 6 times. The figure A1 shows the solution for 6 iterations.

While the algorithm can be used to solve for optimal trajectories of x and u, it can also be applied for estimating time derivatives $\dot{x}(t)$: This is achieved by plugging the optimal decision u_0 into the differential equation (3.17). Using this estimate, avoids tedious recalculation of trajectories throughout the present paper.

The main requirement in these assumptions is the existence of an optimal equilibrium for the infinite horizon problem (3.19)–(3.17). If this equilibrium is known, it can be used as an additional constraint in (3.19), in order to improve the convergence properties. In our solution of the model in Sect. 3.2, and further on, we did not use the terminal condition to solve the model but moved forward with a receding horizon to find the (approximate optimal) trajectories. Thus, without a priory knowledge of this equilibrium this convergence can also be ensured. Though the proofs in earlier work were undertaken for an undiscounted NMPC procedure, however, the main proofs carry over to the discounted case, details of which can be found in Gruene et al. [12].

References

1. Arrow, K. (1962). The economic implications of learning by doing. *Review of Economic Studies, 29*, 155–173.
2. Azariadis, C. (2001). The theory of poverty traps: What have we learned? Mimeo, UCLA.
3. Azariadis, C., & Drazen, A. (1990). Threshold externalities in economic development. *Quarterly Journal of Economics, 105*(2), 501–526.
4. Azariadis, C., & Stachurski, J. (2004). Poverty traps. In P. Aghion & S. Durlauf (Eds.), *Handbook of economic growth*. Elsevier Publisher.
5. Barro, R. J., & Ursua, J. F. (2008). *Macroeconomic crises since 1870*. Working Paper 13940, National Bureau of Economic Research.
6. Barro, R. J. (2006). Rare disasters and asset markets in the twentieth century. *The Quarterly Journal of Economics, 121*(3), 823–866.

7. Burke, M., Hsiang, S. M., & Miguel, E. (2011). Global non-linear effect of temperature on economic production. *Nature,527*, 235–239.
8. Burke, E. J., Perry, R. H. J., & Brown, S. J. (2010). An extreme value analysis of UK drought and projections of change in the future. *Journal of Hydrology*, *388*(2010), 131–143.
9. Dechert, W. D., & Nishimura K. (1983). Complete characterization of optimal growth paths in an aggregated model with a non-concave production function. *Journal of Economic Theory, 31*(2), 332–354.
10. Gabaix, X. (2011). Disasterization: A simple way to fix the asset pricing properties of macroeconomic models. *The American Economic Review*, *101*(3), 406–409.
11. Greiner, A., Semmler, W., & Gong, G. (2005). *The forces of economic growth: A time series perspective*. Princeton: Princeton University Press.
12. Gruene, L., & J. Pannek. (2011). Nonlinear model predictive control, Theory and algorithms. Springer Verlag, London.
13. Gruene, L., Semmler, W., & Stieler, M. (2015). Using nonlinear model predictive control for solving dynamic decisions problems in economics. *Journal of Economic Dynamic and Control*, *60*, 112–133.
14. Grüne, L., & Semmler, W. (2004). Using dynamic programming with adaptive grid scheme for optimal control problems in economics. *Journal of Economic Dynamics and Control*, *28*, 2427–2456.
15. Gumbel, E. (1935). Les valeurs extremes des distributions statistiques. *Annales de l'institut Henri Poincare,5*(2), 115–158.
16. Gumbel, E. (1958). *Statistics of extremes*. Columbia University Press.
17. Hochrainer, S., Mechler, R., & Pflug, G. (2009). *Assessing current and future impacts of climate-related extreme events—The case of Bangladesh*. Interim Report IR-09-030, International Institute for Applied Systems Analysis.
18. Hochrainer-Stigler, S., Mechler, R., Pflug, G., & Williges, K. (2014). Funding public adaptation to climate-related disasters. Estimates for a global fund. *Global Environmental Change*, *25*, 87–96.
19. Independent Evaluation of the Asian Development Bank. (2015). *Global increase in climate-related disasters*.
20. Kovacevic, R. M., & Pflug, G. C. (2011). Does insurance help to escape the poverty trap?—A ruin theoretic approach. *Journal of Risk and Insurance*, *78*(4), 1003–1028.
21. Kremer, M., Onatski, A., & Stock, J. (2001). *Searching for prosperity*. NBER Working Paper Series, Working Paper 8020.
22. Lucas, R. E. (1988). The mechanism of economic development. *Journal of Monetary Economics*, *22*, 3–42.
23. Mikosch, T. (2009). *Non-life insurance mathematics—An introduction with the Poisson process* (2nd ed.). Berlin, Heidelberg: Springer.
24. Mittnik, S., Semmler, W., & Haider, A. (2019). *Climate disaster risks—Empirics and a multi-phase dynamic model*. IMF Working Paper WP/19/145
25. Myrdal, G. (1957). *Economic theory and under-developed regions*. London: Duckworth.
26. Nurske, R. (1953). *Problems of capital formation n underdeveloped countries*. Oxford: Oxford University Press.
27. Nyberg, M., et al. (2016). A simple method to calculate first-passage time densities with arbitrary initial conditions. *New Journal of Physics*, *18*, 063019.
28. Press, W. H., et al. (2007). *Numerical recipes: the art of scientific computing* (3rd ed.). New York: Cambridge University Press.
29. Quah, D. T. (1996). Twin-peaks: Growth and convergence in models of distribution dynamics. *The Economic Journal,106*, S.1045–1055.
30. Rietz, T. A. (1988). The equity risk premium a solution. *Journal of Monetary Economics*, *22*(1), 117–131.
31. Romer, P. M. (1990). Endogenous technological change. *Journal of Political Economy*, *98*, S71–102.

32. Rosenstein-Rodan, P. (1943). Problems of industrialization of Eastern and Southeastern Europe. *Economic Journal*, *53*, 202–211.
33. Rosenstein-Rodan, P. (1961). Notes on the theory of the big push. In H. S. Ellis & H. C. Wallich (Eds.), *Economic development in Latin America*. New York: Macmillian.
34. Scitovsky, T. (1954). Two concepts of external economies. *Journal of Political Economy 62*, 143–151.
35. Semmler, W., & Ofori, M. (2007). On poverty traps, thresholds and take-offs. *Structural Change and Economic Dynamics*, *18*(1), 1–26.
36. Singer, H. W. (1949). Economic progress in underdeveloped countries. *Social Research*, *15*(2).
37. Skiba, A. K. (1978). Optimal growth with a convex-concave production function. *Econometrica*, *46*(3), 527–539.
38. Tarski, A. (1955). A lattice-theoretical fixpoint theorem and its applications. *Pacific Journal of Mathematics*, *5*(2), 285–309.
39. Uzawa, H. (1965). Optimum technical change in an aggregative model of economic growth. *International Economic Review*, *6*, 18–31.
40. Yumashev, D., Hope, K., Schaefer, K., Riemann-Campe, F., Iglesias-Suarez, E., Jafarov, E. J., et al. (2019). Climate policy implications of nonlinear decline of Arctic land permafrost and other cryosphere elements. *Nature Communications*. Appendix: Numerical Solution Procedure.

Chapter 4
Rationally Risking Addiction: A Two-Stage Approach

Michael Kuhn and Stefan Wrzaczek

4.1 Introduction

The seminal rational addiction model by Becker and Murphy [2] assumes that individuals enter a state of addiction (even with a very low level of addictive capital) right from the beginning of the time horizon. This is in contrast with empirical evidence (see Volkow et al. [23]) according to which the typical addiction dynamics typically set in only after the addictive good has been consumed over a certain time horizon and/or after it has been consumed in excess of a certain quantity.[1] In other words, typically a threshold needs to be crossed from the recreational to the compulsive consumption of a substance that characterizes addiction. Moreover, empirical data shows that many people underestimate their potential for addiction, i.e. they start consuming the addictive good without knowing when they might get addicted (see e.g. Auld and Matheson [1]).

To address these two points we extend the rational addiction model to explicitly involve two stages: Upon first consumption of the addictive good the individual enters a first stage in which it is not yet addicted but accumulates addictive capital (accumulated through the past consumption of the addictive good subject to some depreciation). It may subsequently move into a second stage, in which it is addicted in the sense of being subject to the three typical mechanisms of addiction (see e.g. Orphanides and Zervos [17], Cawley and Ruhm [5], Strulik [20]): reinforcement (by

[1] Indeed, Volkow et al. [23] liken the onset of addiction to the onset of a chronic brain disease and provide evidence on some of the factors that affect susceptibility to acquiring "addiction".

This research was supported by the Austrian Science Fund (FWF) under Grant P 30665-G27.

M. Kuhn · S. Wrzaczek (✉)
Wittgenstein Centre (IIASA, OeAW, University of Vienna), Vienna Institute of Demography/OeAW, Vienna, Austria
e-mail: stefan.wrzaczek@oeaw.ac.at

J. L. Haunschmied et al. (eds.), *Dynamic Economic Problems with Regime Switches*, Dynamic Modeling and Econometrics in Economics and Finance 25,
https://doi.org/10.1007/978-3-030-54576-5_4

which the marginal utility of consumption of the addictive good increases with the stock of addiction), tolerance (by which individuals can compensate an increasingly negative impact from the stock of addiction only by consuming larger quantities of the addictive good) and withdrawal (by which the short-run utility loss from abstinence increases with the stock of addiction).[2]

The timing of the onset of addiction (i.e. entry into the second stage) is subject to a random process, which is shaped by the stock of addictive capital (i.e. with the cumulative consumption of the addictive good so far), where a higher stock of addictive capital implies a higher probability of becoming addicted. The randomness itself reflects a certain element of chance in the neurological processes that govern the entry into addiction, as well as the individual's ignorance about its own addiction threshold. Finally, we assume that addiction is an absorbing state in the following sense: While individuals may shed the addiction by stopping the consumption of the addictive good and subsequently running down their stock of addictive capital, we assume that they remain sensitive to the consumption of even minute quantities of the addictive good in the sense that in this case, the addiction dynamics set in immediately (in fact, akin to the original model by Becker and Murphy [2]).[3]

Explicitly considering a two-stage model of rational addiction with a pre-addiction stage and a random transition into addiction, we contribute insights on a number of aspects that so far have received only insufficient attention in the literature. First, our model allows to study in an explicit way the behaviour that is leading into addiction. Importantly, while such behaviour involves risk-taking, it is not yet subject to the neurological processes that drive addiction. The clear distinction between the two lifecycle stages before and after the onset of addiction allows us to study how the behaviours and outcomes the individual rationally anticipates after the onset of addiction affects risky behaviour and, vice versa, how risky (or precautionary) behaviours affect the course and outcomes of addiction. Second, we study how different patterns of addiction (involving escalation towards permanent high-level addiction or quitting) may arise depending on the states (addictive capital and financial assets) at the onset of addiction.

From a mathematical perspective, our analysis marries two innovative fields of applied modelling: First, in order to analyze the random transition between a first lifecycle stage in which the individual engages in the recreational consumption of an addictive good and accumulates addictive capital, and in the second stage involving addiction, we apply a novel transformation method developed by Wrzaczek et al. [24]. Specifically, the transformation of the underlying two-stage optimal control model with a random switching time into an age-structured optimal control model allows us to study the dynamics of addictive and non-addictive consumption, as well as of the stock of addiction and financial assets in a unified way,

[2]See Volkow et al. [23] for how these mechanisms are grounded in neurobiological changes within the brain's stimulus-and-reward system.

[3]Koob [15] argues that long-term neurological changes aimed at maintaining the stability of the stress-and-reward system in the presence of addiction may ultimately be responsible for a former addict's permanent vulnerability to relapse.

showing how pre- and post-addiction behaviours and outcomes link into each other. Furthermore, the transformation allows us to employ the well-established numerical methods by Veliov [21] to study a numerical example.

Second, we extend the analysis of Skiba points[4] within a model of rational addiction by studying how their emergence is shaped by the pre-addiction stage. Intuitively, a Skiba point means a specific value of the state variable (addictive capital), where two different trajectories (leading to different long-run solutions, e.g. transition into long-run addiction as opposed to quitting) imply the same value of the objective function (see Grass et al. [11]). In the context of the Becker-Murphy model the presence of a Skiba point would mean that for a specific value of the initial addiction capital, the individual is indifferent to different trajectories of addiction (see Caulkins et al. [4]). But, given that the addictive capital is the accumulated (and discounted) past consumption of the addictive good, its initial value should always be zero. This, in turn, would imply that a Skiba point is irrelevant. In contrast, the concept becomes meaningful in our framework, where addiction sets in only after a random time span during which addictive capital is accumulated. The Skiba point then turns out to be crucial as it separates the basin of attraction of two different optimal long-run solutions.

Our paper builds on and adopts many core features of the classical model of rational addiction, as pioneered by Becker and Murphy [2] with further analysis presented in, e.g. Caputo [3] and Ferguson [9].[5] The model has been subsequently adapted to study addiction cycles (Dockner and Feichtinger [7]), the role of imperfect information on addiction thresholds (Orphanides and Zervos [17]), myopia and hyperbolic discounting (Orphanides and Zervos [18], Gruber and Köszegi [12]), multiple equilibria (Orphanides and Zervos [17], Gavrila et al. [10], Caulkins et al. [4]) and the nexus between addiction and more conventional health behaviours as drivers of health and longevity (Strulik [20], Jones et al [14]).

By expressly focussing on the role of a prior "experimentation" stage with addictive consumption, our model is most closely related to Orphanides and Zervos [17] who, to our knowledge, are the only authors who explicitly incorporate this important stage into a model of rational addiction. While their model also involves a Skiba point, where the long-run outcome of addiction depends on the level of addictive capital, their emphasis is on a learning process, where individuals are able to infer their type from whether or not they suffer from symptoms of addiction after consuming the addictive good. Thus, the Skiba point turns out to be relevant when the individual first observes an outcome that reveals its propensity to become an addict. If at that point the stock of addictive capital is too high, which is the case if discovery is too late, the individual becomes rationally addicted. One distinguishing feature to our model is that in Orphanides and Zervos [17] individuals observe symptoms of addiction in random order and can escape addiction if they observe mild symptoms.

[4] In other contributions Skiba points are also referred to as Dechert-Nishimura-Skiba (DNS), Dechert-Nishimura-Sethi-Skiba (DNSS) or Sethi-Skiba points.

[5] For an overview of the economic modelling of addiction see Melberg and Rogeberg [16] or Auld and Matheson [1].

This implies that addiction is to some extent reversible. In contrast, following the medical evidence, addiction is an absorbing state in our model. Another distinction lies in the definition of the switching time, and thus, the point at which the Skiba property may become relevant. While in our model, the switching point is defined by the (clinical) onset of addiction and the differences in the "neurological" rules that govern behaviour, in Orphanides and Zervos [17] it is defined by the point in time at which the individual becomes aware of its addiction. Finally, while omitting the learning issue, our approach allows for a more explicit and richer characterization of the optimal allocation. In studying the structure of multiple equilibria, our model also relates to Gavrila et al. [10] and Caulkins et al. [4], the key difference here being our consideration of the first pre-addiction stage of the lifecycle.

The remainder of the paper is structured as follows. The following section introduces the model while Sect. 4.3 derives the optimal allocation and dynamics of the model. Section 4.4 proceeds to determine the steady states as well as the conditions for a Skiba point. Section 4.5 illustrates the model by way of a numerical analysis, and Sect. 4.6 concludes.

4.2 The Model

In this section, we present an extension of the classical Becker-Murphy rational addiction model (from now on BM is used as an abbreviation for Becker-Murphy). The time horizon is assumed to be separated into two stages, which can be characterized as follows:

First stage (no addiction): The individual enjoys utility from the consumption of two goods, of which one is addictive and contributes to the accumulation of a stock of addictive capital. However, at this stage, the individual has not yet built up a level of the stock that is large enough to trigger the typical effects of addiction, i.e. addictive capital does not yet influence the individual's utility and productivity. The individual is *not addicted*.

Second stage (addiction): The stock of addictive capital enters utility, as it does in the BM-model, and triggers the mechanisms of reinforcement, tolerance and withdrawal. Moreover, it is assumed that addiction reduces labour productivity, resulting in a loss of earnings. The individual is *addicted*.

We assume that the individual does not know when it gets addicted, but that it is aware of the rate of becoming addicted (from now on referred to as 'switching rate'), which is influenced by the stock of $S_1(t)$ of "addictive capital" (modelled as a state variable).[6] In mathematical terms this means that the model is separated into two stages at the switching time $s \in [0, \infty)$, which is a non-negative random variable. Let $([0, \infty), \Sigma, \mathbb{P})$ be a probability space and $\mathcal{F}$ the cumulative probability function (with corresponding density $\mathcal{F}'(t)$) that the model has switched by t. Then

[6]Note that state and control variables corresponding to stage i ($i = 1, 2$) are denoted by subscript i.

the switching rate, which is assumed to depend on the first stage stock of addictive capital, can be defined as

$$\eta(S_1(t)) = \frac{\mathcal{F}'(t)}{1-\mathcal{F}(t)}. \tag{4.1}$$

The two stages are defined as follows:

First stage (no addiction):

During the first stage the individual enjoys utility from consuming a quantity $c_1(t)$ of an addictive good that contributes to the accumulation of addictive capital $S_1(t)$ at t, and from consuming a quantity $y_1(t)$ of a non-addictive good that does not contribute to the accumulation of $S_1(t)$. We assume non-negative addictive consumption and a minimal level $y_0 \geq 0$ for non-addictive consumption. The stock $S_1(t)$ evolves according to

$$\dot{S}_1(t) = c_1(t) - \delta S_1(t), \quad S_1(0) = 0, \tag{4.2}$$

where $\delta \geq 0$ denotes depreciation. Following Becker and Murphy [2] (page 678), we do not model investments aimed at reducing the stock of addiction. The intertemporal budget constraint (originally introduced as integral equation, see equation (4) on page 677 in Becker and Murphy [2]), is formulated as a classical state equation,

$$\dot{A}_1(t) = rA_1(t) + w_1 - pc_1(t) - y_1(t), \quad A_1(0) = 0, \lim_{t\to\infty} A_1(t) = 0, \tag{4.3}$$

where the state variable $A_1(t)$ denotes assets at t, where w_1 is the wage rate in the absence of addiction, and where p is the price of the addictive good, with the non addictive good acting as numeraire. Generally, w_1 and p could depend on time, but for simplicity they are assumed to be constant. Note that the assumption of zero assets in the long-run limit is both necessary for obtaining a unique solution and intuitive as it reflects that individuals will not reckon to be infinitely alive in a set-up, for instance, where the discount rate incorporates a (constant) risk of mortality.

The objective function is the aggregated utility over time discounted at a rate ρ. The function $u^1(y_1(t), c_1(t))$ denotes the instantaneous utility from consumption of the non-addictive and addictive goods. Since the individual is not addicted in the first stage, it does not suffer any (dis-)utility from $S_1(t)$. Moreover we assume $u^1(\cdot)$ to be concave with respect to both consumption goods and to be additive separable between them. The time horizon $s \in [0, \infty)$ of the first stage is a random variable, where the switching rate $\eta(S(t))$ is known. This implies that the value of the second period has to be included in the optimization problem in the spirit of a salvage value. Altogether the optimization problem of the first stage reads

$$\begin{aligned}
\max_{\substack{y_1(t)\geq y_0\\ c_1(t)\geq 0}} \quad & \mathbb{E}_s\Big[\int_0^s e^{-\rho t}u^1(y_1(t), c_1(t))\ dt + e^{-\rho s}V^*(S_1(s), A_1(s), s)\Big] \\
\text{s.t. } & \dot{S}_1(t) = c_1(t) - \delta S_1(t), \quad S_1(0) = 0 \\
& \dot{A}_1(t) = rA_1(t) + w_1 - pc_1(t) - y_1(t) \\
& A_1(0) = 0, \lim_{t\to\infty} A_1(t) = 0.
\end{aligned} \tag{4.4}$$

where $V^*(S_1(s), A_1(s), s)$ denotes the value function of stage 2. To put it differently, the individual optimizes the expected value of the first stage including the optimal behaviour of the second stage (depending on the stock of addiction and assets at the switching time).

Second stage (addiction):

At the switching time s, the individual gets addicted. Thus, the second stage is represented by the classical BM-model. Utility is now given by $u^2(y_2(t, s), c_2(t, s), S_2(t, s))$ and depends on both consumption goods (again in an additive separable form) and on the stock of addiction. More specifically, we assume that

$$\frac{\partial u^2}{\partial y_2} > 0, \frac{\partial^2 u^2}{\partial y_2^2} < 0, \frac{\partial u^2}{\partial c_2} > 0, \frac{\partial^2 u^2}{\partial c_2^2} < 0, \frac{\partial u^2}{\partial S_2} < 0, \frac{\partial^2 u^2}{\partial S_2^2} < 0, \frac{\partial^2 u^2}{\partial S_2 \partial c_2} > 0. \tag{4.5}$$

In particular, this implies that while addictive capital, S_2, lowers utility (implying tolerance, where we focus, without loss of generality, on harmful addiction), it raises the marginal utility of addictive consumption, c_2 (implying reinforcement and withdrawal). The dynamics of the state variables are the same as in the first stage. Since the second-stage allocation depends on the switching time (entering the stage by the initial conditions for the states), the state and control variables are not only indexed by t but also by the switching time s. Moreover, we assume that, as a consequence of addiction, the wage rate during the second stage does not exceed the first stage wage and may well fall short, i.e. $w_2 \leq w_1$.

Since we do not consider the possibility that the individual is cured from addiction and reverts to stage 1, the time horizon is set to infinity.[7] The optimal control model of the second stage is then represented by

$$\begin{aligned} \max_{\substack{y_2(t,s) \geq y_0 \\ c_2(t,s) \geq 0}} \quad & \int_s^\infty e^{-\rho t} u^2(y_2(t, s), c_2(t, s), S_2(t, s)) \, dt \\ \text{s.t. } & \dot{S}_2(t, s) = c_2(t, s) - \delta S_2(t, s), \quad S_2(s, s) = \lim_{t \to s-} S_1(t) \\ & \dot{A}_2(t, s) = r A_2(t, s) + w_2 - p c_2(t, s) - y_2(t, s), \\ & A_2(s, s) = \lim_{t \to s-} A_1(t), \lim_{t \to \infty} A_2(t, s) = 0. \end{aligned} \tag{4.6}$$

The optimized value of the objective function of the above second-stage problem is denoted by $V^*(S_1(s), A_1(s), s)$ and enters the optimization problem of the first stage.

Altogether, the model (4.4) s.t. (4.6), is a two-stage optimal control model with random switching time, which will be transformed into an age-structured optimal

[7]This is in line with the typical observation that individuals who have shed their addiction will typically refrain from any (future) consumption of the addictive good, such that $c_2 \equiv 0$ from some point onward. Any small amount $c_2 > 0$ would immediately retrigger addiction, implying that, technically speaking, such an individual remains in stage 2 even if $c_2 \equiv 0$ and $S_2 = 0$ from some point. As we will see below, our model allows for such an allocation.

control model. This approach offers certain advantages such as an improved analytical representation, the scope for a numerical solution and new ways of presenting results, e.g. in relation to the duration of addiction $t - s$ (see Veliov [22] and Wrzaczek et al. [24] for details).

Full model:

From now on, we use the age-structured representation of the above multistage optimal control model with random switching time, which has been obtained by using the transformation proposed in Wrzaczek et al. [24]:

$$\max_{\substack{y_1(t), y_2(t,s) \geq y_0 \\ c_2(t), c_2(t,s) \geq 0}} \int_0^\infty e^{-\rho t} \left(z(t) u^1 (y_1(t), c_1(t)) + Q(t) \right) dt$$

$$\begin{aligned}
\text{s.t. } & \dot{S}_1(t) = c_1(t) - \delta S_1(t), \qquad S_1(t) = 0, \\
& \dot{A}_1(t) = r A_1(t) + w_1 - p c_1(t) - y_1(t), \; A_1(0) = 0, \; \lim_{t \to \infty} A_1(t) = 0, \\
& \dot{z}_1(t) = -\eta(S_1(t)) z_1(t), \qquad z(0) = 1, \\
& \frac{dS_2(t,s)}{dt} = c_2(t,s) - \delta S_2(t,s), \quad t \geq s, \\
& S_2(s,s) = S_1(s), \quad \forall s \geq 0, \\
& \frac{dz_2(t,s)}{dt} = 0, \quad t \geq s, \\
& z_2(s,s) = z_1(s) \eta(S_1(s)), \quad \forall s \geq 0, \\
& \frac{dA_2(t,s)}{dt} = r A_2(t,s) + w_2 - p c_2(t,s) - y_2(t,s), \\
& A_2(s,s) = A_1(s), \; \lim_{t \to \infty} A_2(t,s) = 0 \;\; \forall s \geq 0, \\
& Q(t) = \int_0^t z_2(t,s) u^2 (c_2(t,s), y_2(t,s), S_2(t,s)) \, ds.
\end{aligned} \tag{4.7}$$

To give an intuitive understanding for the transformation, consider that at every t the individual might get addicted at a rate η. Here, the onset of addiction is tantamount to a switch to a different life-regime featuring in our model a change to the utility function and the wage rate and implying a different optimal behaviour. By the transformation, the individual considers these possible switches (including the corresponding lifetime trajectory) and includes them into the optimization in the following way. Remaining non-addicted is weighted by probability $z_1(t)$. To account for the possibility of being already addicted at t, $Q(t)$ is included in the objective function, which aggregates the utilities (weighted correspondingly) that are realized if addiction has set in before t.

From a mathematical viewpoint, one advantage of the transformation to an age-structured form is that the two-stage model (with stochastic switch) can be represented by a single deterministic model, for which all influences (dependence of the switch on the stock of addictive capital, dependence of utility on second-stage decisions, inter-dependence between the first and second stage) can be glanced in a direct and intuitive way from the first order conditions, the shadow prices, as well as the stage-1 and stage-2 dynamics.

4.3 Optimal Allocation and Dynamics

To obtain the optimality conditions and the adjoint equations for problem (4.7), we apply the Maximum Principle for age-structured optimal control models (see Feichtinger et al. [8]). The Hamiltonian is given by

$$\begin{aligned}\mathcal{H}(t, s, \Omega, \Xi, \Psi) = z_1 u + Q + &\\ &\lambda_S(c_1 - \delta S_1) + \lambda_A(r A_1 + w - p c_1 - y_1) + \lambda_z(-\eta z_1) + \\ &\xi_S(c_2 - \delta S_2) + \xi_z \cdot 0 + \xi_A(r A_2 + \bar{w} - p c_2 - y_2)\end{aligned} \tag{4.8}$$

where Ω, Ξ and Ψ are vectors of the control, state and adjoint variables respectively, and where λ_k (for $k = \{S, A, z\}$) and ξ_k (for $k = \{S, A, z\}$) denote the adjoint variable of the states S, A and z corresponding to stages 1 and 2, respectively.

The first order conditions for inner solutions read (t and s are surpressend)

$$\frac{\partial \mathcal{H}}{\partial y_1} = z_1 u^1_{y_1} - \lambda_A = 0, \tag{4.9}$$

$$\frac{\partial \mathcal{H}}{\partial c_1} = z_1 u^1_{c_1} + \lambda_S - \lambda_A p = 0, \tag{4.10}$$

$$\frac{\partial \mathcal{H}}{\partial y_2} = z_2 u^2_{y_2} - \xi_A = 0, \tag{4.11}$$

$$\frac{\partial \mathcal{H}}{\partial c_2} = z_2 u^2_{c_2} + \xi_S - \xi_A p = 0. \tag{4.12}$$

If the above equation cannot be solved (i.e. if there are no inner solutions), the controls are equal to the boundary. For the adjoint equations, we obtain

$$\begin{aligned}\dot{\lambda}_S &= (\rho + \delta)\lambda_S + \eta_{S_1}\lambda_z z_1 - \xi_S(t,t) - \xi_z(t,t) z_1 \eta_{S_1}, \\ \dot{\lambda}_A &= (\rho - r)\lambda_A - \xi_A(t,t), \\ \dot{\lambda}_z &= (\rho + \eta)\lambda_z - u^1 - \xi_z(t,t)\eta, \\ \frac{d\xi_S}{dt} &= (\rho + \delta)\xi_S - z_2 u^2_{S_2}, \\ \frac{d\xi_z}{dt} &= \rho\xi_z - u^2, \\ \frac{d\xi_A}{dt} &= (\rho - r)\xi_A.\end{aligned} \tag{4.13}$$

with the following transversality conditions

$$\lim_{t\to\infty} e^{-\rho t}\lambda_i(t) = 0, \quad i \in \{S, A, z\},$$
$$\lim_{t\to\infty} e^{-\rho t}\xi_i(t, s) = 0, \quad \forall s, i \in \{S, A, z\}. \tag{4.14}$$

Note that all characteristic lines of model (4.7) are isolated. Thus, the transversality conditions for (time-dependent) optimal control models can be applied (see Grass et al. [11]) for every characteristic line.

Recalling our assumption of additive separability between the two consumption goods (as is in line with Becker and Murphy [2]), we obtain from (4.9) and (4.11) the Euler equations with respect to consumption of the non-addictive good

$$\dot{y}_1 = -\frac{u^1_{y_1}}{u^1_{y_1 y_1}}\left[r - \rho - \eta\frac{u^1_{y_1} - u^2_{y_2}}{u^1_{y_1}}\right], \tag{4.15}$$

$$\frac{dy_2}{dt} = -\frac{u^2_{y_2}}{u^2_{y_2 y_2}}(r - \rho). \tag{4.16}$$

While the stage-2 consumption dynamics correspond in a standard manner to the gap between interest rate and subjective discount rate, the stage-1 dynamics contain an additional effect. To the extent that at the point of becoming addicted, the marginal utility from conventional consumption drops below the value in the absence of addiction, i.e. to the extent that $u^1_{y_1} > u^2_{y_2}$, the individual tends to advance conventional consumption in line with the risk of becoming addicted. A reduction in the marginal utility of conventional consumption may arise due to a general numbing of the brain's reward system to both drug and non-drug related stimuli (see Volkow et al. [23]). The converse is true for $u^1_{y_1} < u^2_{y_2}$, which may arise, for instance, if the individual suffers a sharp drop in earnings due to its addiction. In this case, the individual accumulates precautionary savings which are then dissolved over time, as long as addiction does not set in.

The dynamics of addictive consumption are as follows:

$$\dot{c}_1 = -\frac{u^1_{c_1}}{u^1_{c_1 c_1}}\left[\underbrace{\left[(r-\rho) + \eta\frac{u^2_{y_2}}{u^1_{y_1}}\right]\frac{u^1_{y_1}}{u^1_{c_1}}p}_{(i.1)} + \underbrace{(\rho+\delta)\frac{\lambda_S}{z_1 u^1_{c_1}} - \frac{\xi_S}{z_1 u^1_{c_1}} + \eta_{S_1}\frac{\lambda_z - \xi_z}{u^1_{c_1}}}_{(ii.1)} + \underbrace{(-\eta)}_{(iii.1)}\right], \tag{4.17}$$

$$\frac{dc_2}{dt} = -\frac{u^2_{c_2}}{u^2_{c_2 c_2}}\left[\underbrace{(r-\rho)\frac{u^2_{y_2}}{u^2_{c_2}}p}_{(i.2)} + \underbrace{(\rho+\delta)\frac{\xi_S}{z_2 u^2_{c_2}} - \frac{u^2_{S_2}}{u^2_{c_2}}}_{(ii.2)} + \underbrace{\frac{u^2_{c_2 S_2}}{u^2_{c_2}}(c_2 - \delta S_2)}_{(iii.2)}\right]. \tag{4.18}$$

Stage 1—no addiction: According to (4.17), the consumption dynamics of the addictive good in the absence of addiction are driven by three forces:

(i.1) The typical Euler dynamics plus the expected stage-2 value of wealth conditional on transition into addiction. Both are weighted by a factor $\frac{u^1_{y_1}}{u^1_{c_1}}p$, which equals 1 in a setting in which the stock of addictive capital does not matter ($\lambda_S = 0$). In this case, c_1 is chosen just like a non-addictive good. Using (4.9) it is possible to eliminate λ_A from (4.10), which implies $\frac{u^1_{y_1}}{u^1_{c_1}}p < 1$ if $\lambda_S < 0$, i.e. if the sotck of addictive capital has a negative value. This suggests a stifling of dynamics for harmful addictive goods. The converse applies for $\lambda_S > 0$.

(ii.1) A collection of effects relating to the dynamics of the value of addictive capital,[8]

$$\lambda_S(t) = \int_t^\infty e^{-(\rho+\delta)(\tau-t)} \left[\xi_S(\tau,\tau) + z_1(\tau)\eta_{S_1}(\xi_z(\tau,\tau) - \lambda_z(\tau))\right] d\tau,$$

during stage 1. Assuming that the stock of addictive capital has a negative value in both stages 1 and 2, such that both $\lambda_S < 0$ and $\xi_S < 0$, the first two terms in (ii.i) imply an ambiguous effect. On the one hand, addictive consumption tends to be advanced in the presence of strong discounting of the future stock of addictive capital in stage 1; on the other hand, it tends to be postponed in a precautionary way if the stock of addiction carries a high negative value in stage 2. Finally, if the rate of addiction increases in the stock of addictive capital, $\eta_{S_1} > 0$, this tends to imply a precautionary postponement of consumption if the value of remaining without addiction,[9]

$$\lambda_z(t) = \int_t^\infty e^{-\rho(\tau-t)} \frac{z_1(\tau)}{z_1(t)} \left(u^1 + \eta \int_\tau^\infty e^{-\rho(\tau'-\tau)} u^2 d\tau'\right) d\tau,$$

exceeds the value of being addicted,[10]

$$\xi_z(t,s) = \int_t^\infty e^{-\rho(\tau-t)} u^2 d\tau,$$

evaluated at the time of transition into addiction, $s = t$.

(iii.1) Finally, the risk of moving into addiction itself tends to lead to an advancement of addictive consumption. This effect may appear somewhat counterintuitive

---

[8]The value of addictive capital at time t within the pre-addiction stage 1 consists of the discounted stream of (i) the stage-2 value of addictive capital, $\xi_S(\tau,\tau)$ if the transition occurs at τ plus (ii) the expected change in the value of the addictive capital for an incrase in the switching rate, $\eta_{S_1} > 0$, due to the accumulation of addictive capital in period τ.

[9]The value of remaining without addiction at time t consists of the discounted stream of (i) the value of pre-addiction utility, u^1, within period τ of the expected remaining lifetime without addiction plus (ii) the expected discounted stream of continuation utility in addiction, u^2, should a switch occur in period τ.

[10]Intuitively, the value of being addicted at some time $t \geq s$ consists of the discounted stream of stage-2 utility over the remaining lifetime (within addiction).

given that individuals might rather defer addictive consumption for precautionary reasons. Note, however, that this motivation is captured under the effects contained in (ii). The present effect is thus reflecting a *direct* effect, where the individual seeks to advance consumption in line with the risk that the present life-stage may end. Which of the effects, precautionary deferral or advancement of consumption, dominates is an empirical question.

Stage 2—addiction: According to (4.18), the consumption dynamics of the addictive good under addiction are driven by the following forces:

(i.2) The Euler dynamics, which are dampened to the extent that the stock of addictive capital carries a negative value, $\xi_S < 0$.

(ii.2) The dynamics of the value of addictive capital, ξ_S, under addiction. As is readily shown, we have that[11]

$$\xi_S(t, s) = z_1(s)\eta(S_1(s)) \int_t^{\infty} e^{-(\rho+\delta)(\tau-t)} u^2_{S_2}\, d\tau < 0$$

for $u^2_{S_2} < 0$. We then have a tendency towards advanced consumption due to discounting and a tendency towards postponement of consumption in order to lower the direct disutility from being addicted, $u^2_{S_2} < 0$. Both effects relate to the extent of tolerance, as measured by $u^2_{S_2}$. While strong discounting of the future discomfort associated with tolerance speaks for an advancement of addictive consumption, the converse is true in respect to the prevention of tolerance.

(iii.2) An increase in consumption over time that comes with the increase in the marginal utility of the addictive good, when addictive capital is accumulated, such that $\frac{dS_2(t,s)}{dt} = c_2 - \delta S_2 > 0$. This is the reinforcement effect of addiction.

Assuming for the moment that $r = \rho$ so that consumption would remain constant according to the conventional Euler dynamics (i.2), we see that an increase in addictive consumption occurs if and only if the stock of addictive capital is increasing, $\frac{dS_2(t,s)}{dt} > 0$, while, at the same time, its value, ξ_S, is increasing (i.e. getting less negative) or not decreasing by too much.

4.4 Steady-State Analysis and Skiba Points

As is standard for rational addiction models, we assume an infinite time horizon. Therefore, the long-run optimal solution approaches a steady state of the canonical system. This implies that we need to study the steady states for the dynamics of both stages, irrespective of the age-structured representation of our model.

[11] The value of addictive capital at time t, conditional on a transition into addiction at $s \leq t$, consists of the discounted stream of the utility loss from addcitive capital, $u^2_S < 0$, over the remaining lifetime $[t, \infty)$, weighted with the unconditional/ex-ante probability $z_1(s)\eta(S_1(s))$ of having had a transition into addiction at time s.

In order to establish an explicit solution of the steady state, from now on we assume $r = \rho$ and employ specific functions for the switching rate and utility. Specifically, we assume a linear function for the switching rate

$$\eta(S_1) = aS_1,$$

with $a > 0$.

In line with our previous assumption, the utility functions for both stages are separable in both consumption goods. Thus, for the second-stage utility (under addiction) we assume a sub-utility of non-addictive consumption $\frac{y_2^{1-\sigma}}{1-\sigma}$ (increasing and concave if $0 < \sigma < 1$), and a sub-utility for the addictive good that is akin to the one in Becker and Murphy [2]. Overall, we have

$$u^2(y_2, c_2, S_2) = \frac{y_2^{1-\sigma}}{1-\sigma} + \alpha_c c_2 - \frac{\alpha_{cc}}{2}c_2^2 - \alpha_S S_2 - \frac{\alpha_{SS}}{2}S_2^2 + \alpha_{cS}c_2 S_2,$$

with $\sigma \in [0, 1]$ and $\alpha_c, \alpha_{cc}, \alpha_S, \alpha_{SS}, \alpha_{cS} > 0$. Note that the utility should also decrease in S (i.e. $u_S^2 < 0$), which is not fullfilled for all possible choices of S and c. However, in our numerical example $u_S^2 < 0$ holds along the optimal paths in stage 2.

The utility function of the first stage is analogous to the second-stage one, with the exception that the parameters related to S_1 are zero, i.e. $\alpha_S = \alpha_{SS} = \alpha_{cS} = 0$. Thus,

$$u^1(y_1, c_1) = \frac{y_1^{1-\sigma}}{1-\sigma} + \alpha_c c_1 - \frac{\alpha_{cc}}{2}c_1^2,$$

where σ, α_c and α_{cc} are the same as for the second stage.

The following two subsections develop the steady states and their stability for the two stages.

4.4.1 Long-Run Allocation in Stage 1

Within this subsection, we derive the steady states for the first stage.

Interior steady state I-1:

Consider an inner solution in the steady state for both controls. A positive steady state value $\hat{c}_1$ implies a positive steady state value for $\hat{S}_1$, and therefore, (by the dynamics of z_1) $\hat{z}_1 = 0$.[12] The dynamics of S_1 and A_1 then yield $\hat{S}_1$ and $\hat{y}_1$ as a function of $\hat{c}_1$. Moreover, $\dot{\lambda}_S = 0$ implies (by using $\hat{z}_1$) $\hat{\lambda}_S = 0$.

From the first order condition for addictive consumption we obtain the following equation for $\hat{c}_1$,

[12] Steady state values are denoted by a hat.

$$\frac{\alpha_c - \alpha_{cc}\hat{c}_1}{p} = (w^1 - p\hat{c}_1)^{-\sigma}, \tag{4.19}$$

which can be solved numerically. The expressions for the control and state variables can be found in Table 4.1.

The analysis of the Eigenvalues of the Jacobian reveals that the steady state is always a saddle point (two negative, three positive and one zero Eigenvalues).

Boundary steady state NAC-1 (no addictive consumption):

Here, we assume that the addictive consumption is on the lower boundary, i.e. $\hat{c}_1 = 0$. The dynamics of the stock of addictive capital then implies $\hat{S}_1 = 0$. The dynamics of z_1 is negative for positive S_1 and zero otherwise. Thus, the steady state value of z_1 is not unique and lies in the interval [0, 1]. The dynamics of A_1, λ_z and λ_S, as well as the first order condition for non-addictive consumption, imply the steady state values of the remaining control and state variables, which are summarized in Table 4.1.

NAC-1 is feasible only if the first order condition with respect to addictive consumption is non-positive, i.e. only if $\left.\frac{\partial \mathcal{H}}{\partial c_1}\right|_{c_1=0} \leq 0$. After some manipulation, the condition reads (for a detailed derivation we refer to Appendix A)

$$\alpha_c \leq \frac{a\left(\frac{(w^1)^{1-\sigma}}{1-\sigma} - \frac{(w^2)^{1-\sigma}}{1-\sigma}\right)}{\rho(\rho+\delta)} + (w^1)^{-\sigma} p. \tag{4.20}$$

According to this condition, the marginal utility of the first unit of addictive consumption in stage 1 must fall short of the sum of (i) the marginal utility of the consumption of the non-addictive good that would be foregone in stage 1 and (ii) the value of maintaining a zero probability of getting addicted. Here, the valuation corresponds to the difference between the steady-state utility in the non-addiction as opposed to the addiction stage, the latter conditional on having shed the addiction in the steady state. Note here that for the (unlikely) onset of addiction from an (almost) zero addictive stock implies it can be shed (almost) instantaneously.

The analysis of the Jacobian (on the boundary) shows a saddle path property independent of the parameters.

Boundary steady state MNC-1 (minimal non-addictive consumption):

Analogously to NAC-1, we now assume a minimal non-addictive consumption steady state, i.e. $\hat{y}_1 = y_0$. Consequently, the asset dynamics implies a maximal long-term steady state value for the addictive consumption $\hat{c}_1 = \bar{c}_1 := \frac{w^1 - y_0}{p}$. By the dynamics of S_1 and z_1 we are able to derive the steady state values of the remaining state and control variables, which are again listed in Table 4.1.

The feasibility condition for this steady state is obtained by the first order condition for addictive consumption (i.e. $\left.\frac{\partial \mathcal{H}}{\partial c_1}\right|_{c_1=\bar{c}_1} \geq 0$), and turns out to be fulfilled for all parameter values.

Table 4.1 Steady states of the first stage (pre-addiction)

Steady state	States			Controls	
	$\hat{S}_1$	$\hat{A}_1$	$\hat{z}_1$	$\hat{y}_1$	$\hat{c}_1$
I-1	$\frac{\hat{c}_1}{\delta}$	0	0	$w^1 - p\hat{c}_1$	Solve (4.19)
NAC-1	0	0	$\hat{z}_1 \in [0, 1]$	w^1	0
MNC-1	$\frac{\bar{c}_1}{\delta}$	0	0	y_0	$\bar{c}_1$

The Eigenvalues of the Jacobian (on the boundary) show that the equilibrium is always a saddle point independent of the parameters.

Summary of steady states of stage 1:

In contrast to the boundary steady states, which can be derived analytically, the interior steady state can only be derived numerically (solution of (4.19)). For fixed σ, p and w^1 the value of $\hat{c}_1$ only depends on the parameters α_c and α_{cc} of the utility function. In the numerical example considered in Sect. 4.5, we adopt the parameters from Caulkins et al. [4].

4.4.2 Long-Run Allocation in Stage 2

Within this subsection, we derive the steady states of the second stage given that the individual became addicted at s. This implies the initial conditions $S_2(t,t) = S_1(t)$, $A_2(t,t) = A_1(t)$, and $z_2(t,t) = z_1(t)aS_1(t)$.

Interior steady state I-2:

Here, we assume interior solutions for both controls, implying a positive stock of addiction.

The assumption $r = \rho$ and the adjoint equation for ξ_A imply $\xi_A(t,s) = \xi_A(s,s)$ for $t \geq s$. Using this within the first order condition for non-addictive consumption, we obtain constant $y_2(t,s) = y_2(s,s)$ for $t \geq s$, i.e. constant non-addictive consumption during the second stage in the interior region. From the adjoint equation for ξ_S together with the dynamics of S_2 and A_2, as well as from the first order condition for non-addictive consumption, we obtain $\hat{y}_2$, $\hat{c}_2$, as functions of $\hat{S}_2$, as reported in Table 4.2. From the first order condition for c_2 we then get the following equation for $\hat{S}_2$,

$$0 = \frac{1}{\rho+\delta}\left(-\alpha_S - \alpha_{SS}\hat{S}_2 + \alpha_{cS}\delta\hat{S}_2\right) - p\left(w^2 - p\delta\hat{S}_2\right)^{-\sigma} + \left(\alpha_c - \alpha_{cc}\delta\hat{S}_2 + \alpha_{cS}\hat{S}_2\right), \tag{4.21}$$

which can be solved numerically.

Under the assumption $r = \rho$ the four-dimensional canonical system can be reduced to a two-dimensional one. Analysis of the Eigenvalues of this system then implies that the steady state exhibits the following stability properties,[13]

$$\begin{aligned} \rho^2 - 4a + 4b &< 0 : \text{unstable focus}, \\ \rho^2 > \rho^2 - 4a + 4b &> 0 : \text{unstable node}, \\ \rho^2 - 4a + 4b &> \rho^2 : \text{saddle point}, \end{aligned} \tag{4.22}$$

where

$$a = -\delta(\rho + \delta), \tag{4.23}$$

$$b = \frac{\alpha_{SS}}{\alpha_{cc}} - (2\delta + \rho)\frac{\alpha_{cS}}{\alpha_{cc}}. \tag{4.24}$$

Boundary steady state NAC-2 (no addictive consumption):

For this steady state we assume no addictive consumption, i.e. $\hat{c}_2 = 0$. From the dynamics of the stock of addiction we obtain $\hat{S}_2 = 0$ and from the asset dynamics $\hat{y}_2 = w^2$. All steady state values are listed in Table 4.2.

This boundary steady state is only feasibly if the first order condition for c_2 is non-positive (i.e. $\left.\frac{\partial \mathcal{H}}{\partial c_2}\right|_{c_2=0} \leq 0$). This implies (the proof is similar to that of (4.20) in Appendix A)

$$0 \geq -\frac{\alpha_S}{\rho + \delta} - (w^2)^{-\sigma} p + \alpha_c. \tag{4.25}$$

The Eigenvalues of the Jacobian (on the boundary) prove that this steady state is always a saddle point.

Boundary steady state MNC-2 (minimal non-addictive consumption):

In contrast to NAC-2, we now consider minimal non-addictive consumption, i.e. $\hat{y}_2 = y_0$. Consequently the long-term value of addictive consumption is maximal, i.e. $\bar{c}_2 := \frac{1}{p}(w^2 - y_0)$. From the dynamics for S_2 and the adjoint equations, the steady state values for the state variables can be obtained (listed in Table 4.2).

From the first order condition for c_2, we obtain the feasibility condition $(\left.\frac{\partial \mathcal{H}}{\partial c_2}\right|_{c_2=\bar{c}_2} \geq 0)$, which is (the proof is similar to that of (4.20) in Appendix A)

$$0 \leq \frac{1}{\rho + \delta}\left(-\alpha_S - \alpha_{SS}\frac{\bar{c}_2}{\delta} + \alpha_{cS}\bar{c}_2\right) - p y_0^{-\sigma} + \left(\alpha_c - \alpha_{cc}\bar{c}_2 + \alpha_{cS}\frac{\bar{c}_2}{\delta}\right). \tag{4.26}$$

---

[13] The Eigenvalue corresponding to $\frac{d\xi_A}{dt}$ is always equal to zero (since the dynamics is always zero). Since A_2 does not enter the equations for $\frac{dS_2}{dt}$ and $\frac{d\xi_S}{dt}$, the Eigenvalue corresponding to $\frac{dA_2}{dt}$ can be isolated as $r = \rho$.

Table 4.2 Steady states of the second stage (addiction)

Steady state	States		Controls	
	$\hat{S}_2$	$\hat{A}_2$	$\hat{y}_2$	$\hat{c}_2$
I-2	Solve (4.21)	0	$w^2 - p\delta\hat{S}_2$	$\delta\hat{S}_2$
NAC-2	0	0	w^2	0
MNC-2	$\frac{\bar{c}_2}{\delta}$	0	y_0	$\bar{c}_2$

The Eigenvalues of the Jacobian (on the boundary) prove that the steady state is always a saddle point.

Summary of steady states of stage 2:

Analogously to the first stage, the boundary steady states can be derived analytically, whereas for the interior one it is necessary to solve an equation numerically. The Eigenvalues of the interior steady states show that it is unstable (node or focus) rather than a saddle point if $b < a$ holds (obtained by manipulation of (4.22)). Employing (4.23) and (4.24) one can show that

$$b < a \Leftrightarrow \alpha_{cS} > \frac{\alpha_{SS}}{2\delta + \rho} + \frac{\delta(\delta + \rho)}{2\delta + \rho}\alpha_{cc},$$

which equals the condition that has been found in Caulkins et al. [4]. In similarity to the BM-model, this implies that the reinforcement effect (as captured by α_{cS}) be sufficiently strong relative to the tolerance effect (as captured by α_{SS} and α_{cc}) or, put differently, that there is sufficient adjacent complementarity in the spirit of Ryder and Heal [19].

The boundary steady states exist depending on conditions (4.25) and (4.26). This is the classical case for a Skiba point to occur in an optimal control model (see Grass et al. [11]). For the rational addiction model this has been found, e.g. by Caulkins et al. [4] for an easier version of the BM-model without non-addictive consumption and an intertemporal budget constraint. Nevertheless, the model shows for certain parameter constellations that Skiba points exist as part of the optimal solution. If Skiba points govern the optimal solution of the second stage, this has to be included in the optimization of the first stage. This becomes obvious from the first stage Euler equations with respect to non-addictive consumption (4.15) and addictive consumption (4.17).

In the next section the Skiba property of the second stage will be discussed in more detail.

4.4.3 Skiba Property in Stage 2

A Skiba point of an optimal control model is defined as an initial point (i.e. specific values for the initial conditions) for which two different optimal solutions exist, where the indifference point property implies that the two solutions yield the same value for the optimal control problem, and where the threshold point property implies that, in a small neighbourhood of the Skiba point, the two solutions converge to different limit sets. For a detailed definition see Grass et al. [11].

The classical case of a Skiba point involves an unstable steady state (i.e. unstable focus or node) with two saddle points that can be reached by a single unstable manifold of the unstable steady state. The I-2 steady state of the second stage of our model is an unstable focus or node if $b < a$ holds. The two boundary steady states are always saddle points. The above definition then implies that a Skiba point satisfies the following condition

$$V^{NAC2}(S, A) = V^{MNC2}(S, A), \tag{4.27}$$

where $V_{NAC2}(S, A)$ and $V_{MNC2}(S, A)$ are the value functions corresponding to the NAC-2 and the MNC-2 steady state, respectively. Both depend on (S, A), the initial conditions of stage 2, i.e. the values of S and A at the switching time from the first to the second stage. Due to autonomy there is no explicit dependence on the initial time (see Grass et al. [11]), i.e. the switching time in our context. Both value functions consist of the utility accumulated within the inner region and of the utility on the boundary. Defining $V_I^{NAC2}(S, A)$ and $V_B^{NAC2}(S_2(\tau_1))$ as the parts of the NAC-2 value function corresponding to the inner region and the boundary steady state, respectively, we have

$$V^{NAC2}(S, A) = V_I^{NAC2}(S, A) + e^{-\rho\tau_1} V_B^{NAC2}(S_2(\tau_1))$$

with

$$\begin{aligned} V_I^{NAC2}(S, A) = &\frac{\hat{y}_2^{1-\sigma}(1 - e^{-\rho\tau_1})}{(1-\sigma)\rho} \\ &+ \int_0^{\tau_1} e^{-\rho t} \Big(\alpha_c c_2(t) - \frac{\alpha_{cc}}{2}(c_2(t))^2 - \alpha_S S_2(t) \\ &- \frac{\alpha_{SS}}{2}(S_2(t))^2 + \alpha_{cS} c_2(t) S_2(t)\Big)\, dt \\ V_B^{NAC2}(S_2(\tau_1)) = &\frac{1}{\rho}\left[\frac{(w^2)^{1-\sigma}}{1-\sigma} - \frac{\rho\alpha_S}{\rho+\delta} S_2(\tau_1) - \frac{\rho\alpha_{SS}}{2(\rho+2\delta)}(S_2(\tau_1))^2\right]. \end{aligned} \tag{4.28}$$

Here, $V_I^{NAC2}(S, A)$ consists of the present value of the utility stream enjoyed along the (optimal) unstable manifold approaching the boundary $c_2 = 0$, with $c_2(t)$ and $S_2(t)$ being the control and state paths in the interval $t \in [0, \tau_1)$. The time at which the boundary is reached, τ_1, now refering to the time since the onset of addiction is determined by the condition $A_2(\tau_1) = 0$, where switching at another value would contradict the condition $\lim_{t\to\infty} A_2(t) = 0$. An analytic derivation of this case is tedious (if possible at all) and would not give further insight due to very long expressions.

The value $V_B^{NAC2}(S_2(\tau_1))$ denotes the present value of the utility stream enjoyed for the boundary $c_2 = 0$. Noting that it can be derived analytically, it is easy to see that it only depends on $S_2(\tau_1)$, which is the value of the stock of addiction when the optimal trajectory hits the boundary.

Analogously we can define the value function of stage 2 corresponding to the MNC2 steady state.

$$V^{MNC2}(S, A) = V_I^{MNC2}(S, A) + e^{-\rho\tau_2} V_{MNC2}^B(S_2(\tau_2))$$

with

$$\begin{aligned} V_I^{MNC2}(S, A) = &\frac{\hat{y}_2^{1-\sigma}(1 - e^{-\rho\tau_2})}{(1-\sigma)\rho} \\ &+ \int_0^{\tau_2} e^{-\rho t}\Big(\alpha_c c_2(t) - \frac{\alpha_{cc}}{2}(c_2(t))^2 - \alpha_S S_2(t) - \\ &\frac{\alpha_{SS}}{2}(S_2(t))^2 + \alpha_{cS} c_2(t) S_2(t)\Big)\, dt \\ V_B^{MNC2}(S_2(\tau_2)) = &\frac{1}{\rho}\Big[\frac{y_0^{1-\sigma}}{1-\sigma} + \left(\alpha_c - \frac{\alpha_S}{\rho+\delta}\right)\bar{c}_2 - \left(\frac{\alpha_{cc}}{2} + \frac{\alpha_{SS}}{(\rho+\delta)(\rho+2\delta)}\right)\bar{c}_2^2 \\ &-\frac{\rho\alpha_S}{\rho+\delta} S_2(\tau_2) - \frac{\rho\alpha_{SS}}{2(\rho+2\delta)}(S_2(\tau_2))^2 + \\ &\left(\alpha_{cS} - \frac{\rho\alpha_{SS}}{(\rho+\delta)(\rho+2\delta)}\right)\bar{c}_2 S_2(\tau_2)\Big]. \end{aligned} \tag{4.29}$$

In general, condition (4.27) cannot be solved analytically but only numerically. It is possible, however, for the special case of a static budget constraint $w_i - pc_i(\cdot) - y_2(\cdot) = 0$ for $i = 1, 2$, which we obtain by setting $\dot{A}_i(\cdot) = A_i(\cdot) \equiv 0, \quad i = 1, 2$ in (4.4) and (4.6), respectively. Without the asset state, one control can be directly expressed as a function of the other. This simplification corresponds to the model that has been used by Ferguson [9], to provide a profound interpretation of the rational addiction model. Ferguson [9] does not consider the Skiba property, however, but is more interested in the impact of price variations. Caulkins et al. [4] consider another version of the rational addiction model without any budget constraint, neither static nor dynamic. This implies a different condition for the Skiba point. The equilibrium properties of the modified second stage model are given in the following Proposition.

Proposition 1 *Consider the optimal control model of the second stage (4.6) without assets, such that $w^2 = c_2(t)p + y_2(t)$ holds for $\forall t$, and let $0 > \delta(\rho+\delta) + \frac{\partial c}{\partial S}\left(\frac{\alpha_{SS}}{\alpha_{cS}} - \rho\right)$, such that the interior steady state is no saddle point. The following holds. If $\beta_1\beta_2 > 0$ where*

$$\beta_1 = \frac{y_0^{1-\sigma}}{1-\sigma} - \frac{(w^2)^{1-\sigma}}{1-\sigma} + \bar{c}_2\left(\alpha_c - \frac{\alpha_S}{\rho+\delta}\right) - \bar{c}_2^2\left(\frac{\alpha_{cc}}{2} + \frac{\alpha_{SS}}{(\rho+\delta)(\rho+2\delta)}\right)$$
$$\beta_2 = \bar{c}_2\left(\alpha_{cS} - \frac{\rho\alpha_{SS}}{(\rho+\delta)(\rho+2\delta)}\right), \tag{4.30}$$

a Skiba point S^* does not exist. If $\beta_1 > 0$ ($\beta_1 < 0$) MNC-2 dominates (is dominated by) NAC-2. If $\beta_1\beta_2 < 0$, S^* exists and is given by

$$S^* = -\frac{\beta_1}{\beta_2}. \tag{4.31}$$

Proof $0 > \delta(\rho+\delta) + \frac{\partial c}{\partial S}\left(\frac{\alpha_{SS}}{\alpha_{cS}} - \rho\right)$ ($\frac{\partial c}{\partial S} > 0$ can be derived explicitely) implies that the interior steady state is unstable (node or focus). It is obtained by evaluating the Eigenvalues of the Jacobian of the inner equilibrium of the second stage model without dynamic budget constraint. Since this derivation is quite standard, it is omitted.

A static budget constraint equals the case of $A = 0$, which is equivalent to case $\tau_1 = \tau_2 = 0$. Therefore, (4.27) can be reduced to $V_B^{NAC2}(S) = V_B^{MNC2}(S)$. After some manipulation we obtain a linear function $\Lambda(S) = \beta_1 + \beta_2 S$ (β_1 and β_2 are given in the Proposition). Solving $\Lambda = 0$ gives the expression for S^*.

If $\beta_1 > 0$ and $\beta_2 > 0$ holds no solution can be found and MNC-2 dominates NAC-2. If, on the other hand, $\beta_1 < 0$ and $\beta_2 < 0$ holds, NAC-2 dominates MNC-2.

□

The explicit condition contained in the proposition is a particular convenience of our modelling framework and is accessible to intuition. Noting that for $\dot{A}_2(\cdot) = A_2(\cdot) \equiv 0$, any switch to addiction will immediately lead to the boundary allocation such that $\tau_1 = \tau_2 = 0$ and the condition for a Skiba point reads , $V_B^{NAC2}(S, 0) = V_B^{MNC2}(S, 0)$. It can be checked then that $\beta_1 = V_B^{MNC2}(0, 0) - V_B^{NAC2}(0, 0)$ gives the net benefit of continued addictive consumption in the (hypothetical) absence of addictive capital, while β_2 denotes the marginal net benefit from having an addictive stock of capital $S \geq 0$ at the point of entering addiction and continuing addictive consumption. Note that $\beta_2 > (< 0)$ if reinforcement effects play a more prominent role than tolerance. It is then immediate that if both $\beta_1 < (>) 0$ and $\beta_2 < (>) 0$ it never (always) pays to remain in addiction. More interestingly, for $\beta_1 > 0 > \beta_2$ the individual would prefer the continued consumption of the addictive good but only at low levels of addictive capital, to begin with, suggesting that the NAC-2 boundary, i.e. abstinence, is reached if and only if S is sufficiently high at the point of switching. In contrast, for $\beta_1 < 0 < \beta_2$ the individual would prefer to give up addictive consumption when addictive capital plays little role but may be drawn into permanent addiction in the presence of a high stock of addictive capital to begin with. Here, the NAC-2 boundary is reached if and only if S is sufficiently low at entry into addiction.

4.5 Numerical Example

In this section, we solve the model with the parameters listed in Table 4.3. For our example, we employ the same parameter values that were used by Caulkins et al. [4] with the exception of (i) σ, w^1 and w^2, which were not used previously and, therefore, have been fixed by us; and (ii) α_S and α_{cS}, which in Caulkins et al. [4], were used as bifurcation parameters to analyze the different scenarios concerning the Skiba property of the model. Before turning to the results of the numerical example, two remarks are in order:

No calibration: The numerics employed here are meant to illustrate the qualitative behaviour of a two-stage approach to the BM-rational addiction model, in particular with respect to uncertainty, with the exception of Orphanides and Zervos [17], has been neglected so far. It is not our intention within the scope of this exercise to provide a quantitatively realistic calibration, which would need to relate to specific addictive goods. We reserve this for future research.

No bifurcation analysis: We do not provide a bifurcation analysis either. Although interesting from a mathematical point of view, this is beyond the scope of this article, and again, will be the focus of another paper.

In the following, we discuss the results of the numerical example point by point:

Non-addictive consumption (Fig. 4.1 **left panel)**: Consumption of the non-addictive good during the pre-addiction stage 1 takes off from a low level and subsequently converges towards its steady state. From the Euler equation (4.15), it can be glanced that under our assumption $r = \rho$, the reason for this lies with the drop in income, and thus, in the consumption level should the individual become addicted.[14] In order to cushion this, the individual initially engages in precautionary saving which is gradually reduced over time allowing for increasing levels of consumption when addiction does not set in. Note that such adjustment behaviour is not possible if we consider addictive consumption alone (e.g. Caulkins et al. [4]) and/or a static budget constraint (e.g. Ferguson [9]). Consumption of the non-addictive good during the addiction stage 2 is not plotted, as it is merely reflecting the pattern of addictive consumption after the onset of addiction. In any of the scenarios, the individual then consumes the steady state level $\hat{y}_2$. As we will see below, the parameter constellation we employ implies the presence of a Skiba point. It then follows that whenever the stock of addiction is sufficiently low, as is the case for early enough transitions (see left panel of Fig. 4.2), the individual switches to the maximum (NAC-2) level of non-addictive consumption, whereas it switches to minimal (MNC-2) consumption for late transitions at a high stock of addiction.

Addictive consumption (Fig. 4.1 **right panel)**: Consumption of the addictive good during the pre-addiction stage 1 is declining over time towards a steady state. Inspection of the Euler equation (4.17), shows that the forces pulling towards an advancement of consumption, in particular, the increasing risk of becoming

[14] Formally, this implies that $u_y^2 > u_y^1$ in (4.15).

addicted dominate the forces pushing towards a delay, in particular, the tendency to lower precautionary saving. What role is played by the change in the shadow price of addictive capital is ambiguous without further analysis. For the addiction stage 2, the Skiba property can be observed. Here, the particular parametrization of our model implies that a MNC-2 steady state with permanent high-level consumption of the addictive good is reached if and only if the stock of addictive capital upon entry into addiction is sufficiently high. Given the ongoing accumulation of addictive capital during the pre-addiction stage, this implies that if the individual gets addicted quite early (approximately up to $t = 35$), it is optimal to follow NAC-2, i.e. to stop the consumption of the addictive good. If the addiction sets in later, the individual has to "pay up for the gamble" during the first stage, i.e. it is optimal to become heavily addicted and to follow MNC-2. The optimal consumption paths towards the respective boundary are depicted by dashed lines, whereas the boundary steady states are depicted by solid lines. The times, τ_1 and τ_2, at which the boundary steady state is reached can be glanced immediately. It turns out that for our parametrization the time span to shed an addiction (i.e. to reach NAC-2) is somewhat longer than the time span towards becoming maximally addicted (i.e. to reach MNC-2). Furthermore, the speed of approaching either of the boundary steady states is moderately increasing with a later onset of addiction. Thus, the duration of, as one could say perhaps, "moderate" addiction is declining for those individuals who get addicted at a late stage. Interestingly the difference between the levels of both steady states is quite considerable. Going to the NAC-2 steady state (possible in finite time), is near the 'cold turckey'-behaviour, which means that the consumption of the addictive good has to be stopped immediately.

Addictive capital (Fig. 4.2 **left panel)**: Ongoing consumption of the addictive good implies an increasing stock of addictive capital which in the absence of a switch to addiction approaches a steady-state level. The figure also illustrates that it is optimal to shed an addiction if and only if the addiction stock is sufficiently low at the switch, which due to the ongoing accumulation of addictive capital during stage 1 implies that individuals who acquire an addiction early on will subsequently shed it, whereas individuals who acquire an addiction at a late stage will continue to build up addictive capital at an even higher rate.

Assets (Fig. 4.2 **right panel)**: The shape of the assets illustrates the early accumulation of precautionary savings and their subsequent running down. While this holds true even for the case in which the individual remains without addiction, the onset of addiction will induce a fast melt-down of assets and, indeed, the accumulation of some debt regardless of whether or not an addicted individual sheds the addiction. The course of addiction, however, bears on the use of the assets. While individuals who are shedding their addiction employ the assets to increasingly support their non-addictive consumption despite a reduction in their earnings; individuals who are moving to become long-term addicts require funding for their increasing levels of addictive consumption. The figure also shows that the accumulation of debt depends on the timing of addiction. Individuals who fall into addiction at a relatively late stage are more prone to accumulate debt. Notably, however, this is not due to the fact that they are more prone to turn out long-term addicts but rather

Table 4.3 Parameters employed for the numerical example

$\rho = r$	0.05	σ	1.5
a	0.1	α_C	5
p^c	1	α_{CC}	10
δ	0.1	α_S	1
w^1	3	α_{SS}	0.16
w^2	1.5	α_{CS}	1.6

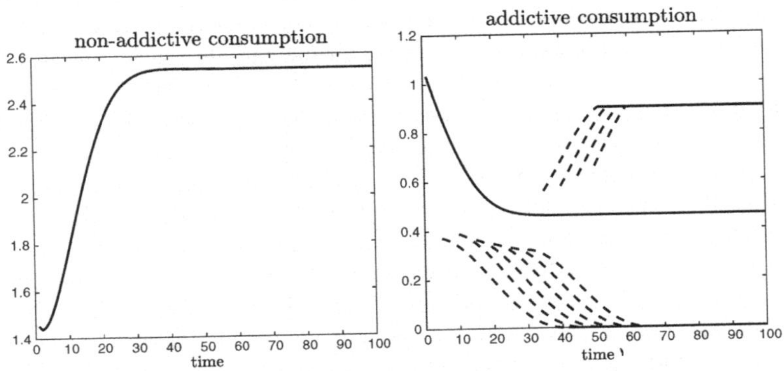

Fig. 4.1 Non-addictive and addictive consumption over time

down to the fact that by the time they become addicted, they have already run-down their assets. In the data, this would suggest a spurious correlation between becoming indebted and the propensity to become a long-run addict.

Probability of remaining without addiction over time (Fig. 4.3): The probability of staying remaining without addiction is a decreasing function over time, which is trivial as long as the addiction stock does not diminish over time. As we have seen, the probability of becoming addicted acts as an additional discount factor within the pre-addiction stage. Indeed, this is reflected in the adjoint Eq. (4.13), for z_1, where η enters in the same way as ρ. As the propensity to remain without addiction is driven by the level of the addiction stock, however, this part of the discount factor is endogenous, and indeed, increasing over time. At this point, it is well worth cautioning that we are not claiming our parametrization to be representative of any particular setting or type of addiction. Thus, it is quite possible, for instance, that for certain addictive goods, a declining pattern of consumption, such as we observe in Fig. 4.1, ultimately leads to a decline in the stock of addictive capital, and thus, to a positive probability never acquiring an addition. Naturally, such a scenario is then also prone to lead to very different dynamics and Skiba properties with respect to the other variables.

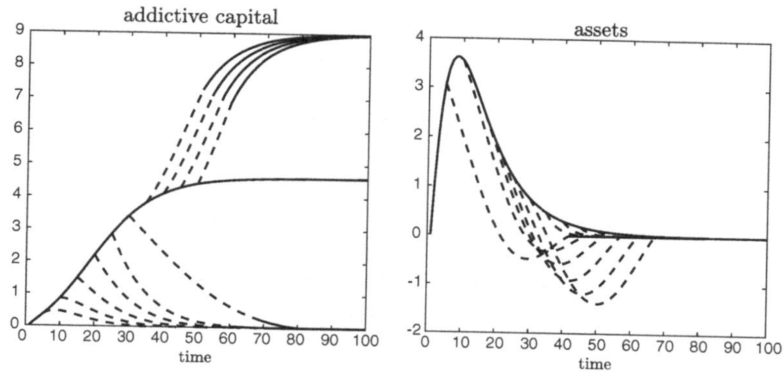

Fig. 4.2 Addictive capital and assets over time

Fig. 4.3 Probability of remaining without addiction over time

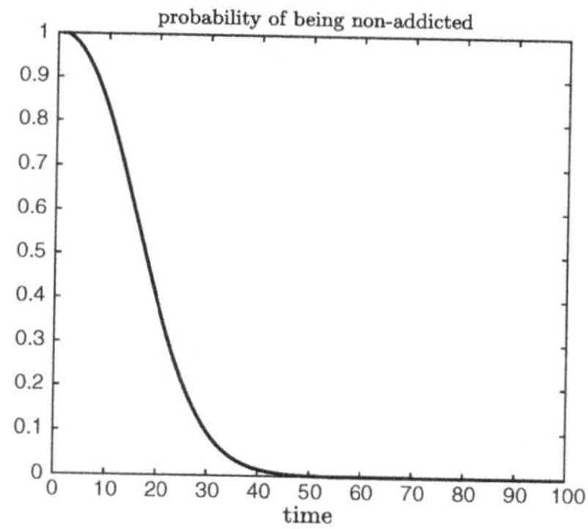

4.6 Conclusions

In this contribution, we have proposed an extension to the classical Becker and Murphy [2] model of rational addiction which introduces a pre-addiction stage during which the individual consumes the addictive good at an increasing risk of becoming addicted but is not yet subject to the pathology of addiction. In modelling addiction akin to a non-communicable disease, the onset of which is random but to some extent preventable, we follow the gist of current neurological research on addiction (see Volkow et al. [23]). Our modelling allows us to study the interaction between the behavioural patterns of addicted individuals, the timing of the onset of addiction and the risky consumption, as well as other behaviours before the set-on of addiction. The mathematical modelling of such a set-up poses challenges in that it involves a two-stage optimal control model with endogenous and random switching. We apply a novel transformation to an age-structured optimal control model by Wrzaczek et al. [24], to provide a solution to the model that allows us to depict in a compact and coherent way how the consumption patterns (and other model dynamics) are linked across the two stages. A final contribution lies in an analysis of how the emergence of divergent optimal behaviour (towards either the termination of addictive consumption or long-run high-level addictive consumption) in the presence of a Skiba point depends on the behavioural patterns before the onset of addiction. For a simple set-up with a static budget constraint we are, thus, able to provide precise conditions showing which path is followed, depending on the stock of addictive capital at the switching point which can then be linked to pre-addictive behaviours. A numerical analysis illustrates this link, where for the parameters under consideration an early transition into addiction (with a relatively low stock of addictive capital) induces the individual to terminate it in finite time, whereas a late transition (with a relatively high stock of addictive capital) induces the individual to stick and reinforce its addictive habit up to a point of maximal addictive consumption.

Such a behavioural pattern is not inconsistent with the gateway theory according to which unchecked recreational consumption of cannabis consumption may open the way into addiction from stronger drugs such as opiates (see e.g. Hall [13] and Curran et al. [6]). We should caution, however, that for the moment one should mostly take our findings for their illustrative character. Whether or not there is a positive correlation between a later onset of addiction and a tendency for it to progress into permanent and heavy addiction is testable, in principle. Before confronting this result against the evidence, however, we would need to ascertain its robustness both in light of the assumption that there is no asset accumulation and in light of the specific parametrization. In future research, we intend to apply the framework to study behavioural patterns before and after the onset of addiction, based on a more careful calibration of the model to data on the habits and outcomes related to particular addictive substances. We will also explore the underlying incentives and how they can be shaped by policymaking. In so doing, one particular focus will be on the role of policymaking with respect to the prevention of risky behaviours. This sets our work apart from most of the previous research based on the Becker and Murphy model, which implicitly or explicitly assumes the immediate presence of addiction, and thus, does not allow for a proper conception of the idea of "prevention".

Acknowledgements The authors are grateful for very helpful and constructive comments by two anonymous referees.

A Derivation of Equation (4.20)

Starting point is $\left.\frac{\partial \mathcal{H}}{\partial c_1}\right|_{c_1=0} \leq 0$, i.e.

$$z_1 u^1_{c_1} + \lambda_S - \lambda_A p \leq 0. \tag{4.32}$$

Using (4.9) we obtain

$$z_1 u^1_{c_1} + \lambda_S - p z_1 u^1_{y_1} \leq 0. \tag{4.33}$$

Using the functional specification and $\hat{c}_1 = 0$ we arrive (after rearrangement) at

$$\alpha_c \leq p(w_1)^{-\sigma} - \frac{\lambda_S}{z_1}. \tag{4.34}$$

Solving and inserting the steady state value of λ_S, which is $\hat{\lambda}_S = \frac{\alpha z_1}{\rho(\rho+\delta)}\left(\hat{u}^2 - \hat{u}^1\right)$, we obtain (4.20). For the derivation of λ_S we used the result by Caulkins et al. [4] stating that the long-run optimal solution converges to the non-addictive consumption steady state if S is sufficiently low.

Equation (4.25) and (4.26) can be proven analogously.

References

1. Auld, M. C., & Matheson, J. A. (2014). Addiction. In A. E. Culyer (Ed.), *Encyclopedia of health economics* (Vol. 1, pp. 19–25). North Holland: Elsevier.
2. Becker, G. S., & Murphy, K. M. (1988). A theory of rational addiction. *Journal of Political Economy*, *96*(4), 675–700.
3. Caputo, M. R. (2017) Uncovering the behavioral implications of the rational addiction assertion. *The B.E. Journal of Theoretical Economics* 20160120.
4. Caulkins, J. P., Feichtinger, G., Hartl, R. F., Kort, P. M., Novak, A. J., & Seidl, A. (2013). Multiple equilibria and indifference-threshold points in a rational addiction model. *Central European Journal of Operations Research*, *21*(3), 507–522.
5. Cawley, J., & Ruhm, C. (2012). The economics of risky health behaviors. In M. V. Pauly, T. G. McGuire, & P. P. Barros (Eds.), *Handbook of health economics* (Vol. 2, pp. 95–199). New York: Elsevier.
6. Curran, H. V., Freeman, T. P., Mokrysz, C., Lewis, D. A., Morgan, C. J., & Parsons, L. H. (2016). Keep off the grass? Cannabis, cognition and addiction. *Nature Reviews Neuroscience*, *17*(5), 293–306.
7. Dockner, E. J., & Feichtinger, G. (1993). Cyclical consumption patterns and rational addiction. *American Economic Review*, *83*(1), 256–263.
8. Feichtinger, G., Tragler, G., & Veliov, V. M. (2003). Optimality conditions for age-structured control systems. *Journal of Mathematical Analysis and Applications*, *288*, 47–68.
9. Ferguson, B. S. (2000). Interpreting the rational addiction model. *Health Economics*, *9*, 587–598.
10. Gavrila, C., Feichtinger, G., Tragler, G., Hartl, R. F., & Kort, P. M. (2005). History-dependence in a rational addiction model. *Mathematical Social Sciences*, *49*(3), 273–293.
11. Grass, D., Caulkins, J. P., Feichtinger, G., Tragler, G., & Behrens, D. (2008). *Optimal control of nonlinear processes: With applications in drugs.*, Corruption and terror Berlin: Springer.
12. Gruber, J., & Köszegi, B. (2001). Is addiction rational? Theory and evidence. *Quarterly Journal of Economics*, *116*, 1261–1303.
13. Hall, W. (2015). What has research over the past two decades revealed about the adverse health effects of recreational cannabis use? *Addiction*, *110*(1), 19–35.
14. Jones, A. M., Laporte, A., Rice, N., & Zucchelli, E. (2019). Dynamic panel data estimation of an integrated Grossman and Becker-Murphy model of health and addiction. *Empirical Economics*, *56*(2), 703–733.
15. Koob, G. F. (2011). Neurobiology of addiction. *Focus*, *9*(1), 55–65.
16. Melberg, H. O., & Rogeberg, O. J. (2010). Rational addiction theory: A survey opinions. *Journal of Drug Policy Analysis*, *3*(1), 5.

17. Orphanides, A., & Zervos, D. (1995). Rational addiction with learning and regret. *Journal of Political Economy*, *103*, 739–758.
18. Orphanides, A., & Zervos, D. (1995). Myopia and addictive behaviour. *Economic Journal*, *108*, 75–91.
19. Ryder, H. E, Jr., & Heal, G. M. (1973). Optimum growth with intertemporally dependent preferences. *Review of Economic Studies*, *40*, 1–33.
20. Strulik, H. (2018). Smoking kills: An economic theory of addiction, health deficit accumulation, and longevity. *Journal of Health Economics*, *61*, 1–12.
21. Veliov, V. M. (2003). Newton's method for problems of optimal control of heterogeneous systems. *Optimization Methods and Software*, *18*(6), 689–703.
22. Veliov, V. M. (2008). Optimal control of heterogeneous systems: Basic theory. *Journal of Mathematical Analysis and Applications*, *346*, 227–242.
23. Volkow, N. D., Koob, G. F., & McLellan, A. T. (2016). Neurobiologic advances from the brain disease model of addiction. *New England Journal of Medicine*, *374*, 363–371.
24. Wrzaczek, S., Kuhn, M., & Frankovic, I. (2020). Using age structure for a multi-stage optimal control model with random switching time. *Journal of Optimization Theory and Optimization*, *184*, 1065–1082.

Chapter 5
Modeling Social Status and Fertility Decisions Under Differential Mortality

Sergey Orlov, Elena Rovenskaya, Matthew Cantele, Marcin Stonawski, and Vegard Skirbekk

5.1 Introduction

Investigations into the behavior of social animals have revealed a strong correlation between higher fertility and a higher social status at the individual level. For example, the closest animal relatives of modern-day humans, chimpanzees, and bonobos have been found to exhibit a direct positive relationship between their hierarchy status and fertility. Here, it is particularly interesting that this relationship seems to hold regardless of the sex, as bonobos' societies are matriarchal and chimpanzees' societies are patriarchal societies [34, 37, 45, 49, 50]. In societies of these social animals, the dominant male (in case of chimpanzees) and male offspring of the dominant female (in case of bonobos) have greater mating opportunities, often limiting the reproductive success of other members.

S. Orlov (✉) · E. Rovenskaya · M. Cantele
International Institute for Applied Systems Analysis, Laxenburg 2361, Austria
e-mail: orlov@iiasa.ac.at

S. Orlov · E. Rovenskaya
Faculty of Computational Mathematics and Cybernetics, Lomonosov Moscow State University, Moscow 119991, Russia

M. Cantele
School of BioSciences, University of Melbourne, Parkville, Victoria, Australia

M. Stonawski
Center for Advanced Studies of Population and Religion (CASPAR), Cracow University of Economics, 31-510 Cracow, Poland

V. Skirbekk
Centre for Fertility and Health, Norwegian Institute of Public Health, Nydalen, PO Box 4404, 0403 Oslo, Norway

Columbia Aging Center, Mailman School of Public Health, Columbia University, New York, NY 10032, USA

J. L. Haunschmied et al. (eds.), *Dynamic Economic Problems with Regime Switches*, Dynamic Modeling and Econometrics in Economics and Finance 25,
https://doi.org/10.1007/978-3-030-54576-5_5

Empirical analyses find that the fertility between humans and their social status (defined based on education, occupation, and income/wealth) varies by context. A number of studies, particularly from more recent periods, indicate a relatively low fertility for individuals with a higher social status, especially for women [19, 28, 36, 43], while some studies find that higher social status of humans relates to higher fertility [20, 41] as with the majority of non-human social animals. Both situations are observed in different environments and under different social conditions even within the same country [2]. A negative relation between social status and fertility in humans presents a stark contrast to the positive relation observed for social animals. This discrepancy has been used by scholars to critique the use of socio- and evolutionary biology in understanding human behavior [48].

We attempt to explain this discrepancy by applying notions of evolutionary life history theory to our understanding of the modern demographic trends. This theory posits that individuals face a resource allocation problem in their attempt to achieve higher biological fitness [33]; see e.g., [26, 27]. Within certain environments, having a lower number of offspring, but with a higher investment in their well-being, may be a viable strategy to increase fitness [30, 40]. Longitudinal studies are scarce but some indicate that preferring status to fertility over the relatively short time frame (ca. 5 generations) reduces long-term fitness [24].

Natural disasters, including droughts, floods, infectious disease outbreaks, hurricanes, extreme temperature swings, and tsunamis often have long-term demographic implications [16, 17, 25, 39], contributing to high mortality and a lower population growth before the demographic transition (UN [32, 46]. As a consequence, the fertility level necessary to maintain a given population size in the context of fluctuating mortality varied [8, 32]. European societies have over historic periods applied mechanisms that regulated fertility up or down, maintaining a certain population equilibrium given resource availability and mortality, also in anticipation of natural events and resource constraints [7, 51]. Indeed, one's exposure to both climatic and famine-related risk has been found to correlate with the fertility behavior [35]. In several case studies, the regions with a greater frequency of extreme weather events have been found to exhibit relatively high fertility levels, and also a high preference to fertility [9, 12, 42].

Some researchers have theorized that social status plays an important role for survival and eventually higher fitness in the context of frequent and severe high-mortality events [11]. Education as one component of the social status has also been linked explicitly to a lower disaster vulnerability [18, 22, 38] as well as lower (particularly female) fertility [43]. Baker et al. [3] offer a meta-analysis of how education affects adult mortality; they show that lower education is associated with a higher likelihood of death.

Thus, the fact that the human population has been frequently subjected to high-mortality events throughout the vast majority of human evolutionary history may explain the drive toward higher social status, even at a cost to fertility. In light of the extant theoretical research, we pose the following research question: Within which environments could social-status-seeking at a cost to fertility have evolved as an adaptation for survival and ultimately higher biological fitness over the long term?

5.2 The Model

In order to address our research question, we propose a model that explicitly accounts for the trade-off between social status and fertility in different economic, environmental, and social situations.

We assume that at time moments $t = 1, 2, 3 \ldots$ there appears and lives generation t consisting of N_t adult individuals, homogeneous in all their attributes. Every individual from generation t inherits capital B_{t-1} from the previous generation and earns wage w_t over their entire life. These together make up capital $K_t = B_{t-1} + w_t$ she has available, which she can use for on consumption, rearing children, and bequest. Every individual in generation t is assumed to choose, independently from others, the number of children to have (denoted by n_t), on the amount of resources to allocate into their children's education (denoted by $\lambda\alpha_t$, where λ is the cost of full education and $\alpha_t \in [0, 1]$ is the education level of offspring), and on the share of the reminder $K_t - \kappa n_t - \lambda\alpha_t n_t$ to be passed on to the next generation (denoted by b_t; here κ is the amount of resources required to rear one child).

We assume wage w_t to be positively related to the individual's education level α_t as follows

$$w_t = s + r\lambda\alpha_{t-1},$$

where $s > 0$ is a minimum wage[1] that an individual gets over their entire life without optional education, r is a coefficient defining how the education level pays off in terms of the total wage. In case of $r > 1$, education pays off and an individual earns more ($r\lambda\alpha_t$) than their parents directly invested in their education ($\lambda\alpha_t$). In case of $r = 1$, education is budget-neutral. In case of $0 < r < 1$, optional education allows an individual to earn extra money, but it costs more to their parents than an individual actually gets. In case of $r = 0$, education has no influence on the wage of an individual. Case $r < 0$ describes the situation, in which people with higher education receive a lower salary than people without education over their whole life. This can occur, for example, because they spend a significant part of their life on obtaining education, while non-educated people start working much earlier.

Each generation t and their offspring are assumed to suffer from natural disasters, which can lead to premature deaths, namely, mortality events are happening before children become adults and make their own decisions; we assume that those remove fraction of children $\mu(\alpha_t)$ given their education level α_t. Consistent with the results revealing a negative correlation between the number of deaths caused by natural disasters and the education level of subjects, in our model, we assume differential mortality and use the power-law function for the probability of death of every child; i.e., we assume $\mu(\alpha_t) = 1 - \alpha_t^{\beta}$. Here parameter $\beta \geq 0$ characterizes the strength of mortality events affecting offspring of generation t. After suffering from natural disasters, individuals of generation t fade away and those children who

[1]It is also assumed here that the minimal level of individual's consumption necessary to survive is already subtracted from the value of s.

survive become adults of new generation $t+1$, make their decisions and the process repeats. Thus, the dynamics of the population size and the total capital per individual are the following:

$$N_{t+1} = N_t[1 - \mu(\alpha_t)]n_t,\ N_1 = \bar{N}_1, \tag{5.1}$$

$$K_{t+1} = b_t(K_t - \kappa n_t - \lambda\alpha_t n_t) + s + r\lambda\alpha_t,\ K_1 = \bar{K}_1. \tag{5.2}$$

Here $\bar{N}_1$ is the population size of the first generation, $\bar{K}_1$ is the total capital per individual in the first generation, and b_t is the fraction of $(K_t - \kappa n_t - \lambda\alpha_t n_t)$ that is bequeathed to offspring of an individual in generation t.

Being consistent with a life history theory approach of optimal resource allocation, we describe an individual's decision-making problem as follows. Each generation t is assumed to derive their utility J_t from (a) their consumption (which is a standard assumption (e.g., [5, 6]), (b) the expected number of their survived children, (c) their children's education (equivalent to the resources allocated into children's education), and (d) the bequest left to their children (following the idea of the "warm-glow" type of altruism in bequeathing [21, 44, 53]). We follow a well-established tradition of using logarithmic utilities [13, 15, 23, 52] and convolute the competing goals (a), (b), (c), and (d) into a single weighted-sum utility function

$$\begin{aligned} J_t = {} & \omega_C \ln[(1 - b_t)(K_t - \kappa n_t - \lambda\alpha_t n_t)] + \omega_R \ln[(1 - \mu(\alpha_t))n_t] \\ & + \omega_E \ln[(1 - \mu(\alpha_t))\lambda\alpha_t n_t] + \omega_B \ln[b_t(K_t - \kappa n_t - \lambda\alpha_t n_t)]. \end{aligned} \tag{5.3}$$

In (5.3), weights ω_C, ω_R, ω_E, and ω_B reflect preferences of an individual in generation t to Consume (C), Reproduce (R), to invest in children's Education (E), and to Bequeath (B), respectively. Without loss of generality, we assume $\omega_C + \omega_R + \omega_E + \omega_B = 1$.

Let us point out that our approach of explicitly including the bequest and children's education in the utility function differs from other models presented in literature, which rather consider children's consumption [5, 6] and human capital [13, 21, 31]. The form of utility used in this study allows for describing the trade-off between having more children and higher education of fewer offspring in a clear and transparent way.

Let us define the population growth rate (PGR_t) at time moment t as the ratio between the number N_{t+1} of survived individuals in generation $t+1$ and the number N_t of survived individuals in generation t:

$$\text{PGR}_t = \frac{N_{t+1}}{N_t}.$$

We define the long-term population growth rate as follows:

$$\mathrm{PGR} = \lim_{t\to\infty} \mathrm{PGR}_t = \lim_{t\to\infty} \frac{N_{t+1}}{N_t}.$$

The long-term population growth rate PGR characterizes the long-term dynamics of the population size.

Let us call a dynasty the infinite sequence of generations with a fixed "dynastic" set of preferences $\boldsymbol{\omega} = (\omega_C, \omega_R, \omega_E, \omega_B)$, which is inherited by individuals' offspring. Then the higher the PGR of a dynasty is, the more offspring the dynasty has over the sufficiently long term. Thus, we use PGR as a proxy of biological fitness. Also, we use the combination of education and income as a proxy of social status [1, 10, 29]. Fertility is characterized by a number of offspring of an individual.

To address the research question posed in the introduction, we would like to compare long-term population growth rates PGR for different strategies, i.e., different sets of preferences (weights) $\boldsymbol{\omega}$, which define decisions regarding consumption, the number of children, their education and bequest. By this comparison, we would like to reveal strategies that are optimal in terms of maximizing biological fitness under different economic and environmental conditions as well as to delineate the parameter sets, for which the dynasty grows and shrinks. Hence, we consider the following problem

$$\begin{aligned}
&\underset{\boldsymbol{\omega}=(\omega_C,\omega_R,\omega_E,\omega_B)}{\text{Maximize}} \quad \mathrm{PGR}\\
&\text{S.t.}\, \omega_C + \omega_R + \omega_E + \omega_B = 1,\\
&\omega_i \geq 0,\ i = C, R, E, B,\\
&N_{t+1} = N_t[1-\mu(\alpha_t)]n_t,\ N_1 = \bar{N}_1,\\
&K_{t+1} = b_t(K_t - \kappa n_t - \lambda\alpha_t n_t) + s + r\lambda\alpha_t,\ K_1 = \bar{K}_1
\end{aligned} \tag{5.4}$$

where α_t, n_t, b_t are optimal solutions in the individual's decision-making problem

$$\begin{aligned}
&\underset{\alpha_t, n_t, b_t}{\text{Maximize}}\, J_t\\
&\text{S.t.}\ K_t - \kappa n_t - \lambda\alpha_t n_t \geq 0,\\
&\alpha_t \in [0,1],\, n_t \geq 0,\, b_t \in [0,1],
\end{aligned} \tag{5.5}$$

in which K_t and $\boldsymbol{\omega}$ are given.

Importantly, we assume that our model (5.4), (5.5) is applicable at a local scale, that is that the environmental and economic conditions can be treated as external factors, and decisions of individuals have no influence on the environment and the economy. Hence, we assume that all economic and environmental parameters, namely, the independent part of the wage, s, the cost of full education, λ, the coefficient of the return of the investment in education, r, the cost of rearing one child, κ, and the strength of mortality event, β, do not change over time. Solving the two-stage problem (5.4)–(5.5) under these assumptions allows for investigation of strategies $\boldsymbol{\omega} = (\omega_C, \omega_R, \omega_E, \omega_B)$ that are optimal over the long term under fixed economic and environmental conditions. Explaining the processes related to the dynamic fertility at the global scale would require considering feedbacks between globally implemented decisions of individuals and the environment and the economy, which is outside of the scope of this study.

5.2.1 Solution to the Individual's Decision-Making Problem (5)

We take advantage of the separation of variables, which is possible in (5.3), and represent the utility function J_t in the following form:

$$J_t = u_t(\alpha_t, n_t) + v_t(b_t)$$

where

$$\begin{aligned} u_t(\alpha_t, n_t) = {} & (\omega_C + \omega_B)\ln(K_t - \kappa n_t - \lambda\alpha_t n_t) \\ & + \omega_R \ln[(1 - \mu_t(\alpha_t))n_t] + \omega_E \ln[(1 - \mu_t(\alpha_t))\lambda\alpha_t n_t] \end{aligned} \tag{5.6}$$

and

$$v_t(b_t) = \omega_C \ln(1 - b_t) + \omega_B \ln b_t.$$

Therefore, problem (5.5) can be split into two independent problems:

$$\begin{aligned} & \text{Maximize}_{\alpha_t, n_t}\ u_t(\alpha_t, n_t) \\ & \text{S.t.}\quad K_t - \kappa n_t - \lambda\alpha_t n_t \geq 0, \\ & \qquad\quad \alpha_t \in [0, 1], n_t \geq 0; \end{aligned} \tag{5.7}$$

and

$$\begin{aligned} & \text{Maximize}_{b_t}\ v_t(b_t) \\ & \text{S.t.}\quad b_t \in [0, 1]. \end{aligned} \tag{5.8}$$

Theorem 5.1 *Problem* (5.7) *has a solution* (α_t^*, n_t^*) *as follows:*

1. *If* $\omega_R + \omega_E > 0$ *then the solution is unique and, additionally, if (a)* $\frac{\omega_E + \beta(\omega_R + \omega_E)}{\lambda} \leq \frac{\omega_R - \beta(\omega_R + \omega_E)}{\kappa}$, *then*

$$\alpha_t^* = \frac{\omega_E + \beta(\omega_R + \omega_E)}{\omega_R - \beta(\omega_R + \omega_E)} \frac{\kappa}{\lambda},$$

$$n_t^* = (\omega_R - \beta(\omega_R + \omega_E))\frac{K_t}{\kappa};$$

if (b) $\frac{\omega_E+\beta(\omega_R+\omega_E)}{\lambda} > \frac{\omega_R-\beta(\omega_R+\omega_E)}{\kappa}$, *then*

$$\alpha_t^* = 1, n_t^* = (\omega_R + \omega_E)\frac{K_t}{\kappa + \lambda};$$

2. *If* $\omega_R = \omega_E = 0$, *then*

$$\alpha_t^* \in [0, 1], n_t^* = 0.$$

Proof Let us first consider case 1. We rewrite utility function (5.6) in the form

$$u_t(\alpha_t, n_t) = (\omega_C + \omega_B)\ln(K_t - \kappa n_t - \lambda\alpha_t n_t) + \omega_R \ln n_t + \omega_E \ln \lambda\alpha_t n_t + (\omega_R + \omega_E)\ln \alpha_t^\beta.$$

First, we calculate the derivatives

$$\frac{\partial u_t(\alpha_t,n_t)}{\partial \alpha_t} = -\frac{(\omega_C+\omega_B)\lambda n_t}{K_t-\kappa n_t-\lambda\alpha_t n_t} + \frac{\omega_E+\beta(\omega_R+\omega_E)}{\alpha_t},$$
$$\frac{\partial u_t(\alpha_t,n_t)}{\partial n_t} = -\frac{(\omega_C+\omega_B)(\kappa+\lambda\alpha_t)}{K_t-\kappa n_t-\lambda\alpha_t n_t} + \frac{\omega_R+\omega_E}{n_t}$$

and find stationary points $(\bar{\alpha}_t, \bar{n}_t)$ of $u_t(\alpha_t, n_t)$ by solving the system of equations $\frac{\partial u_t(\alpha_t,n_t)}{\partial \alpha_t} = 0$, $\frac{\partial u_t(\alpha_t,n_t)}{\partial n_t} = 0$; we obtain a unique stationary point as follows

$$\bar{\alpha}_t = \frac{\omega_E + \beta(\omega_R + \omega_E)}{\omega_R - \beta(\omega_R + \omega_E)}\frac{\kappa}{\lambda}, \bar{n}_t = (\omega_R - \beta(\omega_R + \omega_E))\frac{K_t}{\kappa}.$$

By applying the second partial derivative test at $(\bar{\alpha}_t, \bar{n}_t)$ (i.e., checking the negative definiteness of the Hessian matrix of $u_t(\alpha_t, n_t)$), we obtain

$$\left.\frac{\partial^2 u_t(\alpha_t, n_t)}{\partial \alpha_t^2}\right|_{\substack{\alpha_t = \bar{\alpha}_t \\ n_t = \bar{n}_t}} = -\frac{(\omega_C + \omega_B + \omega_E + \beta(\omega_R + \omega_E))(\omega_R - \beta(\omega_R + \omega_E))^2\lambda^2}{(\omega_C + \omega_B)(\omega_E + \beta(\omega_R + \omega_E))\kappa^2} < 0,$$

$$\left.\frac{\partial^2 u_t(\alpha_t, n_t)}{\partial \alpha_t^2}\right|_{\substack{\alpha_t = \bar{\alpha}_t \\ n_t = \bar{n}_t}} \left.\frac{\partial^2 u_t(\alpha_t, n_t)}{\partial n_t^2}\right|_{\substack{\alpha_t = \bar{\alpha}_t \\ n_t = \bar{n}_t}} - \left(\left.\frac{\partial^2 u_t(\alpha_t, n_t)}{\partial \alpha_t \partial n_t}\right|_{\substack{\alpha_t = \bar{\alpha}_t \\ n_t = \bar{n}_t}}\right)^2 = \frac{(\omega_R - \beta(\omega_R + \omega_E))\lambda^2}{(\omega_C + \omega_B)(\omega_E + \beta(\omega_R + \omega_E))K_t^2} > 0.$$

Thus, function $u_t(\alpha_t, n_t)$ is concave in the totality of its two arguments α_t and n_t, and its unique maximum point $(\bar{\alpha}_t, \bar{n}_t)$ is its global maximum point.

Note that the inequalities of case 1 (*a*) ensure that expression $\omega_R - \beta(\omega_R + \omega_E)$ is positive. Hence, $\bar{n}_t \geq 0$ and, moreover, $\bar{\alpha}_t \in [0, 1]$. Also, the first inequality of problem (5.7) is satisfied at this point:

$$K_t - \kappa\bar{n}_t - \lambda\bar{\alpha}_t\bar{n}_t = (\omega_C + \omega_B)K_t > 0.$$

Therefore, the global maximum point $(\bar{\alpha}_t, \bar{n}_t)$ satisfies all the constraints of problem (5.7) and, hence, $\left(\alpha_t^*, n_t^*\right) = (\bar{\alpha}_t, \bar{n}_t)$ is the solution of problem (5.7) in case 1 (*a*).

Under inequalities of case 1 (*b*) value $\bar{\alpha}_t$ either does not exist or lies outside segment [0, 1]. Thus, the stationary point $(\bar{\alpha}_t, \bar{n}_t)$ does not satisfy the constraints of problem (5.7). In order to find the conditional maximum in problem (5.7) in this case, let us consider domain D generated by the constraints of problem (5.7) and within which the utility function $u_t(\alpha_t, n_t)$ is to be maximized:

$$D = \left\{(\alpha_t, n_t) : 0 \leq \alpha_t \leq 1, 0 \leq n_t \leq \frac{K_t}{\kappa + \lambda\alpha_t}\right\}.$$

Let us consider the parametrization of domain D by the set of non-intersecting curves

$$n_t = \xi_t \frac{K_t}{\kappa + \lambda\alpha_t}, \ \xi_t \in [0, 1],$$

and consider the function

$$\begin{aligned}\bar{u}(\alpha_t, \xi_t) =& u(\alpha_t, n_t)|_{n_t = \xi_t \frac{K_t}{\kappa + \lambda\alpha_t}} = (\omega_C + \omega_B)\ln((1 - \xi_t)K_t) \\ &+ \omega_R \ln \xi_t \frac{K_t}{\kappa + \lambda\alpha_t} + \omega_E \ln \lambda\alpha_t\xi_t \frac{K_t}{\kappa + \lambda\alpha_t} + (\omega_R + \omega_E)\ln \alpha_t^{\beta}.\end{aligned}$$

For any $\xi_t \in (0, 1)$

$$\frac{\partial\bar{u}(\alpha_t, \xi_t)}{\partial\alpha_t} = \lambda\kappa \frac{\frac{\omega_E + \beta(\omega_R + \omega_E)}{\lambda} - \alpha_t \frac{\omega_R - \beta(\omega_R + \omega_E)}{\kappa}}{\alpha_t(\kappa + \lambda\alpha_t)}.$$

Thanks to the inequality of case 1 (*b*) and $\alpha_t \in [0, 1]$ in domain D we have the following set of relations

$$\begin{aligned}&\frac{\omega_E + \beta(\omega_R + \omega_E)}{\lambda} - \alpha_t \frac{\omega_R - \beta(\omega_R + \omega_E)}{\kappa} \\ &\quad \geq \min\left(\frac{\omega_E + \beta(\omega_R + \omega_E)}{\lambda} - \frac{\omega_R - \beta(\omega_R + \omega_E)}{\kappa}, \frac{\omega_E + \beta(\omega_R + \omega_E)}{\lambda}\right) > 0\end{aligned}$$

Therefore, $\frac{\partial \bar{u}(\alpha_t, \xi_t)}{\partial \alpha_t} > 0$ and function $\bar{u}(\cdot, \xi_t)$ increases with respect to $\alpha_t \in (0, 1]$ for any $\xi_t \in (0, 1)$ and, hence, has its maximum point $\alpha_t = 1$. Then, by solving equation

$$\frac{\partial}{\partial \xi_t}\left(\bar{u}(\alpha_t, \xi_t)|_{\alpha_t=1}\right) = -\frac{\omega_C + \omega_B}{1 - \xi_t} + \frac{\omega_R + \omega_E}{\xi_t} = 0,$$

we find a unique stationary point $\bar{\xi}_t = \omega_R + \omega_E \in (0, 1)$. The negative second derivative

$$\frac{\partial^2}{\partial \xi^2}\left(\bar{u}(\alpha_t, \xi_t)|_{\alpha_t=1}\right) = -\frac{\omega_C + \omega_B}{(1 - \xi_t)^2} - \frac{\omega_R + \omega_E}{\xi_t^2} < 0$$

confirms that point $\xi_t = \bar{\xi}_t$ is a global maximizer of $\bar{u}(\alpha_t, \xi_t)|_{\alpha_t=1}$ on $[0, 1]$. Returning to the original variables, we conclude, that point $(\alpha_t^*, n_t^*) = \left(1, (\omega_R + \omega_E)\frac{K_t}{\kappa+\lambda}\right)$ is a solution of problem (5.7) in case 1 (*b*).

In case 2, utility function (5.6) becomes $u_t(\alpha_t, n_t) = (\omega_C + \omega_B)\ln(K_t - \kappa n_t - \lambda\alpha_t n_t)$. Due to the fact that it monotonically decreases with respect to α_{t+1} and n_t, the global maximizer of problem (5.7) in case 2 is such that $n_t^* = 0$ and α_t^* is any point from segment $[0, 1]$.

The proof is complete. ∎

Theorem 5.2 *Problem* (5.8) *has a solution* b_t^* *as follows:*

1. *If* $\omega_C + \omega_B > 0$, *then the solution is unique and*

$$b_t^* = \frac{\omega_B}{\omega_C + \omega_B}$$

2. *If* $\omega_C = \omega_B = 0$, *then*

$$b_t^* \in [0, 1].$$

Proof Let us find stationary points $\bar{b}_t$ of function $v_t(b_t)$ by solving equation

$$v_t'(b_t) = -\frac{\omega_C}{1 - b_t} + \frac{\omega_B}{b_t} = 0.$$

We obtain

$$\bar{b}_t = \frac{\omega_B}{\omega_C + \omega_B} \in [0, 1]$$

only if $\omega_C + \omega_B > 0$. The second derivative evaluated at the stationary point $\bar{b}_t$ is

$$v_t''(\bar{b}_t) = -\frac{\omega_C}{(1-\bar{b}_t)^2} - \frac{\omega_B}{\bar{b}_t^2} \le 0.$$

Thus, $b_t^* = \bar{b}_t$ is the maximizer of function $v_t(b_t)$ given $\omega_C + \omega_B > 0$. If $\omega_C = \omega_B = 0$, then any point of segment [0, 1] is the maximizer in problem (5.8).

The proof is complete. ∎

We obtain that, besides the degenerate case 2 of the two weights ω_B, ω_C being zero, the fraction of the bequeathed wealth, $b_t^* = \frac{\omega_B}{\omega_C+\omega_B}$, depends on the proportion between weights ω_C, ω_B. This means that fraction b_t^* depends on the "ratio" between an individual's preferences about their own consumption and bequest. Particularly, in the limit case when $\omega_C = 0$, the available resources are allocated solely to the bequest if individual behaves optimally, and vice versa, in case $\omega_B = 0$ all the available resource is consumed and nothing is bequeathed. In the degenerate case of $\omega_C = \omega_B = 0$, an individual cares neither about consumption, nor about bequests, so it follows from the Theorem 1 that in this case ($\omega_R + \omega_E = 1$) there will be no amount money to split and the value b_t^* does not play any role and can be any number between zero and one.

5.2.2 Dynamics and the Explicit Formula for the Long-Term Population Growth Rate

Here, we consequently investigate model dynamics and find the explicit form of long-term population growth rate as follows.

Let us consider dynamic Eqs. (5.1), (5.2) of the model. Due to Theorems 5.1 and 5.2, every individual in generation t chooses to have n_t^* children, to give them education of level α_t^*, and to leave bequest as fraction b_t^* of the remainder of their capital. Given that, formulas (5.1), (5.2) become

$$N_{t+1} = N_t[1 - \mu(\alpha_t^*)]n_t^*, \tag{5.9}$$

$$K_{t+1} = b_t^*(K_t - \kappa n_t^* - \lambda\alpha_t^* n_t^*) + s + r\lambda\alpha_t^* = \omega_B K_t + s + r\lambda\alpha_t^*. \tag{5.10}$$

Theorem 5.1 implies the following optimal level of education constant over time for each generation in the non-degenerate case

$$\alpha_t^* = \alpha_* = \begin{cases} \frac{\omega_E+\beta(\omega_R+\omega_E)}{\omega_R-\beta(\omega_R+\omega_E)}\frac{\kappa}{\lambda} & \text{if } \frac{\omega_E+\beta(\omega_R+\omega_E)}{\lambda} \le \frac{\omega_R-\beta(\omega_R+\omega_E)}{\kappa}, \\ 1 & \text{if } \frac{\omega_E+\beta(\omega_R+\omega_E)}{\lambda} > \frac{\omega_R-\beta(\omega_R+\omega_E)}{\kappa}; \end{cases} \tag{5.11}$$

the optimal wage then becomes

$$w_t = w_* = s + r\lambda\alpha_*. \tag{5.12}$$

Theorem 5.2 implies the following constant over time optimal bequeathed fraction for each generation in the non-degenerate case

$$b_t^* = b_* = \frac{\omega_B}{\omega_C + \omega_B}.$$

The degenerate case solution remains the same. Hereafter, we consider only non-degenerate cases, i.e., we assume that $\omega_R + \omega_E > 0$; case when $\omega_R = \omega_E = 0$ implies the absence of any offspring at all and thus zero population growth rate which is clearly not optimal in terms of maximizing the long-term population growth rate.

Lemma *Let α_* and w_* be defined by relations* (5.11) *and* (5.12), *respectively. Then the long-term population growth rate* PGR *of the dynasty with the set of preferences $\boldsymbol{\omega} = (\omega_C, \omega_R, \omega_E, \omega_B)$ takes the following form*

$$\mathrm{PGR}(\omega_C, \omega_R, \omega_E, \omega_B) = \begin{cases} \frac{\alpha_*^\beta(\omega_R - \beta(\omega_R+\omega_E))}{1-\omega_B}\frac{w_*}{\kappa} & \text{if } \alpha_* < 1, \\ \frac{\omega_R+\omega_E}{1-\omega_B}\frac{w_*}{\kappa+\lambda} & \text{if } \alpha_* = 1. \end{cases}$$

Proof Given formulas (5.11), (5.12), capital (5.10) can be represented by the formula $K_{t+1} = \omega_B K_t + w_*$, or, by solving it, $K_t = \omega_B^{t-1}K_1 + \frac{1-\omega_B^{t-1}}{1-\omega_B}w_*$, where $\omega_B \in [0, 1)$ and K_1 is the initial value of capital for generation 1. Then, $\lim_{t\to\infty} K_t = \frac{w_*}{1-\omega_B}$ as ω_B^{t-1} tends to zero while t tends to infinity.

The population growth rate PGR_t thanks to formula (5.9) and the results of Theorem 1 becomes

$$\mathrm{PGR}_t = [1 - \mu(\alpha_t^*)]n_t^* = \begin{cases} \alpha_*^\beta(\omega_R - \beta(\omega_R + \omega_E))\frac{K_t}{\kappa} & \text{if } \alpha_* < 1, \\ (\omega_R + \omega_E)\frac{K_t}{\kappa+\lambda} & \text{if } \alpha_* = 1. \end{cases}$$

Hence, using the fact that $\lim_{t\to\infty} B_t = \frac{w_*}{1-\omega_B}$ and the definition of PGR we obtain the formula in the statement of Lemma.

The proof is complete.

■

Theorem 5.3 *Optimal preferences ω_R^*, ω_E^*, ω_B^*, (while $\omega_C^* = 0$) optimal level of education α_*, and optimal long-term population growth rate* PGR_* *in problem* (5.4) *can be represented in the following form:*

(1) $\omega_E^* = 0, \omega_R^* + \omega_B^* = 1, \omega_R^* > 0, \alpha_* = \frac{\beta}{1-\beta}\frac{\kappa}{\lambda}, \mathrm{PGR}_* = \left(\frac{\beta}{1-\beta}\frac{\kappa}{\lambda}\right)^\beta ([1-\beta]\frac{s}{\kappa} + \beta r)$ *if*

$$(a)\ \ \beta \in \left[0, \frac{\lambda}{\lambda+\kappa}\right), r \le 0; \quad or \quad (b)\ \ \beta = 0, r \in \left(0, \frac{s}{\kappa}\right);$$

$$\begin{aligned}(2)\quad &\omega_E^* = \bar{\omega}_{ER}^* \omega_R^* > 0,\ \bar{\omega}_{ER}^* = \tfrac{1}{2\beta \frac{s}{\kappa}}\Big(\big[(1-\beta)\tfrac{s}{\kappa} - (1+\beta)r\big] \\ &-\sqrt{\big(\tfrac{s}{\kappa} - r\big)\big[(1-\beta)^2\tfrac{s}{\kappa} - (1+\beta)^2 r\big]}\Big),\ \omega_B^* = 1 - \big(\omega_R^* + \omega_E^*\big),\ \alpha_* \\ &= \tfrac{\bar{\omega}_{ER}^* + \beta(1+\bar{\omega}_{ER}^*)}{1-\beta(1+\bar{\omega}_{ER}^*)}\tfrac{\kappa}{\lambda},\ \ \mathrm{PGR}_* = \mathrm{PGR}\big(0, \omega_R^*, \omega_E^*, \omega_B^*\big), if\end{aligned}$$

$$\beta \in \big(0, \bar{\beta}_*\big), \beta < \frac{\lambda}{\lambda+\kappa}, r \in \left(0, \frac{s}{\kappa}\right);$$

$$(3)\quad \omega_E^* \geq \frac{\frac{\lambda}{\kappa+\lambda} - \beta}{\frac{\kappa}{\kappa+\lambda} + \beta}\omega_R^* > 0, \omega_B^* = 1 - \big(\omega_R^* + \omega_E^*\big), \alpha_* = 1, \mathrm{PGR}_* = \frac{s+r\lambda}{\kappa+\lambda}\ if$$

$$(a)\ \beta \in \left(0, \frac{\lambda}{\lambda+\kappa}\right), r \geq \frac{s}{\kappa};\ or\ (b)\ \beta \in \left[\bar{\beta}_*, \frac{\lambda}{\lambda+\kappa}\right), r \in \left(0, \frac{s}{\kappa}\right);\ or\ (c)\ \beta = 0, r > \frac{s}{\kappa};$$

$$(4)\quad \omega_B^* = 1 - \big(\omega_R^* + \omega_E^*\big) < 1, \alpha_* = 1, \mathrm{PGR}_* = \frac{s+r\lambda}{\kappa+\lambda}\ if$$

$$\beta \geq \frac{\lambda}{\lambda+\kappa};$$

$$(5)\quad \omega_B^* = 1 - \big(\omega_R^* + \omega_E^*\big) < 1, \alpha_* = \min\left(\frac{\omega_E^*}{\omega_R^*}\frac{\kappa}{\lambda}, 1\right), \mathrm{PGR}_* = r\ if$$

$$\beta = 0, r = \frac{s}{\kappa}.$$

Value $\bar{\beta}_*$ *is a unique root to equation* $\mathrm{PGR}\big(0, 1, \bar{\omega}_{ER}^*, \omega_B^*\big) = \frac{s+r\lambda}{\kappa+\lambda}$.

The proof of Theorem 5.3 is presented in the Appendix; qualitatively different cases are summarized in Tables 5.1 and 5.2 and discussed thereafter.

Each of the five cases presented in Tables 5.1 and 5.2 (marked by different colors) describes a society with certain economic parameters (namely, r describing the distribution of income between educated and non-educated people, λ describing the

Table 5.1 Optimal strategies maximizing long-term fitness

Income gap \ Mortality strength	No $\beta = 0$	Low $\beta \in (0, \bar{\beta}_*)$	Medium $\beta \in \left[\bar{\beta}_*, \frac{\lambda}{\lambda+\kappa}\right)$	High $\beta \geq \frac{\lambda}{\lambda+\kappa}$
Negative $r \leq 0$	Maximize offspring	Maximize offspring	Maximize offspring	Maximize allocation in social status to survive
Low $0 < r < \frac{s}{\kappa}$	Maximize offspring	Balance reproduction and social status	Prioritize social status	Maximize allocation in social status to survive
Balanced $r = \frac{s}{\kappa}$	No influence of social status on population growth	Prioritize social status	Prioritize social status	Maximize allocation in social status to survive
High $r > \frac{s}{\kappa}$	Prioritize social status	Prioritize social status	Prioritize social status	Maximize allocation in social status to survive

Table 5.2 Brief description of optimal strategies in problem (5.4). Here, optimal preference ω_C^* is zero and optimal preference $\omega_B^* = 1 - \omega_R^* - \omega_E^*$

Name	Optimal ratio $\omega_{ER}^* = \frac{\omega_E^*}{\omega_R^*}$ between the preferences towards social status and reproduction	Optimal allocation in education	Long-term population growth rate
Maximize offspring	$\omega_{ER}^* = 0$ minimal	$\alpha_* = \frac{\beta}{1-\beta}\frac{\kappa}{\lambda}$ roughly proportional to the mortality event strength	$PGR_* = \alpha_*^\beta \left[(1-\beta)\frac{s}{\kappa} + \beta r\right]$ negatively related to the mortality event strength
Balance reproduction and social status	$\omega_{ER}^* = \bar{\omega}_{ER}^*$ positively related to the mortality event strength	$\alpha_* = \frac{\beta + (1+\beta)\bar{\omega}_{ER}^*}{(1-\beta) - \beta\bar{\omega}_{ER}^*}\frac{\kappa}{\lambda}$ roughly proportional to optimal ratio ω_{ER}^*	$PGR_* = \alpha_*^\beta \left[r + \left(\frac{s}{\kappa} - r\right)\left(\frac{1}{1+\bar{\omega}_{ER}^*} - \beta\right)\right]$ negatively related to the mortality event strength
Prioritize social status	$\omega_{ER}^* \geq \frac{\frac{\lambda}{\kappa+\lambda} - \beta}{\frac{\kappa}{\kappa+\lambda} + \beta}$ any that is sufficiently large compared with the ratio between full education cost and child rearing cost reduced roughly proportionally to the mortality event strength	$\alpha_* = 1$ maximal	$PGR_* = \frac{s + r\lambda}{\kappa + \lambda}$ positively related to the income gap
Maximize allocation in social status to survive	$\omega_{ER}^* \geq 0$ any	$\alpha_* = 1$ maximal	$PGR_* = \frac{s + r\lambda}{\kappa + \lambda}$ positively related to the income gap
No influence of social status on population growth	$\omega_{ER}^* \geq 0$ any	$\alpha_* = \min\left(\omega_{ER}^* \frac{\kappa}{\lambda}, 1\right)$ proportional to the ratio between the chosen preference towards social status and the preference towards reproduction	$PGR_* = r = \frac{s}{\kappa}$ does not depend on education level and its cost

education cost, s describing the minimum salary individual gets over their lifespan, and κ describing the child-rearing cost) and environmental parameter (namely, β describing the strength of mortality events). A combination of these parameters implies a certain optimal strategy, i.e., an optimal set of preferences (weights) $\omega_C^*, \omega_R^*, \omega_E^*, \omega_B^*$. Multiple sets of preferences correspond to each optimal strategy. It is the ratio between the preference toward social status attainment and the preference toward reproduction that defines which strategy is optimal.

We distinguish several qualitatively different situations. From the economic side, we consider situations with a negative, low-, and high-income gap between people with full education and people with basic education only, corresponding to $r \leq 0$, $0 < r < \frac{s}{\kappa}$, and $r \geq \frac{s}{\kappa}$, respectively. From the environmental side, we consider situations of no, low-, medium-, and high-mortality events corresponding to $\beta = 0$, $\beta \in (0, \bar{\beta}_*)$, $\beta \in [\bar{\beta}_*, \frac{\lambda}{\lambda+\kappa})$ and $\beta \geq \frac{\lambda}{\lambda+\kappa}$, respectively. Here, value $\bar{\beta}_*$ depends on all four economic parameters (one can find the exact equation for $\bar{\beta}_*$ in Appendix B). Importantly, different economic situations in the society imply different bounds for the classification of the strength of mortality events.

In what follows, we describe what strategies optimize the long-term population growth in different societies and reveal in what cases seeking for higher social status becomes optimal for survival. The strategy "Maximize offspring" is optimal in societies in which the gain in the survival potential due to a higher social status is not significant enough to motivate compromising the number of offspring against investment into their education. In the absence of premature mortality, social status does not affect the survival probability; parents optimize their utility to which the number of offspring makes a greater contribution. This motivates them to choose the "Maximize offspring" strategy. Another situation, which also promotes the choice of this strategy, is the case of a negative income gap. A need to sacrifice a part of the income in order to obtain a higher social status demotivates a rational individual from investing in education, but this consideration trades-off with a potential gain in the number of survived offspring due to a higher status. Correspondingly, in the presence of low- and medium-strength mortality events, it remains optimal to maximize offspring in our model. However, if a society experiences high-strength mortality events, optimal behavior will change. In this case, indifferently to their preferences, the allocation for the education of children to offer them the best opportunities to survive must be maximal. We refer to this case as "Maximize allocation in social status to survive" strategy. This strategy is optimal regardless of the relation between incomes of educated and non-educated.

The strategy "Prioritize social status" is similar to the latter case. It recommends setting a sufficiently high weight on social status while retaining some weight on reproduction too. In this scenario, despite a balanced approach to set the preferences, the optimal allocation for social status must be maximal. This strategy becomes optimal in a number of cases when mortality event strength is between zero and medium and the income gap is between low and high. The higher the income gap and mortality event strength, the more motivated the dynasty is toward allocating resources into social status.

"Balance reproduction and social status" is a strategy that prescribes a certain optimal ratio between the weight toward social status and the weight toward reproduction. The ratio depends positively on the mortality event strength; the optimal allocation for social status is roughly proportional to this ratio. Let us point out that in this case, as well as in the other cases, the long-term population growth rate is positively related both to the maximal number of non-educated children an individual can afford and to the income inequality (gap).

Our model also demonstrates a special case in which the income gap is strictly equal to the ratio of the basic income and the cost of rearing a child in the absence of premature mortality. In this case, which we call "No influence of social status on population growth", any weight combination between reproduction and social status is optimal; the ratio of the chosen weights defines the optimal allocation for social status; however, the long-term population growth rate is unrelated to the choice of weights and equals to the income gap.

Considering situations, in which enhancing the social status significantly promotes survival (i.e., strategies "Prioritize social status" and "Maximize allocation in social status to survive") with $\mathrm{PGR}_* = \frac{s+r\lambda}{\kappa+\lambda}$, we conclude that the total income of an individual over their life span, $w = s + r\lambda$, must be larger than the amount of resources $\kappa + \lambda$ to be spent for rearing one full-educated child in order for this dynasty to survive over the long-term.

5.3 Transitions Between the Boundaries: Possibility of Regime Switches of Optimal Strategies Maximizing Biological Fitness

Theorem 5.3 describes sets of strategies (preferences $\omega_C^*, \omega_R^*, \omega_E^*, \omega_B^*$) optimal in the long term under certain fixed economic and environmental conditions. They were derived under the assumption that the basic wage, s, the cost of full education, λ, the coefficient of the return of the investment in education, r, the cost of rearing one child, κ, and the strength of mortality event, β, as well as preferences $\omega_C, \omega_R, \omega_E, \omega_B$ do not change over time.

Interestingly, the transition between the boundaries of the different cases is not always continuous in terms of the optimal sets of preferences. Such a discontinuity can be interpreted as a regime switch of an optimal strategy. Let us consider, for instance, cases 1 and 5 of Theorem 5.3. The Hausdorff distance between the corresponding sets of optimal preferences $\Omega_1 = \{\omega_C^* = \omega_E^* = 0, \omega_R^* + \omega_B^* = 1, \omega_R^* > 0\}$ and $\Omega_5 = \{\omega_C^* = 0, \omega_B^* = 1 - (\omega_R^* + \omega_E^*) < 1\}$ is non-zero (equal to $\sqrt{3/2}$), which indicates the discontinuity of the transition between these cases. Let us consider a situation with a non-positive income gap, $r \leq 0$, and high-mortality events, $\beta \geq \lambda/(\lambda + \kappa)$. In such a situation, $\bar{\omega}^* = (\omega_C^* = 0, \omega_R^* = \omega_E^* = \omega_B^* = \frac{1}{3})$ from Ω_5 is one possible optimal strategy (see case 5 of Theorem 5.3). If parameter β changes to $\beta < \frac{\lambda}{\lambda+\kappa}$ (note that this change can be very small if the initial value of β lies close to the boundary between the cases), strategy $\bar{\omega}^*$ becomes non-optimal, because in case 1 (a) the optimal preference is $\omega_E^* = 0$. This means that the affected individual needs to switch from $\bar{\omega}^*$ to a strategy from the optimal set Ω_1. The distance (in the Euclidian metric) between even the closest strategy from this set $(\omega_C^* = \omega_E^* = 0, \omega_R^* = \omega_B^* = \frac{1}{2})$ and the previously optimal strategy $\bar{\omega}^*$ is $\sqrt{1/6}$—due to an arbitrary small change in β.

5.4 Long-Term Survival Under Different Economic and Environmental Conditions

Inequalities $\text{PGR}_* > 1$, $\text{PGR}_* = 1$, and $\text{PGR}_* < 1$ describe the situations, in which a dynasty following the optimal strategy grows, remains stable, or dies out over the long term correspondingly. Originally, the environmental and economic conditions are defined by the set of five parameters, $s, \lambda, \kappa, r, \beta$ in the model; in fact the number of parameters can be reduced to four: $s/\kappa, \lambda/\kappa, r, \beta$; all are naturally dimensionless.

In this section, we investigate how the long-term population growth rate depends on the model parameters. We delineate parameter sets for which $\mathrm{PGR}_* > 1$, $\mathrm{PGR}_* = 1$, and $\mathrm{PGR}_* < 1$. The visualization of the obtained results is presented in Fig. 5.1.

Each panel in Fig. 5.1 represents PGR_* in axes β and r depending also on different values of parameters s/κ and λ/κ, varying along large horizontal and vertical arrows correspondingly. The value of PGR_* is marked by color. The black line corresponds to $\mathrm{PGR}_* = 1$ separating the parameter set, in which a dynasty survives and grows ($\mathrm{PGR}_* > 1$) from the parameter set, in which it dies out ($\mathrm{PGR}_* < 1$).

We can see that if the both fractions s/κ and λ/κ are small, which means that individuals do not have enough means for child-rearing, then for any β and for r such that $r < \frac{1+\lambda/\kappa - s/\kappa}{\lambda/\kappa}$ the long-term population growth rate $\mathrm{PGR}_* < 1$ and the dynasty dies out. Let us point out that for smaller fractions λ/κ, the set of such r becomes larger. Increasing each fraction leads to an increase in the long-term population growth rate, which can eventually become greater than 1 for small enough r and β.

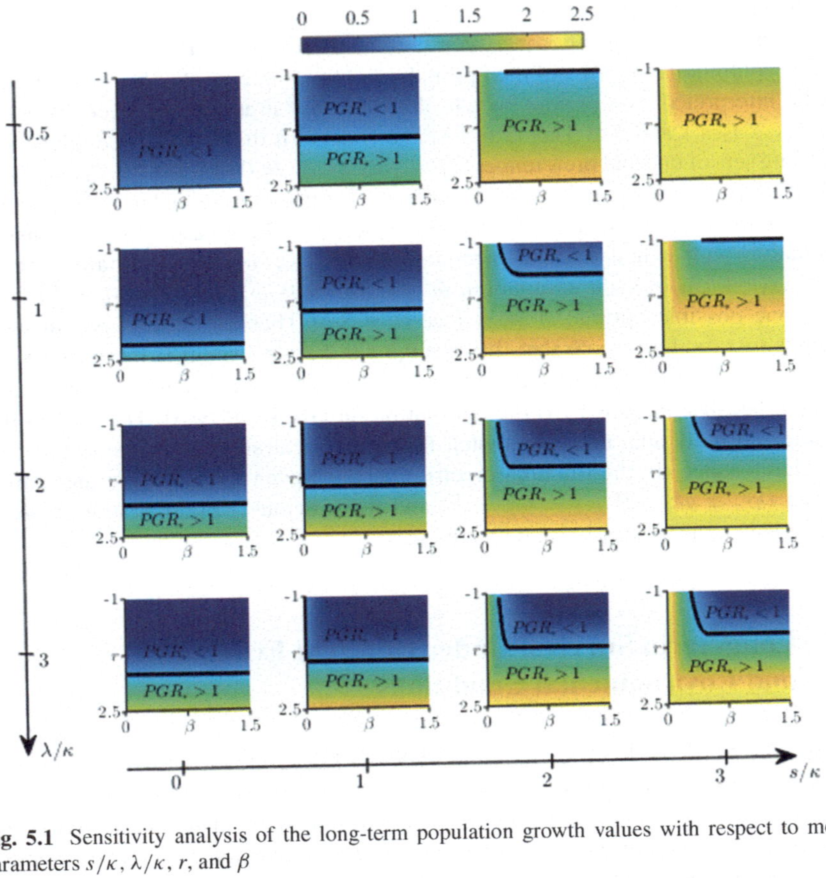

Fig. 5.1 Sensitivity analysis of the long-term population growth values with respect to model parameters s/κ, λ/κ, r, and β

For a sufficiently large minimum salary with respect to child-rearing, s/κ, the long-term population growth rate can be greater than 1 for any mortality events strength β and any non-negative income gap r. Also, one can see that long-term population growth is non-decreasing in r given all other parameters fixed; the dependence on β can exhibit different monotonicities (i.e., it may increase, decrease, and be constant).

As individual preferences defining the optimal solution of resource allocation problem (2.1) and thus maximizing biological fitness are sticky over generations (in this context the individual preferences are changed through phenotypic plasticity which has eventual costs and limits preventing from instantaneous change to those preferences maximizing fitness [14]), Fig. 5.1 alludes to the fact that a slight but sudden change in environmental or economic conditions may move a dynasty from a growing to a shrinking trend. As optimal weights in changed environment emerge as a result of a long-lasting social evolutionary process and depending on how big the drop in PGR is, the dynasty may not be in the position to adapt to the new conditions swiftly enough and will die out or even if it manages to quickly adapt the highest PGR can become less than one and thus dynasty may die out without introducing fast changes in the socio-economic system to prevent extinction.

Let us, for instance, consider a society with a near-zero income gap in the environment with mortality events of low strength with a minimum salary which is a twice more than child-rearing costs and full education is three times more expensive than child-rearing (see the panel in the fourth row and the third column). In the proposed situation, $\mathrm{PGR}_* > 1$ but it is close to the black line defining $\mathrm{PGR}_* = 1$ from the left. Due to a sudden switch of the strength of mortality events to a higher level, which may happen even in the near future due to climate change [4, 47], the point on the plot would jump over the line $\mathrm{PGR}_* = 1$ to the right and will find itself in the area with $\mathrm{PGR}_* < 1$, where the dynasty will turn out to be on the verge of extinction.

5.5 Conclusion

Our analysis here demonstrates the optimal employment of alternative strategies in order to optimize biological fitness with respect to the trade-off between having more children and giving each child a higher social status. The first strategy prescribes allocating available resources toward maximizing the number of children, with a minimal investment in social status. Following this strategy improves an individual's biological fitness by ensuring the maximum number of survived offspring and is optimal when the severity of mortality events is moderate. However, this strategy does not optimize fitness when mortality events become more frequent or severe. Under environmental conditions of elevated mortality, the second strategy involving allocation resources to ensure a high social status of offspring is optimal; here the number of children decreases as each child "costs" more to rear and educate.

Our study provides insights, in a highly stylized fashion, into how major economic and environmental parameters influence long-term population dynamics. For example, the model shows that a higher income inequality promotes higher investment in social status, particularly under high-mortality conditions. Building upon previous literature, this study provides a quantitative rationale for allocating resources into social status at a cost to immediate biological fitness. Within an environment largely characterized by the absence of high-mortality events, social status-seeking at a cost to reproduction may be the result of evolutionary lag, or a mismatch between phenotypes not yet adapted to life within low mortality contexts. This study provides the first version of a model explaining how seeking social status at a cost to fertility may become optimal in terms of maximizing biological fitness in societies experiencing premature mortality where mortality is disproportionately high among low-status individuals.

The model developed here allows for comparison of the long-term population growth rate under different environmental and economic situations, as well as for comparison of optimal strategy sets. We distinguish the model parameters sets, for which the dynasty exhibits qualitatively different long-term behavior ranging from a growing to a declining trend. We consider possible sudden shifts in the model parameters, which may lead to jumps of the long-term population growth rate between these parameter sets, which may lead to a qualitatively different long-term survival future of the dynasty. Also, we show that in some cases, a slight change in environmental conditions can lead to a regime switch of an optimal strategy maximizing biological fitness.

Acknowledgements We thank Heine Strand, Martin Fieder, Warren Sanderson, Brian Fath, and Arkady Kryazhimskiy for their comments through making this project.

Funding statement S. Orlov acknowledges the support of the Russian Academy of Sciences acting as a Russian National Member Organization at IIASA, which allowed him to participate in the IIASA's Young Scientist Summer Program (YSSP) in 2014 and carry out a major part of the research presented in this paper. We acknowledge support by a Starting Grant of the European Research Council ERC Grant Agreement No. 241003-COHORT (provided to V. Skirbekk).

Appendix

Proof of Theorem 5.3 First, let us show that the optimal preference $\omega_C^* = 0$; for that let us assume the opposite. Let set $(\hat{\omega}_C, \hat{\omega}_R, \hat{\omega}_E, \hat{\omega}_B)$, where $\hat{\omega}_C > 0$, be optimal. Then, as one can see from the formula for PGR in Lemma, set $\left(\frac{\hat{\omega}_C}{2}, \hat{\omega}_R, \hat{\omega}_E, \hat{\omega}_B + \frac{\hat{\omega}_C}{2}\right)$ delivers even bigger value to PGR, as both denominators are greater in the latter case. This contradicts with the optimality of $(\hat{\omega}_C, \hat{\omega}_R, \hat{\omega}_E, \hat{\omega}_B)$ and, therefore, $\omega_C^* = 0$ is optimal. Hereafter we assume, that $\omega_C = 0$, and, in turn, $\omega_B = 1 - (\omega_R + \omega_E)$.

Another observation is that PGR$(\omega_C, \omega_R, \omega_E, \omega_B)$ is not defined for $\omega_B = 1$, or, given the previous assumption, for $\omega_R = \omega_E = 0$. So, below we suppose, that $\omega_R + \omega_E > 0$. Also, below, when we define optimal sets of preferences, we suppose that $\omega_R^* \geq 0$, $\omega_E^* \geq 0$, $\omega_B^* \geq 0$, and $\omega_R^* + \omega_E^* + \omega_B^* = 1$.

Let us rewrite the long-term population growth rate more explicitly than in Lemma using equalities (5.11), (5.12) as follows:

$$\text{PGR} = \begin{cases} \left(\frac{\omega_E+\beta(\omega_R+\omega_E)}{\omega_R-\beta(\omega_R+\omega_E)}\frac{\kappa}{\lambda}\right)^{\beta} \frac{\omega_R-\beta(\omega_R+\omega_E)}{1-\omega_B} \frac{s+r\lambda\frac{\omega_E+\beta(\omega_R+\omega_E)}{\omega_R-\beta(\omega_R+\omega_E)}\frac{\kappa}{\lambda}}{\kappa} & \text{if } \frac{\omega_E+\beta(\omega_R+\omega_E)}{\lambda} \le \frac{\omega_R-\beta(\omega_R+\omega_E)}{\kappa}, \\ \frac{\omega_R+\omega_E}{1-\omega_B}\frac{s+r\lambda}{\kappa+\lambda} & \text{if } \frac{\omega_E+\beta(\omega_R+\omega_E)}{\lambda} > \frac{\omega_R-\beta(\omega_R+\omega_E)}{\kappa}. \end{cases}$$

Substituting $\omega_B = 1-(\omega_R+\omega_E)$, we simplify and obtain

$$\text{PGR} = \text{PGR}(\omega_R,\omega_E) = \begin{cases} \left(\frac{\omega_E+\beta(\omega_R+\omega_E)}{\omega_R-\beta(\omega_R+\omega_E)}\frac{\kappa}{\lambda}\right)^{\beta} \frac{[\omega_R-\beta(\omega_R+\omega_E)]\frac{s}{\kappa}+[\omega_E+\beta(\omega_R+\omega_E)]r}{\omega_R+\omega_E} & \text{if } \frac{\omega_E+\beta(\omega_R+\omega_E)}{\lambda} \le \frac{\omega_R-\beta(\omega_R+\omega_E)}{\kappa}, \\ \frac{s+r\lambda}{\kappa+\lambda} & \text{if } \frac{\omega_E+\beta(\omega_R+\omega_E)}{\lambda} > \frac{\omega_R-\beta(\omega_R+\omega_E)}{\kappa}. \end{cases}$$

One can easily show that function $\text{PGR}(\cdot,\cdot)$ is continuous with respect to ω_R, ω_E in domain $\omega_R \ge 0, \omega_E \ge 0, 0 < \omega_R+\omega_E \le 1$ for any non-negative β, positive s, κ, λ, and an arbitrary r.

In what follows, we derive a maximizer of $\text{PGR}(\omega_R,\omega_E)$ in the following three different cases: $\beta \ge \frac{\lambda}{\lambda+\kappa}$, $\beta = 0$, and $\beta \in \left(0, \frac{\lambda}{\lambda+\kappa}\right)$.

If $\beta \ge \frac{\lambda}{\lambda+\kappa}$ then, due to $\frac{\lambda}{\lambda+\kappa} \ge \frac{\lambda}{\lambda+\kappa}\frac{\omega_R}{\omega_R+\omega_E} \ge \frac{\lambda\omega_R-\kappa\omega_E}{(\lambda+\kappa)(\omega_R+\omega_E)}$, we obtain that $\beta \ge \frac{\lambda\omega_R-\kappa\omega_E}{(\lambda+\kappa)(\omega_R+\omega_E)}$, which is equivalent to $\frac{\omega_E+\beta(\omega_R+\omega_E)}{\lambda} \ge \frac{\omega_R-\beta(\omega_R+\omega_E)}{\kappa}$. In this case $\text{PGR}(\omega_R,\omega_E) = \text{PGR}_* \equiv \frac{s+r\lambda}{\kappa+\lambda}$ is constant. Hence, any preferences $\omega_R^*, \omega_E^*, \omega_B^*$ are optimal and the optimal level of education is $\alpha_* = 1$. This proves case 4.

If $\beta = 0$, then the formula for $\text{PGR}(\omega_R,\omega_E)$ has a simpler form which can be obtained by taking the corresponding limit as follows:

$$\text{PGR}(\omega_R,\omega_E) = \begin{cases} \frac{\omega_R\frac{s}{\kappa}+\omega_E r}{\omega_R+\omega_E} & \text{if } \frac{\omega_E}{\lambda} \le \frac{\omega_R}{\kappa}, \\ \frac{s+r\lambda}{\kappa+\lambda} & \text{if } \frac{\omega_E}{\lambda} > \frac{\omega_R}{\kappa}. \end{cases}$$

Here, we consider three different subcases: $r = \frac{s}{\kappa}$, $r > \frac{s}{\kappa}$, $r < \frac{s}{\kappa}$.

If $r = \frac{s}{\kappa}$ then $\text{PGR}(\omega_R,\omega_E) \equiv \text{PGR}_* = r$ is constant. Hence, any preferences $\omega_R^*, \omega_E^*, \omega_B^*$ are optimal and the optimal level of education in this case is $\alpha_* = \min\left(\frac{\omega_E^*}{\omega_R^*}\frac{\kappa}{\lambda}, 1\right)$. This proves case 5.

Then, for technical reasons, we fix $\omega_R+\omega_E = c \in (0,1]$. Then

$$\text{PGR}(\omega_R) = \begin{cases} \frac{\omega_R}{c}\left(\frac{s}{\kappa}-r\right)+r & \text{if } \frac{\kappa c}{\lambda+\kappa} \le \omega_R \le c, \\ \frac{s+r\lambda}{\kappa+\lambda} & \text{if } 0 \le \omega_R < \frac{\kappa c}{\lambda+\kappa} \end{cases}$$

is a piecewise linear function of ω_R.

If $r < \frac{s}{\kappa}$, then the maximum value of the long-term population growth rate is $\text{PGR}(\omega_R^*) = \text{PGR}_* = \frac{s}{\kappa}$, where $\omega_R^* = c$. Then $\omega_E^* = c - \omega_R^* = 0$. Thus, thanks to the independence of the maximum value of PGR$(\cdot)$ from the parameter $c \in (0, 1]$, the optimal preferences are such that $\omega_E^* = 0$ and the optimal level of education is $\alpha_* = 0$. This proves case 1 (a) and case 1 (b) given $\beta = 0$.

If $r > \frac{s}{\kappa}$, then the maximum value of the long-term population growth rate is $\text{PGR}(\omega_R^*) = \text{PGR}_* = \frac{s+r\lambda}{\kappa+\lambda}$, where $\omega_R^* \in \left[0, \frac{\kappa c}{\lambda+\kappa}\right]$. Then $\omega_E^* = c - \omega_R^* \in \left[\frac{\lambda c}{\lambda+\kappa}, 1\right]$. By returning to the formula of PGR(ω_R, ω_E) and thanks to the independence of the maximum value of PGR$(\cdot)$ from parameter $c \in (0, 1]$, we conclude that the optimal preferences are such that $\omega_E^* \geq \frac{\lambda}{\kappa}\omega_R^*$ and the optimal level of education is $\alpha_* = 1$. This proves case 3 (c).

Let us consider case $\beta \in \left(0, \frac{\lambda}{\lambda+\kappa}\right)$ and let $\omega_R > 0$. Then, we can rewrite the formula for the long-term population growth rate as follows:

$$
\begin{aligned}
&\text{PGR}(\omega_R, \omega_E)\\
&= \begin{cases} \left(\dfrac{\frac{\omega_E}{\omega_R}+\beta\left(1+\frac{\omega_E}{\omega_R}\right)}{1-\beta\left(1+\frac{\omega_E}{\omega_R}\right)}\dfrac{\kappa}{\lambda}\right)^{\beta} \dfrac{\left[1-\beta\left(1+\frac{\omega_E}{\omega_R}\right)\right]\frac{s}{\kappa}+\left[\frac{\omega_E}{\omega_R}+\beta\left(1+\frac{\omega_E}{\omega_R}\right)\right]r}{\left(1+\frac{\omega_E}{\omega_R}\right)} & \text{if } \dfrac{\frac{\omega_E}{\omega_R}+\beta\left(1+\frac{\omega_E}{\omega_R}\right)}{\lambda} \leq \dfrac{1-\beta\left(1+\frac{\omega_E}{\omega_R}\right)}{\kappa},\\ \dfrac{s+r\lambda}{\kappa+\lambda} & \text{if } \dfrac{\frac{\omega_E}{\omega_R}+\beta\left(1+\frac{\omega_E}{\omega_R}\right)}{\lambda} > \dfrac{1-\beta\left(1+\frac{\omega_E}{\omega_R}\right)}{\kappa}. \end{cases}
\end{aligned}
$$

Hence, PGR appears to be dependent on $\frac{\omega_E}{\omega_R}$ only. Denoting $\omega_{ER} = \frac{\omega_E}{\omega_R}$, we have

$$
\begin{aligned}
&\text{PGR}(\omega_{ER})\\
&= \begin{cases} \left(\dfrac{\omega_{ER}+\beta(1+\omega_{ER})}{1-\beta(1+\omega_{ER})}\dfrac{\kappa}{\lambda}\right)^{\beta} \dfrac{[1-\beta(1+\omega_{ER})]\frac{s}{\kappa}+[\omega_{ER}+\beta(1+\omega_{ER})]r}{(1+\omega_{ER})} & \text{if } 0 \leq \omega_{ER} \leq \dfrac{(1-\beta)\lambda-\beta\kappa}{(1+\beta)\kappa+\beta\lambda},\\ \dfrac{s+r\lambda}{\kappa+\lambda} & \text{if } \omega_{ER} > \dfrac{(1-\beta)\lambda-\beta\kappa}{(1+\beta)\kappa+\beta\lambda}. \end{cases}
\end{aligned}
$$

Due to the continuity of function PGR$(\cdot)$ and being constant while $\omega_{ER} > \frac{(1-\beta)\lambda-\beta\kappa}{(1+\beta)\kappa+\beta\lambda}$, it is sufficient to consider the problem of its maximization only on segment $0 \leq \omega_{ER} \leq \frac{(1-\beta)\lambda-\beta\kappa}{(1+\beta)\kappa+\beta\lambda}$. The derivative becomes

$$
\begin{aligned}
&\frac{\partial}{\partial\omega_{ER}}\text{PGR}(\omega_{ER})\\
&= \left(\frac{\omega_{ER}+\beta(1+\omega_{ER})}{1-\beta(1+\omega_{ER})}\frac{\kappa}{\lambda}\right)^{\beta} \frac{\beta\frac{s}{\kappa}(\omega_{ER})^2+\left[(1+\beta)r-(1-\beta)\frac{s}{\kappa}\right]\omega_{ER}+\beta r}{(\omega_{ER}+\beta(1+\omega_{ER}))(1-\beta(1+\omega_{ER}))(1+\omega_{ER})^2}.
\end{aligned}
\tag{5.13}
$$

Here, we again will consider three different subcases $r \geq \frac{s}{\kappa}$, $r \leq 0$ and $r \in \left(0, \frac{s}{\kappa}\right)$.

If $r \geq \frac{s}{\kappa}$, then $(1+\beta)r - (1-\beta)\frac{s}{\kappa} > 0$ and other multipliers and summands in formula (5.13) are also positive. Hence, $\frac{\partial}{\partial\omega_{ER}}\text{PGR}(\omega_{ER}) > 0$ and so PGR(ω_{ER}) increases monotonically in ω_{ER}. Therefore, $\text{PGR}_* = \text{PGR}(\omega_{ER}^*) = \frac{s+r\lambda}{\kappa+\lambda}$, where $\omega_{ER}^* \geq \frac{(1-\beta)\lambda-\beta\kappa}{(1+\beta)\kappa+\beta\lambda}$; the optimal preferences are such that $\omega_E^* \geq \frac{(1-\beta)\lambda-\beta\kappa}{(1+\beta)\kappa+\beta\lambda}\omega_R^*$, and the optimal level of education is $\alpha_* = 1$. This proves case 3 (a).

Let us note, that stationary points of PGR(ω_{ER}) are given by equation:

$$\beta\frac{s}{\kappa}(\omega_{ER})^2+\left[(1+\beta)r-(1-\beta)\frac{s}{\kappa}\right]\omega_{ER}+\beta r=0, \tag{5.14}$$

whose roots, if real, are

$$\omega_{ER}^{1,2}=\frac{1}{2\beta\frac{s}{\kappa}}\left(\left[(1-\beta)\frac{s}{\kappa}-(1+\beta)r\right]\pm\sqrt{\left(\frac{s}{\kappa}-r\right)\left[(1-\beta)^2\frac{s}{\kappa}-(1+\beta)^2r\right]}\right). \tag{5.15}$$

(subscript 1 corresponds to "−", subscript 2 corresponds to "+"). As the left-hand side of (5.14) is quadratic with respect to ω_{ER}, and other multipliers in (5.13) are positive, we conclude that $\mathrm{PGR}(\omega_{ER})$ increases for $\omega_{ER}\le\omega_{ER}^1$ and decreases for $\omega_{ER}^1\le\omega_{ER}\le\omega_{ER}^2$ and then it again increases for $\omega_{ER}\ge\omega_{ER}^2$.

If $r\le 0$, then $\omega_{ER}^1\le 0$ and $\omega_{ER}^2\ge 1$ and, hence, $\mathrm{PGR}(\omega_{ER})$ decreases on segment $\left[0,\frac{(1-\beta)\lambda-\beta\kappa}{(1+\beta)\kappa+\beta\lambda}\right]$. Therefore, $\mathrm{PGR}_*=\mathrm{PGR}\left(\omega_{ER}^*\right)=\left(\frac{\beta}{1-\beta}\frac{\kappa}{\lambda}\right)^\beta\left([1-\beta]\frac{s}{\kappa}+\beta r\right)$, where $\omega_{ER}^*=0$, with preferences such that $\omega_E^*=0$ being optimal, and the optimal level of education is $\alpha_*=\frac{\beta}{1-\beta}\frac{\kappa}{\lambda}$. This proves case 1 (*a*).

Finally, consider $r\in\left(0,\frac{s}{\kappa}\right)$. Consider the expression under the square root in (5.15): $(1-\beta)^2\frac{s}{\kappa}-(1+\beta)^2r$. Its roots are given by $\beta_{1,2}=\frac{\left(\sqrt{\frac{s}{\kappa}}\pm\sqrt{r}\right)^2}{\frac{s}{\kappa}-r}$. One can prove that $\beta_1=\frac{\left(\sqrt{\frac{s}{\kappa}}+\sqrt{r}\right)^2}{\frac{s}{\kappa}-r}>1$ and $\beta_2=\frac{\left(\sqrt{\frac{s}{\kappa}}-\sqrt{r}\right)^2}{\frac{s}{\kappa}-r}<1$.

If $\beta\in\left[\beta_2,\frac{\lambda}{\lambda+\kappa}\right)$, then the roots in (5.15) are either imaginary or real and coincide. Hence, due to monotone increasing of $\mathrm{PGR}(\cdot)$ in this case, $\mathrm{PGR}_*=\mathrm{PGR}\left(\omega_{ER}^*\right)=\frac{s+r\lambda}{\kappa+\lambda}$, where $\omega_{ER}^*\ge\frac{(1-\beta)\lambda-\beta\kappa}{(1+\beta)\kappa+\beta\lambda}$; the optimal preferences are such that $\omega_E^*\ge\frac{(1-\beta)\lambda-\beta\kappa}{(1+\beta)\kappa+\beta\lambda}\omega_R^*$, and the optimal level of education is $\alpha_*=1$. This proves case 3 (*b*) for $\beta\in\left[\beta_2,\frac{\lambda}{\lambda+\kappa}\right)$.

Now let us consider $\beta\in(0,\beta_2)$. In this case $\bar{\omega}_{ER}^*=\omega_{ER}^1>1$ and function $\mathrm{PGR}(\cdot)$ achieves its maximum on segment $\left[0,\frac{(1-\beta)\lambda-\beta\kappa}{(1+\beta)\kappa+\beta\lambda}\right]$ either on the border (at $\omega_{ER}^*(\beta)=0$ or at $\omega_{ER}^*(\beta)=\frac{(1-\beta)\lambda-\beta\kappa}{(1+\beta)\kappa+\beta\lambda}$) or at the local maximum point $\omega_{ER}^*(\beta)=\bar{\omega}_{ER}^*(\beta)$ (if $\bar{\omega}_{ER}^*(\beta)\in\left[0,\frac{(1-\beta)\lambda-\beta\kappa}{(1+\beta)\kappa+\beta\lambda}\right]$). Note that $\omega_{ER}^*(\beta)\ne 0$ because $\mathrm{PGR}(\cdot)$ increases for $0\le\omega_{ER}\le\bar{\omega}_{ER}^*$. Let us write the values of $\mathrm{PGR}(\omega_{ER})$ at the other two points:

$$\mathrm{PGR}\left(\frac{(1-\beta)\lambda-\beta\kappa}{(1+\beta)\kappa+\beta\lambda}\right)=\frac{s+r\lambda}{\kappa+\lambda}=d,$$

$$\begin{aligned}&\mathrm{PGR}\left(\bar{\omega}_{ER}^*(\beta)\right)\\&=\left(\frac{\bar{\omega}_{ER}^*+\beta\left(1+\bar{\omega}_{ER}^*\right)}{1-\beta\left(1+\bar{\omega}_{ER}^*\right)}\frac{\kappa}{\lambda}\right)^\beta\frac{\left[1-\beta\left(1+\bar{\omega}_{ER}^*\right)\right]\frac{s}{\kappa}+\left[\bar{\omega}_{ER}^*+\beta\left(1+\bar{\omega}_{ER}^*\right)\right]r}{\left(1+\bar{\omega}_{ER}^*\right)}\\&=f(\beta).\end{aligned}$$

Let us consider two subcases: $\lambda > \sqrt{\frac{sK}{r}}$ and $\lambda \leq \sqrt{\frac{sK}{r}}$.

If $\lambda > \sqrt{\frac{sK}{r}}$, one can prove that the following inequalities hold: $f(0) - d > 0$, $f(\beta_2) - d < 0$ and $f(\cdot)$ is a monotone decreasing function for $\beta \in (0, \beta_2)$. Then there exists a unique $\bar{\beta}_* \in (0, \beta_2)$ such that $f(\bar{\beta}_*) = d$.

Therefore, if $\beta \in (0, \bar{\beta}_*)$, then $f(\beta) > d$ and $\text{PGR}_* = \text{PGR}(\omega^*_{ER}) = \text{PGR}(\bar{\omega}^*_{ER})$, where $\omega^*_{ER} = \bar{\omega}^*_{ER}$; the optimal preferences are such that $\omega^*_E = \bar{\omega}^*_{ER}\omega^*_R$, and the optimal level of education is $\alpha_* = \frac{\bar{\omega}^*_{ER}+\beta(1+\bar{\omega}^*_{ER})}{1-\beta(1+\bar{\omega}^*_{ER})}\frac{\kappa}{\lambda}$. This proves case 2 for $\lambda > \sqrt{\frac{sK}{r}}$.

If $\beta \in [\bar{\beta}_*, \beta_2)$, then $f(\beta) < d$ and $\text{PGR}_* = \text{PGR}(\omega^*_{ER}) \equiv \frac{s+r\lambda}{\kappa+\lambda}$, where $\omega^*_{ER} \geq \frac{(1-\beta)\lambda-\beta\kappa}{(1+\beta)\kappa+\beta\lambda}$; the optimal preferences are such that $\omega^*_E \geq \frac{(1-\beta)\lambda-\beta\kappa}{(1+\beta)\kappa+\beta\lambda}\omega^*_R$ and the optimal level of education is $\alpha_* = 1$. This proves case 3 (*b*) for $\beta \in [\bar{\beta}_*, \beta_2)$ and $\lambda > \sqrt{\frac{sK}{r}}$.

For the case $\lambda \leq \sqrt{\frac{sK}{r}}$ let us put $\bar{\beta}_* = \frac{\lambda(s-r\kappa)}{r\lambda^2+(s+r\kappa)\lambda+s\kappa} < \beta_2$. One can prove that $f(0) - d > 0$, $f(\bar{\beta}_*) = d$ and the function $f(\beta)$ is monotone decreasing for $\beta \in (0, \bar{\beta}_*)$.

Hence, if $\beta \in (0, \bar{\beta}_*)$, then $\text{PGR}_* = \text{PGR}(\omega^*_{ER}) = \text{PGR}(\bar{\omega}^*_{ER})$ where $\omega^*_{ER} = \bar{\omega}^*_{ER}$; the optimal preferences are such that $\omega^*_E = \bar{\omega}^*_{ER}\omega^*_R$, and the optimal level of education is $\alpha_* = \frac{\bar{\omega}^*_{ER}+\beta(1+\bar{\omega}^*_{ER})}{1-\beta(1+\bar{\omega}^*_{ER})}\frac{\kappa}{\lambda}$. This proves case 2 for $\lambda \leq \sqrt{\frac{sK}{r}}$.

If $\beta \in [\bar{\beta}_*, \beta_2)$, then $\bar{\omega}^*_{ER}(\beta) \geq \bar{\omega}^*_{ER}(\bar{\beta}^*) = \frac{\lambda r}{s} = \frac{(1-\bar{\beta}^*)\lambda-\bar{\beta}^*\kappa}{(1+\bar{\beta}^*)\kappa+\bar{\beta}^*\lambda} \geq \frac{(1-\beta)\lambda-\beta\kappa}{(1+\beta)\kappa+\beta\lambda}$ which means that local maximum point is outside the considered segment and, hence, $\text{PGR}_* = \text{PGR}(\omega^*_{ER}) \equiv \frac{s+r\lambda}{\kappa+\lambda}$, where $\omega^*_{ER} \geq \frac{(1-\beta)\lambda-\beta\kappa}{(1+\beta)\kappa+\beta\lambda}$; the optimal preferences are such that $\omega^*_E \geq \frac{(1-\beta)\lambda-\beta\kappa}{(1+\beta)\kappa+\beta\lambda}\omega^*_R$, and the optimal level of education is $\alpha_* = 1$. This proves case 3 (*b*) for $\beta \in [\bar{\beta}_*, \beta_2)$ and $\lambda \leq \sqrt{\frac{sK}{r}}$.

The proof is complete. ■

References

1. Adler, N. E., Epel, E. S., Castellazzo, G., & Ickovics, J. R. (2000). Relationship of subjective and objective social status with psychological and physiological functioning: Preliminary data in healthy, White women. *Health Psychology, 19*, 586–592. https://doi.org/10.1037/0278-6133.19.6.586.
2. Alvergne, A., & Lummaa, V. (2014). Ecological variation in wealth–fertility relationships in Mongolia: the 'central theoretical problem of sociobiology' not a problem after all? *Proceedings of the Royal Society of London. Series B: Biological Sciences, 281*, 20141733. https://doi.org/10.1098/rspb.2014.1733.
3. Baker, D. P., Leon, J., Smith Greenaway, E. G., et al. (2011). The education effect on population health: A reassessment. *Popul Dev Rev, 37*, 307–332. https://doi.org/10.1111/j.1728-4457.2011.00412.x.
4. Banholzer, S., Kossin, J., & Donner, S. (2014). The impact of climate change on natural disasters. In A. Singh & Z. Zommers (Eds.), *Reducing disaster: Early warning systems for climate change* (pp. 21–49). Netherlands, Dordrecht: Springer.

5. Barro, R. J., & Becker, G. S. (1989). Fertility choice in a model of economic growth. *Econometrica, 57,* 481. https://doi.org/10.2307/1912563.
6. Becker, G. S., & Barro, R. J. (1986). A reformulation of the economic theory of fertility. *National Bureau of Economic Research.*
7. Bengtsson, T., Campbell, C., & Lee, J. Z. (2009). *Life under pressure: Mortality and living standards in Europe and Asia, 1700–1900.* London: The MIT Press, Cambridge, MA.
8. Bengtsson, T., & Dribe, M. (2006). Deliberate control in a natural fertility population: Southern Sweden, 1766–1864. *Demography, 43,* 727–746. https://doi.org/10.1353/dem.2006.0030.
9. Birdsall, N., Kelley, A. C., & Sinding, S. (2003). *Population matters: Demographic change, economic growth, and poverty in the developing world.* Oxford University Press.
10. Bollen, K. A., Glanville, J. L., & Stecklov, G. (2002). Economic status proxies in studies of fertility in developing countries: Does the measure matter? *Population Study, 56,* 81–96. https://doi.org/10.1080/00324720213796.
11. Boone, J. L., & Kessler, K. L. (1999). More status or more children? Social status, fertility reduction, and long-term fitness. *Evolution and Human Behavior, 20,* 257–277.
12. Brown, D. L., Schafft, K. A. (2011). Rural people and communities in the 21st century: Resilience and transformation. *Polity.*
13. de la Croix, D., & Doepke, M. (2003). Inequality and growth: Why differential fertility matters. *American Economic Review, 93,* 1091–1113. https://doi.org/10.1257/000282803769206214.
14. DeWitt, T. J., Sih, A., & Wilson, D. S. (1998). Costs and limits of phenotypic plasticity. *Trends in Ecology & Evolution, 13,* 77–81. https://doi.org/10.1016/S0169-5347(97)01274-3.
15. Doepke, M., & Zilibotti, F. (2007). Occupational choice and the spirit of capitalism. *National Bureau of Economic Research.*
16. Dutton. (2012). *Arid land resources & their mana.* Routledge.
17. Dyson, T. (2011). Population and development. *The demographic transition, 37,* 395–397. https://doi.org/10.1111/j.1728-4457.2011.00417.x.
18. Eide, E. R., & Showalter, M. H. (2011). Estimating the relation between health and education: What do we know and what do we need to know? *Economics of Education Review, 30,* 778–791.
19. Fieder, M., & Huber, S. (2007). The effects of sex and childlessness on the association between status and reproductive output in modern society. *Evolation of Human Behavior, 28,* 392–398. https://doi.org/10.1016/j.evolhumbehav.2007.05.004.
20. Fieder, M., Huber, S., Bookstein, F. L., et al. (2005). Status and reproduction in humans: new evidence for the validity of evolutionary explanations on basis of a university sample. *Ethology, 111,* 940–950.
21. Fink, G., & Redaelli, S. (2005). *Understanding bequest motives an empirical analysis of intergenerational transfers.* Netherlands Central Bank, Research Department.
22. Frankenberg, E., Sikoki, B., & Sumantri, C., et al. (2013). Education, vulnerability, and resilience after a natural disaster. *Ecology and Society, 18.* https://doi.org/10.5751/ES-05377-180216.
23. Galor, O., & Weil, D. N. (1996). The gender gap, fertility, and growth. *American Economic Review, 86,* 374–387.
24. Goodman, A., Koupil, I., & Lawson, D. W. (2012). Low fertility increases descendant socioeconomic position but reduces long-term fitness in a modern post-industrial society. *Proceedings of the Royal Society B: Biological Sciences, 279,* 4342–4351. https://doi.org/10.1098/rspb.2012.1415.
25. Gráda, C. Ó. (2011). Famines past, famine's future. *Development and Changes, 42,* 49–69. https://doi.org/10.1111/j.1467-7660.2010.01677.x.
26. Hamilton, W. D. (1964). The genetical evolution of social behaviour. I. *The Journal of Theoretical Biology, 7,* 1–16. https://doi.org/10.1016/0022-5193(64)90038-4.
27. Hamilton, W. D. (1964). The genetical evolution of social behaviour. II. *The Journal of Theoretical Biology, 7,* 17–52. https://doi.org/10.1016/0022-5193(64)90039-6.

28. Hopcroft, R. L. (2006). Sex, status, and reproductive success in the contemporary United States. *Evol Hum Behav, 27,* 104–120.
29. Huber, S., Bookstein, F. L., & Fieder, M. (2010). Socioeconomic status, education, and reproduction in modern women: an evolutionary perspective. *American Journal of Human Biology Official J Human Biology Council, 22,* 578–587. https://doi.org/10.1002/ajhb.21048.
30. Kaplan, H. S., Lancaster, J. B., Johnson, S. E., & Bock, J. A. (1995). Does observed fertility maximize fitness among New Mexican men? *Human Nature, 6,* 325–360.
31. Lagerlöf, N.-P. (2006). The Galor-Weil model revisited: A quantitative exercise. *Review of Economic Dynamics, 9,* 116–142. https://doi.org/10.1016/j.red.2005.07.002.
32. Livi Bacci, M. (1991). *Population and nutrition: An essay on European demographic history, english translation.* Cambridge University Press, Cambridge.
33. Mayr, E. (1961). Cause and effect in biology kinds of causes, predictability, and teleology are viewed by a practicing biologist. *Science, 134,* 1501–1506.
34. Mitani, J. C., Watts, D. P., & Muller, M. N. (2002). Recent developments in the study of wild chimpanzee behavior. *Evolutionary Anthropology Issues News and Reviews, 11,* 9–25.
35. Mukharjee, A. (2012). *Food security in Asia.* SAGE Publications India.
36. Nettle, D., & Pollet, T. V. (2008). Natural selection on male wealth in humans. *American Naturalist, 172,* 658–666.
37. Newton-Fisher, N. E. (2004). Hierarchy and social status in Budongo chimpanzees. *Primates, 45,* 81–87.
38. Nobles, J., Frankenberg, E., & Thomas, D. (2015). The effects of mortality on fertility: population dynamics after a natural disaster. *Demography, 52,* 15–38. https://doi.org/10.1007/s13524-014-0362-1.
39. Pachauri, R. K., & Reisinger, A. (2007). *IPCC fourth assessment report.* IPCC Fourth Assess Rep.
40. Penn, D. J., & Smith, K. R. (2007). Differential fitness costs of reproduction between the sexes. *Proceedings of National Academy of Sciences, 104,* 553–558.
41. von Rueden, C., Gurven, M., & Kaplan, H. (2011). Why do men seek status? Fitness payoffs to dominance and prestige. *Proceedings of the Royal Society of London. Series B: Biological Sciences, 278,* 2223–2232. https://doi.org/10.1098/rspb.2010.2145.
42. Sinding, S. (2009). Population, poverty and economic development. *Philosophical Transaction of the Royal Society B, 364,* 3023–3030.
43. Skirbekk, V. (2008). Fertility trends by social status. *Demographic Research, 18,* 145–180. https://doi.org/10.4054/DemRes.2008.18.5.
44. Smetters, K. (1999). Ricardian equivalence: long-run Leviathan. *Journal of Public Economics, 73,* 395–421. https://doi.org/10.1016/S0047-2727(99)00011-0.
45. Surbeck, M., Mundry, R., & Hohmann, G. (2011). Mothers matter! Maternal support, dominance status and mating success in male bonobos (Pan paniscus). *Proceedings of the Royal Society of B Biological Sciences, 278,* 590–598.
46. UN (1973). *The determinants and consequences of population trends.* United Nations, New York, NY.
47. van Aalst, M. K. (2006). The impacts of climate change on the risk of natural disasters. *Disasters, 30,* 5–18. https://doi.org/10.1111/j.1467-9523.2006.00303.x.
48. Vining, D. R. (1986). Social versus reproductive success: The central theoretical problem of human sociobiology. *Behavioral and Brain Sciences, 9,* 167–187.
49. de Waal, F. (2007). *Chimpanzee politics: power and sex among Apes, 25th anniversary.* Baltimore, Md: Johns Hopkins University Press.
50. de Waal, F., & Lanting, F. (1998). *Bonobo: The forgotten ape* (1st ed.). Berkeley: University of California Press.
51. Wilson, C., & Airey, P. (1999). How can a homeostatic perspective enhance demographic transition theory? *Popul Stud, 53,* 117–128. https://doi.org/10.2307/2584671.

52. Zhang, J. (1995). Social security and endogenous growth. *Journal of Public Economics, 58,* 185–213.
53. Zhang, J., & Zhang, J. (2001). Bequest motives, social security, and economic growth. *Economic Inquiry, 39,* 453–466. https://doi.org/10.1093/ei/39.3.453.

Chapter 6
Financing Climate Change Policies: A Multi-phase Integrated Assessment Model for Mitigation and Adaptation

Willi Semmler, Helmut Maurer, and Tony Bonen

JEL Classification: C61 · Q54 · Q58 · H5

6.1 Introduction

This paper extends the novel numerical modelling of optimal climate change policy responses developed in Semmler et al. [15]. In that paper, the dynamic decision problem was solved under a single regime of fixed exogenous parameters. We extend that framework here to consider multiple regimes. Under updated parameterizations, we compare the single-regime model against a multi-regime specification in which a new climate financing mechanism—"green bonds"—are introduced to the policy set. The regimes are exhaustive and sequential meaning, there is no possibility of the model returning to an earlier phase. In order to solve the multi-regime model a new numerical algorithm is applied: the arc parameterization method (APM). APM ensures continuous trajectories of the model's state variables, making the numerical

This paper was developed while Willi Semmler was a visiting scholar at the IIASA in July and August 2016. He would like to thank the IIASA for its hospitality. A related version of this paper was worked out while Willi Semmler was a visiting scholar at the IMF Research Department in April 2016. The authors would like to thank Prakash Loungani and his colleagues for many helpful insights.

W. Semmler
New School for Social Research, Henry Arnold Professor of Economics, New York, NY, USA

H. Maurer
Institute for Analysis and Numerics, University of Münster, Müntser, Germany

T. Bonen (✉)
Labour Market Information Council, Ottawa, Canada
e-mail: tony.bonen@lmic-cimt.ca

J. L. Haunschmied et al. (eds.), *Dynamic Economic Problems with Regime Switches*, Dynamic Modeling and Econometrics in Economics and Finance 25,
https://doi.org/10.1007/978-3-030-54576-5_6

solution more realistic. We find that the multi-regime model is Pareto-superior to the single-regime version.

The multi-regime model builds on the structure in Semmler et al. [15], which develops an integrated assessment model (IAM) of climate change's impact on social welfare. In that paper, the finite-horizon model continually solves for the optimal financing allocation to three climate change policy responses: (i) mitigation of increased CO_2 emissions; (ii) adaptation to the unavoidable consequences of climate change and (iii) investment in carbon-neutral, productivity-enhancing infrastructure. Under various sensitivity tests, the consistent optimal solution was found to prioritize funding of the latter category—productivity-enhancing infrastructure—to a much greater extent than the other two policy areas (i.e. over 90% of the total budget).

This policy-focused IAM is extended here to include a new public finance mechanism—so-called "green bonds"—designed to support climate change policy action. The funds raised by green bonds must be allocated to environmental efforts (i.e. among the three allocation options) and are repaid over very long horizons. As argued by Sachs [14], Flaherty et al. [6] and Heine et al. [7], such green bond financing generates more equitable intergenerational outcomes. Intergenerational equity is improved because repayment of the green bonds is a cost faced by future generations who will reap the benefits of an economy spared the worst of climate change's impacts as a result of policies undertaken by earlier generations. It follows that at least two model regimes are required: a period of green bond issuance and a period of their repayment. We also include an antecedent regime in which green bonds do not exist; this represents the current policy world and is similar to the single-regime model in Semmler et al. [15]. We show that the introduction of green bonds reduces total emissions, increases private and public capital, and results in higher overall welfare.

The remainder of the paper is organized as follows. Section 6.2 describes the general form of the IAM. Section 6.3 describes the numerical solution techniques. Sections 6.4 and 6.5 present the results for the single-regime model and its multi-regime extension, respectively. Section 6.6 reports on sensitivity analyses applied to the multi-regime model. Section 6.7 concludes.

6.2 Model Description

Climate change economic models are complex dynamic systems that typically do not lend themselves to standard, closed-form solutions. Common work arounds include linearizing the dynamic system or generating exogenous macroeconomic trajectories that are later integrated with climate dynamics.[1] In contrast, our integrated assessment model (IAM) numerically determines optimal control solutions for the full dynamic system. The model described below extends the IAM developed in

[1] See Bonen et al. [2] for a further discussion.

Semmler et al. [15]. Note that in spite of the use of capital letters, all variables are in per capita terms.

The dynamic system is driven by the state variables $X = (K, R, M, b, g) \in \mathbf{R}^5$ as defined by the following equations:

$$\dot{K} = Y \cdot (\nu_1 g)^{\beta}(1 - \tau_k) - C - e_P - (\delta_K + n)K - u\,\psi R^{-\zeta}, \tag{6.1}$$

$$\dot{R} = -u, \tag{6.2}$$

$$\dot{M} = \gamma u - \mu(M - \kappa\widetilde{M}) - \theta(\nu_3 \cdot g)^{\phi}, \tag{6.3}$$

$$\dot{b} = (r_t - n)b - \alpha_4 e_P - Y \cdot (\nu_1 g)^{\beta}\tau_k + \varsigma_k g, \tag{6.4}$$

$$\dot{g} = \alpha_1 e_P - (\delta_g + n)g + \varsigma_k g, \tag{6.5}$$

where K is the stock of private capital, R is the stock of the non-renewable resource, M is the atmospheric concentration of CO_2, b is the public debt level and g is the stock of public capital. Note that it is from g that climate policy actions are funded.[2]

The dynamic system in (6.1)–(6.5) extends Semmler et al. [15] by introducing $k > 1$ regimes. Specifically, τ_k and ς_k are regime-specific parameters that define three regimes ($k = 1, 2, 3$). When $\tau_k = \varsigma_k = 0$ there are no green bonds in circulation. For $\varsigma_k > 0$, green bonds are issued and the funds allocated to the stock of public capital used for climate change action, g. When $\tau_k > 0$, a special income tax is levied to pay down public debt, b.

The accumulation rate of private capital $\dot{K}$ is driven by, among other factors, output generated under a CES production function in which K and the extracted non-renewable resource u are inputs,

$$Y(K, u) := A(A_K K + A_u u)^{\alpha}. \tag{6.6}$$

Here A is multifactor productivity, A_K and A_u are efficiency indices of the inputs K and u, respectively. In (6.1) private-sector output Y is modified by the infrastructure share allocated to productivity enhancement $\nu_1 g$, for $\nu_1 \in [0, 1]$. This public-private interaction generates gross output $Y(\nu_1 g)^{\beta}$.[3] When green bonds are being repaid, $\tau_k > 0$ reduces net output to $Y(\nu_1 g)^{\beta}(1 - \tau_k)$, from which the economy consumes C, pays a lump sum tax e_P, and is subject to physical δ_K and demographic n depreciation. The last term in (6.1) is the opportunity cost of extracting the non-renewable resource u, where ψ and ζ are the scale and shape parameters that tie the marginal cost of u to the remaining stock of the resource à la Hotelling [9].

The dynamics $\dot{R}$ and $\dot{M}$ specify the environmental drivers of climate change. Equation (6.2) is the stock of the non-renewable resource R depleted by u units in each period. The depletion rate is constrained such that $0 \leq u(t) \leq 0.1$, $\forall t$. Equation (6.3) defines the change of CO_2, which is affected nonlinearly by mitigation efforts,

---

[2] All variables are in per capita terms.

[3] The exponent β is the output elasticity of public infrastructure, $\nu_1 g$.

$\nu_3 g$.[4] The non-renewable resource emits carbon dioxide and thus increases the atmospheric concentration of CO_2 at the rate γ. The environmentally stable level of CO_2 concentration is $\kappa > 1$ times the pre-industrial level $\widetilde{M}$. CO_2 levels at or below this level are naturally re-absorbed into the ecosystem (e.g. oceanic reservoirs) at the rate μ. The last term in (6.3) is the reduction of per-period emissions $\dot{M}$ due to the allocation of $\nu_3 \in [0, 1]$ of public infrastructure g to mitigation projects.

The dynamics $\dot{b}$ and $\dot{g}$ specify the government's fiscal stance. Revenue is generated from the lump sum tax e_P and the regime-specific income tax used to repay green bonds, τ_k. The latter flows directly to debt repayment. The former is allocated among shares $\sum_{i=1}^{4} \alpha_i = 1$: capital accumulation α_1, social transfers α_2 and administrative overhead $\alpha_3 > 0$. The remainder $\alpha_4 = 1 - \alpha_1 - \alpha_2 - \alpha_3$, pays down the stock of debt. Further, we constrain the lump sum tax to be $0 \leq e_P \leq 1, \forall t$.

In addition to repayments $\alpha_4 e_P$ and τ_k, public sector surplus/deficit $\dot{b}$ in (6.4) is driven by the time-varying interest rate r_t, and green bond issuances $\varsigma_k g$. The growth of public capital, Eq. (6.5), evolves according to the revenue stream $\alpha_1 e_P$ and funds raised from green bond issuance $\varsigma_k g$, but depreciates at the population-adjusted rate of $\delta_g + n$.

The objective function is the economy's per capita social welfare. Welfare W is maximized over a given planning horizon $[0, T]$, where $T > 0$ denotes the terminal time. Using a CES welfare function, welfare is a function of T, state variables $X \in \mathbf{R}^5$ and control variables U:

$$W(T, X, U) = \int_0^T e^{-(\rho-n)t} \frac{\left(C \cdot (\alpha_2 e_P)^{\eta} \left(M - \widetilde{M}\right)^{-\epsilon} (\nu_2 g)^{\omega}\right)^{1-\sigma} - 1}{1-\sigma} dt. \quad (6.7)$$

Private consumption C is augmented by three factors: (i) the share $\alpha_2 \in [0, 1]$ of tax revenue e_P used for direct welfare enhancement (e.g. healthcare, social services); (ii) the amount by which atmospheric concentration of CO_2 M is above the pre-industrial level $\widetilde{M}$ and (iii) the share $\nu_2 \in [0, 1]$ of public infrastructure g allocated to climate change adaptation. Exponents $\eta, \epsilon, \omega > 0$ ensure social expenditures and adaptation are welfare enhancing, whereas carbon emissions produce a disutility.[5] Finally, the pure discount rate ρ is adjusted by the population growth rate n.

The policymaker maximizes (6.7) subject to (6.1)–(6.5) via the control vector

$$U = (C, e_P, u, \nu_1, \nu_2, \nu_3) \in \mathbf{R}^6. \quad (6.8)$$

---

[4]Use of R emits carbon dioxide increasing M at the rate γ. The stable level of CO_2 emissions is $\kappa > 1$ of the pre-industrial level $\widetilde{M}$. Some CO_2 is absorbed into oceanic reservoirs at the rate μ.

[5]Note that instead of an independent damage function mapping climate change into output reductions, (6.7) treats climate change as a direct welfare loss. Adaptation efforts are modelled in a similarly direct fashion. We adopt this approach because the welfare impacts of climate change are not limited to lost productivity. For example, loss of life will increase from changing disease vectors and more intense heat waves.

Table 6.1 Definition of climate change policy funding shares

Variable	Definition	In equations
ν_1	Investment in carbon-neutral private infrastructure	$\dot{K}$ (6.1) and $\dot{b}$ (6.4)
ν_2	Climate change adaptation. Funds used to increase the population's resilience to climate change	W (6.7)
ν_3	Climate change mitigation. Funds used to reduce CO_2 emissions	$\dot{M}$ (6.3)

As noted, C is per capita consumption, e_P is a tax on capital gains and the rate of non-renewable resources extracted per period is u. The control variables ν_1, ν_2 and ν_3 determine the allocation of public capital g to carbon-neutral private capital, climate change adaption and climate change mitigation efforts, respectively (see Table 6.1).[6] As shares of g, $\nu_i \in [0, 1]$, $i = 1, 2, 3$ are constrained by $\sum_{i=1}^{3} \nu_i = 1$, total climate change policy funding levels are therefore $\nu_1 g$, $\nu_2 g$ and $\nu_3 g$.

All parameters for the model defined in Eqs. (6.1) through (6.7) are listed in Table 6.2.

6.3 Numerical Solution Techniques

The control problem is discretized on a fine grid, generating a large-scale nonlinear programming problem which is formulated with the Mathematical Programming Language AMPL; see Fourer et al. (1993). In AMPL we employ the interior-point optimization solver IPOPT (see Wächter and Biegler, 2006), that furnishes the control and state variables as well as the adjoint (co-state) variables. In this way, we are able to check whether we have found an *extremal solution* satisfying the necessary optimality conditions. To this end, we first derive an analytical expression of the control as a function of state and adjoint variables via the Maximum Principle. Inserting the computed values of state and adjoint variables into this analytical expression, the values must agree with the directly computed control values with a given tolerance. To verify sufficient conditions for local optimality, we could apply the second-order sufficient conditions presented in Augustin and Maurer [1]. This test amounts to verifying that an associated matrix Riccati equation has a bounded solution. However, since this is a rather elaborate procedure, we refrain from performing this test.

The IPOPT solver is sufficient for single-regime models and was implemented for the solutions reported in Semmler et al. [15]. For the multi-regime extension we introduce a novel application of the arc parameterization method (APM). In Sect. 6.5.3 we show that the APM approach produces realistic trajectories that are Pareto-superior to those of the single-regime model. Below we provide a brief description of the arc parameterization method. A full explanation of APM is offered in Appendix.

[6]Under a single-regime setting, Semmler et al. [15] demonstrate incorporating the ν_i as controls in U improves welfare outcomes versus treating them as fixed parameters.

Table 6.2 Parameter values

Variable	Value	Definition
ρ	0.03	Pure discount rate
n	0.015	Population growth rate
η	0.1	Elasticity of transfers and public spending in utility
ϵ	1.1	Elasticity of CO_2-eq concentration in (dis)utility
ω	0.05	Elasticity of public capital used for adaptation in utility
σ	1.1	Intertemporal elasticity of instantaneous utility
A	1	Total factor productivity
A_K	1	Efficiency index of private capital
A_u	100	Efficiency index of the non-renewable resource
α	0.05	Output elasticity of privately owned inputs, $(A_k K + A_u u)^\alpha$
β	0.5	Output elasticity of public infrastructure, $\nu_1 g$
ψ	0.1	Scaling factor in marginal cost of resource extraction
ζ	2	Exponential factor in marginal cost of resource extraction
δ_K	0.075	Depreciation rate of private capital
δ_g	0.05	Depreciation rate of public capital
α_1	0.1	Proportion of tax revenue allocated to new public capital
α_2	0.7	Proportion of tax revenue allocated to transfers and public consumption
α_3	0.1	Proportion of tax revenue allocated to administrative costs
r_t	0.07	Interest rate charged on debt, fixed $\forall t$
$\widetilde{M}$	1	Pre-industrial atmospheric concentration of greenhouse gases
γ	0.9	Fraction of greenhouse gas emissions not absorbed by the ocean
μ	0.01	Decay rate of greenhouse gases in atmosphere
κ	2	Atmospheric concentration stabilization ratio (relative to $\widetilde{M}$)
θ	0.01	Effectiveness of mitigation measures
ϕ	1	Exponent in mitigation term $(\nu_3 g)^\phi$

6.3.1 Application of Arc Parameterization

Multi-process (multi-phase) optimal control problems have been studied by Clarke and Vinter [4, 5] and later by Augustin and Maurer [1]. To solve our optimal multi-phase control problem, we implement the arc parameterization presented in Maurer et al. [12], Loxton et al. [11] and Lin et al. [10], along with discretization and nonlinear programming methods used previously. Although Maurer et al. [12] apply the arc-parametrization method (APM) only to bang-bang control problems, the APM is extended here to a continuous, multi-phase control problems (see Appendix for further details).

Following this work, we define the multi-process control problem on a grid of $t \in [0, T]$ with $s+1$ phases (regimes) occurring at switch points $t_k \in (0, T)$, $k = 1, ..., s$. Discretization is achieved via a uniform grid of mesh points

$$\tau_i = i \cdot h, \quad h = \frac{1}{N}, \quad i = 0, 1, \ldots, N.$$

The mesh size N must be a multiple of the number $s+1$ of intervals. This condition ensures that the phase boundaries $(k+1)/(s+1)$, $k = 0, \ldots, s+1$, appear as knot points in the applied integration schemes. The resulting discretized control problem is then formulated as a large-scale nonlinear programming problem (NLP) in AMPL.

We define multiple regimes on the specified mesh grid as $s+1$ ordinary differential equations. Let the dynamics of the economic process in the interval $[t_k, t_{t+1}]$ be given by

$$\dot{x}(t) = f_k\big(x(t), u(t)\big), \quad t_k \le t \le t_{k+1} \quad (k = 0, 1, \ldots, s), \tag{6.9}$$

where the right-hand side of the ODE is a C^1 function.

A *continuous* state trajectory $x(t)$ on the entire interval $[0, T]$ is obtained by imposing the continuity condition

$$x(t_k) = x(t_k-), \qquad k = 1, \ldots, s. \tag{6.10}$$

Note that the continuity of the state variables in (6.10) does not automatically ensure the continuity of the control variables; in fact, these can jump when the system transitions between policy phases.

6.4 Single-Phase Model

The model defined in Eqs. (6.1)–(6.7) is discretized on a fine grid $t \in [0, 70]$. While each integer step can be said to represent one year, discrete steps are set at $\Delta t = 0.05$ for the single-phase model, ie., we use $N = 1400$ grid points. For the multi-phase model, subsequently discussed, we also use $N = 1400$ or a refined grid. The finer grid helps the multi-phase pathways be smoother than otherwise. Importantly, the single-phase model does not allow for green bonds or a new tax for repayment. Therefore, the parameters are set at $\varsigma_k = \tau_k = 0$ in Eqs. (6.1), (6.4) and (6.5) for the single-phase model.

The numerical solution technique employed in AMPL requires initial values for each state variable. In addition, model stability is ensured by placing terminal value boundaries (maxima or minima) on certain state variables. The initial values and

Table 6.3 Initial values and terminal constraints

Variable	Initial value $t = 0$	Terminal value $t = 70$
$K(t)$	10	≥ 15
$M(t)$	2.5	≤ 2.5
$R(t)$	1.5	≥ 0
$g(t)$	0.5	Unconstrained
$b(t)$	0.8	Unconstrained

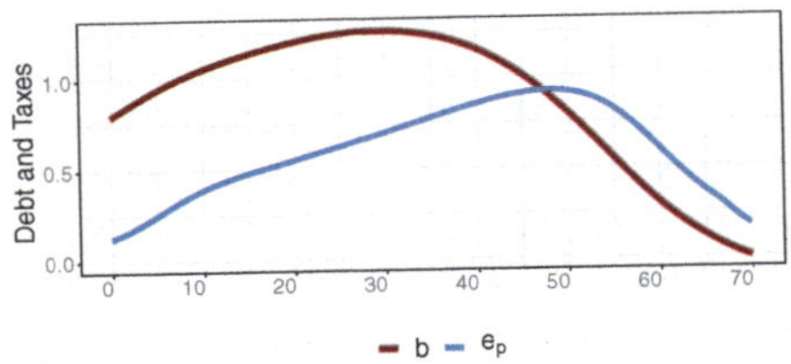

(a) Government debt, b and tax, e_P

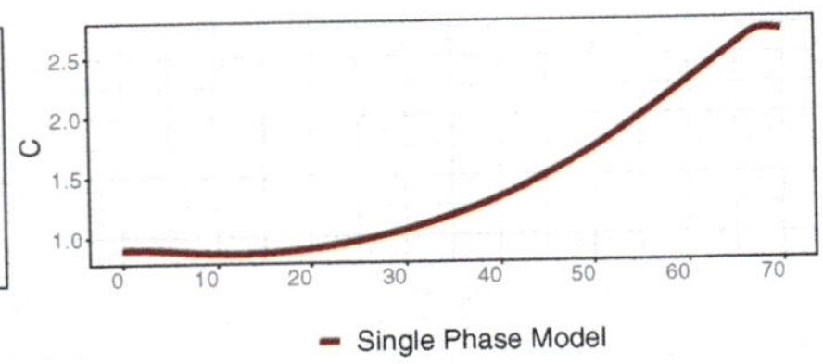

(b) Consumption, C

Fig. 6.1 Government debt, taxes and consumption in single-regime model

terminal constraints are listed in Table 6.3. Note that the terminal constraints ensure that there is at least a 50% increase in private capital over the 70-year period and that CO_2 concentration will only increase slightly over the entire period. In other words, the parameterizations presume climate action is successful. This differs from the setup in Semmler et al. [15] which considered climate policy optimization against a baseline of "business as usual" (i.e. no investment in climate change policies).

Figure 6.1 shows the trajectories of government debt b, taxes e_P and consumption C over the full horizon. Government debt at first rises as various investments are made, but as the terminal point $t = T$ approaches, public debt is driven toward zero so that there are no outstanding obligations at the end of the finite horizon (i.e. no Ponzi schemes are allowed). This downward trend to $b(T) = 0$ is driven by increase is the tax rate e_P which peaks at $t = 48$. Throughout this trajectory consumption increases as it is the major component of welfare in (6.7).

The positive consumption growth rate in the face of increasing taxation comes at the cost of lower investment in capital—both private and public. Figure 6.2 shows the trajectory of private capital stock K. After an initial, brief contraction K increases by approximately 60%. But, in the final 11 periods of the model, the tax burden takes its toll and private capital falls rapidly, and then recovers slightly toward its terminal constraint $K(T) = 15$. Public capital follows a similar pattern, albeit without the slightly recovery at the terminal point. It increases from $g(0) = 0.5$ to $g(56) = 1.2$, only to fall back slightly in the final years of model. The retrenchment is more dramatic for g than it is for K, with public capital stock reaching 0.89 at the terminal point.

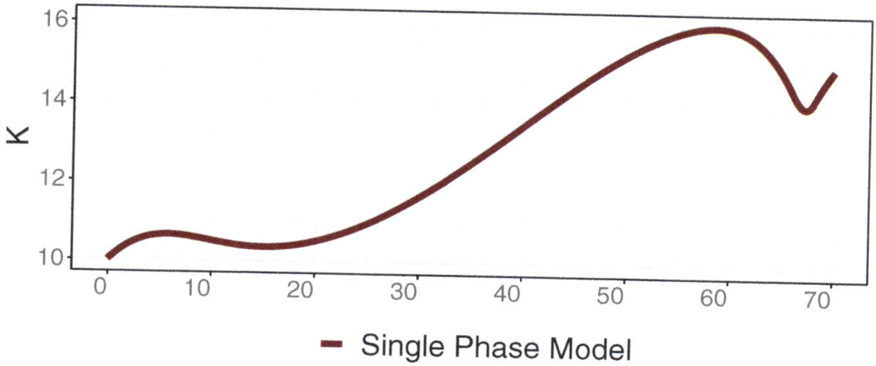

Fig. 6.2 Private capital stock in single-regime model

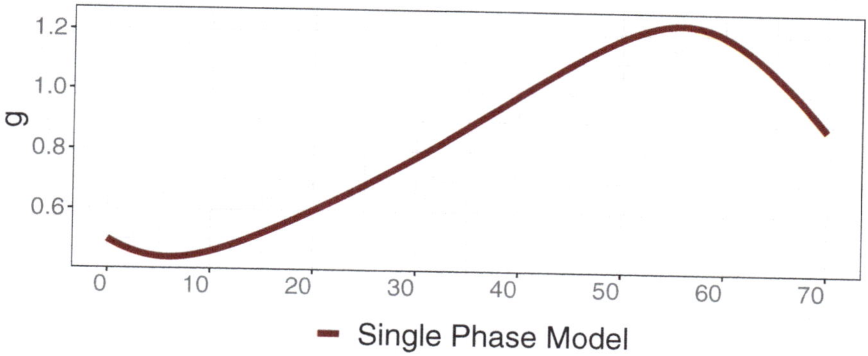

Fig. 6.3 Public capital stock in single-regime model

In addition to reducing the private and public capital stocks, the increased tax revenue needed to pay off government debt has another perverse impact: increasing CO_2 emissions. Recall that private capital K is a carbon-neutral input for the production of Y, with the extracted non-renewable resource u being the alternative input.[7] We find that $u(t) \approx 0$ for $t < 67$, but becomes positive thereafter. At $t = 67$ the extraction rate u rapidly increases (see Fig. 6.4). This behaviour in the model's final decade is driven by the shift in production toward fossil fuel inputs that allow the economy to achieve the no-Ponzi condition $b(T) = 0$ without negatively impacting consumption (Fig. 6.3).

[7] See Eq. (6.6) and accompanying text. We also want to note the long period of no fossil fuel energy extraction comes from the fact that the carbon stock is presumed to start at a low initial level as well as the low efficiency of the fossil fuel-based energy assumed, i.e. $A_u = 100$ (see Sect. 6.6 below).

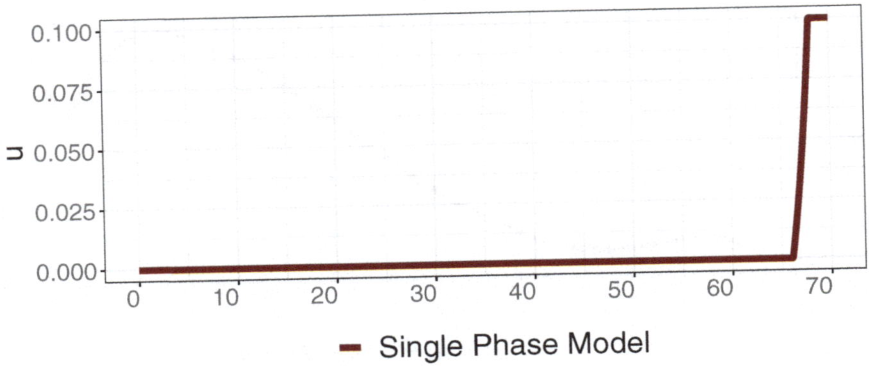

Fig. 6.4 Non-renewable resource extraction rate in single-regime model

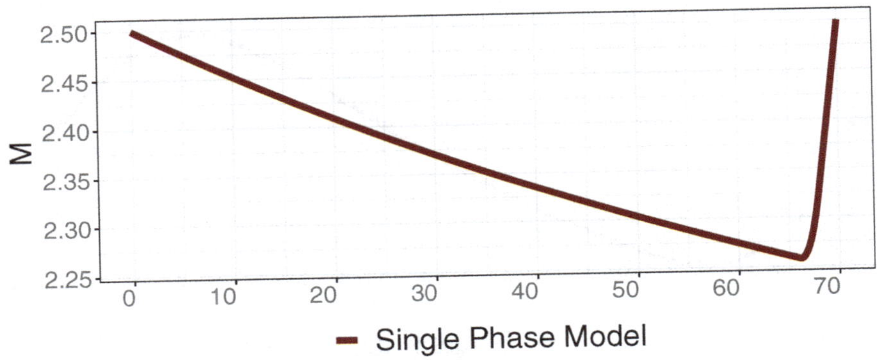

Fig. 6.5 CO_2 emissions in single-regime model

The sudden jump in u translates into additional carbon emissions M. Figure 6.5 shows that reliance on K in production corresponds with a steady decline in CO_2 emissions. This decline is completely reversed by the re-introduction of u in production. Evidently, the somewhat generous terminal condition applied to $M(T)$ allows for this fossil fuel-intensive behaviour.

Finally, Fig. 6.6 shows the relative allocation of g to infrastructure ν_1, adaptation efforts ν_2 and emissions mitigation ν_3. The results are consistent with those reported in Semmler et al. [15]. Namely, approximately 96% of g is allocated to ν_1, 4% to ν_2 and essentially 0% to ν_3 during the first 57 periods. For $t > 60$, ν_1 declines slightly as K falls to it terminal value and adaptation efforts ν_2 increase slightly as non-renewable extraction rates increase (see Fig. 6.4).

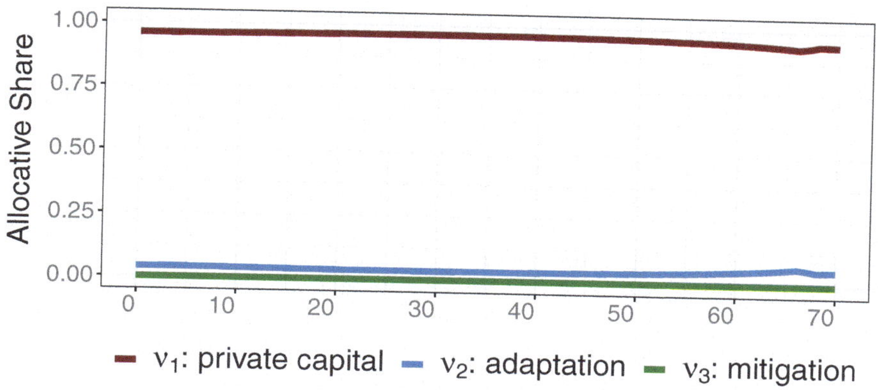

Fig. 6.6 Optimal distribution of public capital in single-regime model

6.5 Multi-phase Extension of the Model

The multi-phase model uses the same parameterization listed in Table 6.2, as well as the same initial values and terminal constraints in Table 6.3.[8] As discussed in Sect. 6.5 and Appendix, the nonlinear multi-phase problem is solved using the arc parameterization method (APM). In accordance with the continuity condition in (6.10), the state variables transition continuously between regimes. However, the model solution generates discontinuous jumps in the controls variables. This is a natural result of the changed dynamic system introduced as the model shifts from one policy environment to another.[9]

The multi-regime model presented here assumes fixed switching points. While it is computationally feasible for the transition times to be endogenously and optimally determined as part of the dynamic system's solution [12], we do not allow for this extra complexity here. Beyond the greater computational expense, preliminary testing indicates that time spent in the second regime (when green bonds are issued) is maximized. This is not surprising as this regime is not constrained by the no Ponzi condition. However, such a result is neither realistic nor helpful as the dynamics of the two other, extremely short regimes become impossible to discern. The three regimes are thus fixed on the intervals $t \in [0, 20]$, $t \in (20, 40]$ and $t \in (40, 70]$.

---

[8] Of course, the initial values apply only to the starting values in phase 1 and the terminal constraints bind only at $T = 70$, at the end of the third phase.

[9] DSGE models have also recently allowed for regime switches (see [8], Sect. 7.2), but not address the issue of discontinuities in the control variables.

6.5.1 *Regimes of the Multi-phase Model*

The first phase, "no green bonds", corresponds exactly to the single-phase model discussed in Sect. 6.4. In particular, $\varsigma_k = \tau_k = 0$ for $k = 1$. In the second phase, $k = 2$, green bonds are introduced as a financing option $\varsigma_2 > 0$. We choose the specific value $\varsigma_2 = 0.05$. The long-term nature of the green bonds means no repayments are made in $k = 2$, such that $\tau_1 = \tau_2 = 0$. In the third and final phase the green bonds come due and the government ceases issuance, $\varsigma_3 = 0$. Repayment is conducted through a special income tax set at 3%, $\tau_3 = 0.03$. Although the accumulated green bond debt feeds into the same overall public debt level, b, the special tax τ_k provides the policymaker with a new mechanism by which to raise revenue. The policymaker can of course continue to control the value of capital taxation e_P throughout the model's three phases.

The introduction of a new asset—green bonds—has implications for financial markets. In general, the issuance of the new green bonds—especially when scaled up—will have price and rebalancing effects on the asset market.[10] In our context we assume that wealth-holders have many assets in their portfolio, such as cash, equity, real estate and bonds. Further, we assume all green bonds are sold at issuance, which implies that wealth-holders bid prices down to a market clearing level. This affects the relative price and return across assets, but can simply be thought of as causing a rebalancing of different types of assets in the portfolio. A similar but more static approach is considered by Tobin [16] in macroeconomic portfolio theory where asset accumulation and portfolio allocation decisions interact with the real side of the economy. A dynamic extension of Tobin's approach is developed in Chiarella et al. [3] in which there are simultaneous asset accumulation and dynamic portfolio decisions.[11]

In the graphical results that follow, the first phase in which no green bonds have been issued is plotted in red. The second phase—green bond issuance—is coloured green, and the third phase, repayment, is blue.

6.5.2 *Multi-phase Model Results*

As in the single-regime model, government debt b builds up rapidly before being driven to the no-Ponzi condition $b(T) = 0$. However, the trajectory of b is somewhat

---

[10]In addition, when green bonds portend reductions in CO_2-emitting energy sources, their issuance might lead to significant devaluation of assets representing fossil fuels if this is expected to increase the risk of these assets becoming (so-called) "stranded" assets. Formally introducing this effect is left for later work.

[11]Note that in the present context the Ricardian equivalence theorem, which says that the real side of the economy will not be affected by deficit spending financed through issuing of bonds, is not applicable in this context. This is because green bonds are used to reduce future damages to GDP, and thus carry some future returns from their investments (in particular from public infrastructure). For details, see Orlov et al. [13].

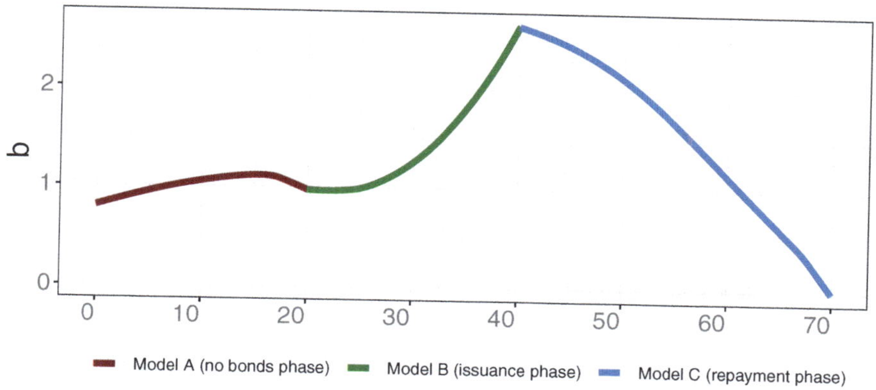

Fig. 6.7 Public debt in 3-regime model

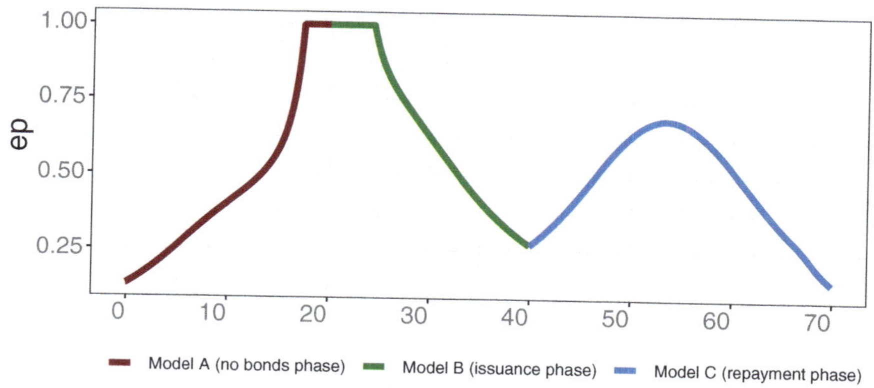

Fig. 6.8 Capital taxation level in 3-regime model

different and specific to the three regimes (see Fig. 6.7). In the first phase, government debt b increases only mildly and even decreases mildly ahead of the green bond issuance phase. Unsurprisingly, the introduction of green bonds increases the stock of public debt threefold to 2.6. The debt level reverses course only as the repayment phase is introduced at $t = 40$.

Interestingly, the standard taxation rate chosen by the policymaker follows a rather different path than before. Figure 6.8 shows e_P rising rapidly at the end of phase 1 and hits the constraint $e_P \leq 1$. The introduction of green bonds allows for the rapid reduction of taxation rates. In the final phase, standard taxation rates increase so as to help reduce the overall tax burden.

Consumption follows a similar upward trajectory in the first and second regimes, but then retrenches slightly in the third phase (see Fig. 6.9). The fall in per capita consumption in $k = 3$ is due to the reduction in output Y stemming from the special tax τ_k. Yet, at its peak per capita consumption reaches $C = 2.97$, surpassing the maximal

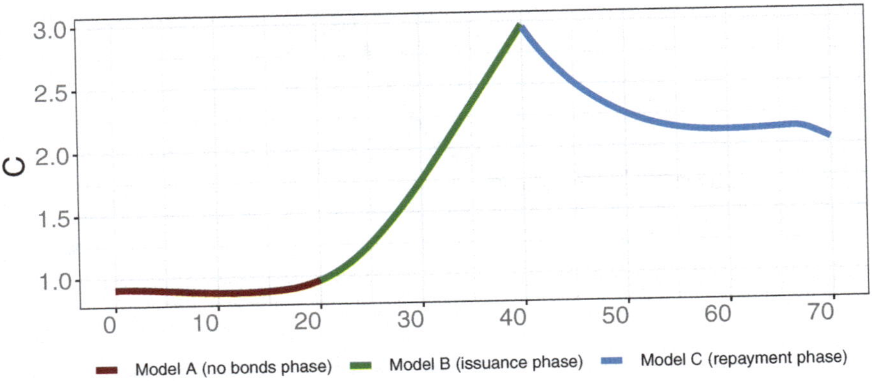

Fig. 6.9 Consumption in 3-regime model

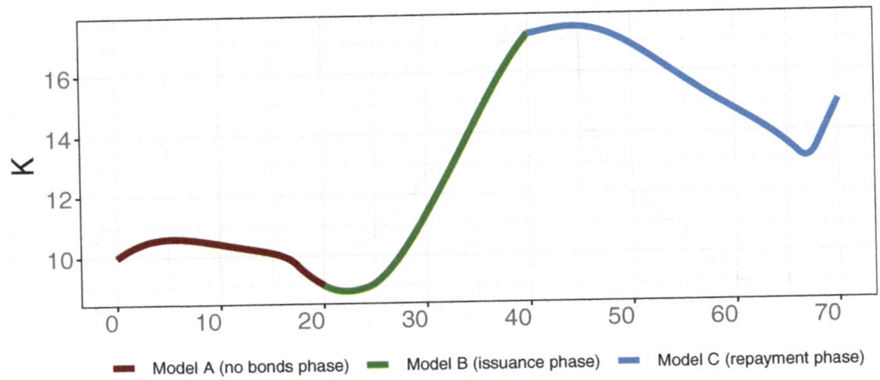

Fig. 6.10 Private capital stock in 3-regime model

observed in the single-phase model, $C = 2.70$. This higher maximal consumption contributes to the overall welfare improvement of the multi-phase model relative to the single-phase version (see Sect. 6.5.3).

Both private and public capital follow as similar pattern as seen in the single-regime model: a rapid increase over the first two-thirds of the finite horizon is reversed with a retrenchment in both as $t \to T$. As before, the retrenchment of private capital K is partially reversed in the final stages of the model (Fig. 6.10), whereas public capital declines monotonically during the third phase (Fig. 6.11).

Carbon dioxide emissions M follow a nearly identical pattern trajectory in the three-regime model in Fig. 6.12 as in the single-regime model (see Fig. 6.5). As before, the terminal value $M(T) = 2.5$ is the reason for the sudden jump in emissions. This constraint can be thought of as a politically determined contract to limit emissions at a higher than optimal level. Where a lower political bound for emissions agreed upon, the final level of emissions modelled would of course be lower as well.

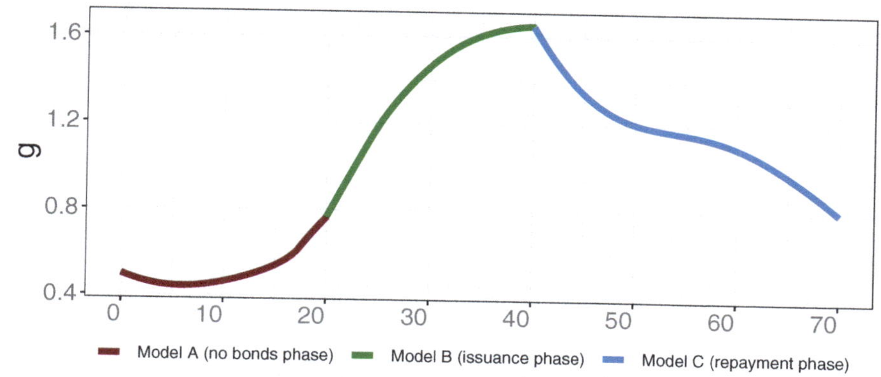

Fig. 6.11 Public capital stock in 3-regime model

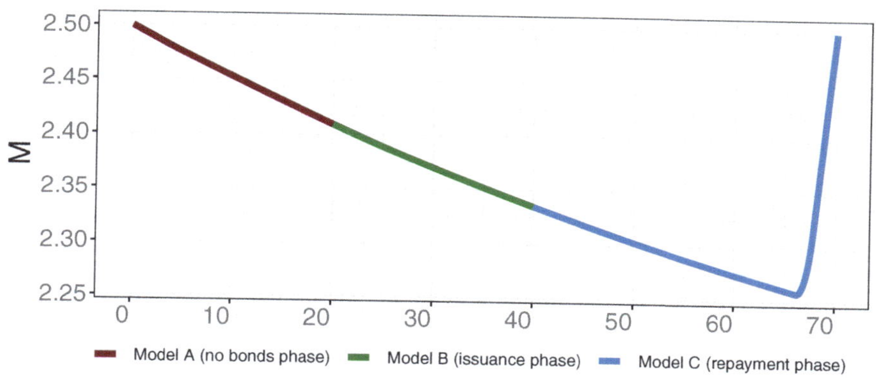

Fig. 6.12 Carbon dioxide emissions in 3-regime model

Figure 6.13 displays some of the discontinuities of the ν_i control variables as the system enters a new regime. In particular, the third regime exhibits additional investments in adaptation efforts, which then slowly reduces as $t \to T$. Mitigation efforts remain steady and around zero throughout the three regimes. The effect of these shifts in allocations is to reduce the rate of investment of public capital g into green infrastructure K relative to the single-phase model (see Fig. 6.6). The average allocation to K is slightly lower at approximately 94% in the multi-regime model versus 95% in the single-regime model.

6.5.3 *Comparison of Welfare Results*

The single and multi-regime models share similar overall pathways, but the slight differences accumulate to large total welfare differentials (see Table 6.4). Average

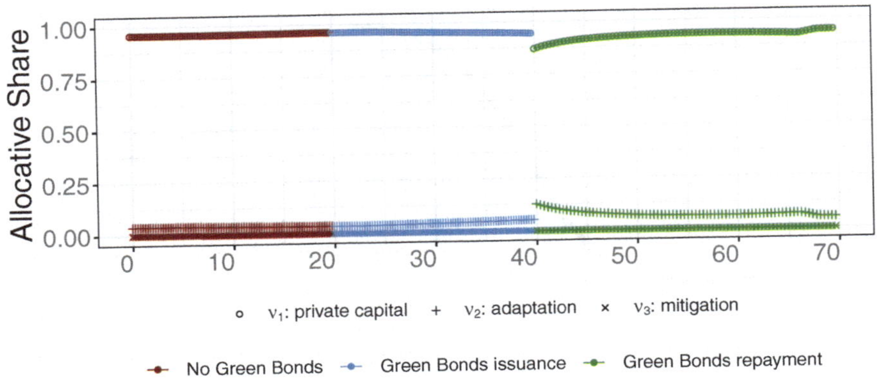

Fig. 6.13 Optimal allocation of public capital

Table 6.4 Single and multi-phase model key variable comparison

		Average	Terminal	Maximum
Single regime				
Consumption	C	1.42	2.69	2.70
Private capital	K	12.82	15.00	16.04
Public capital	g	0.85	0.89	1.23
Carbon emissions	M	2.36	2.50	2.50
Multi-regime				
Consumption	C	1.66	2.05	2.97
Private capital	K	12.63	15.00	17.48
Public capital	g	1.02	0.80	1.65
Carbon emissions	M	2.38	2.50	2.50
Social welfare		*Total*		
Single regime	W	−21.48		
Multi-regime	W	−12.34		

consumption over time was higher in the multi-regime model (1.7 versus 1.4), as is the single-period maximum value (3.0 vs. 2.7). The stock of private and public capital shifted from the single to the multi-regime model. Average K declined slightly in the multi-regime model, whereas average g increased when green bonds were introduced. As expected, from the trajectories observed above, CO_2 concentration levels are virtually identical in the two models, which the multi-regime model exhibiting marginally higher average M (2.38 vs. 2.36).

The higher average pathway of C generated a significantly higher overall social welfare value for the multi-regime model. The values calculated from Eq. (6.7), show that the multi-regime model is superior with at $W_{multi}(T; X; U) = -12.3$ as compared to the single-regime model's total welfare value of $W_{single}(T; X; U) = -21.5$.

6.6 Sensitivity Analysis

Finally, we present here some sensitivity analyses of the multi-phase IAM. The focus is on the efficiency index of A_u, which sets the relative productivity level of the non-renewable resource in production. The model results discussed above set $A_u = 100$. We compare the trajectories for several key variables with higher non-renewable input productivity, for $A_u = 300$ and $A_u = 500$. This is a particularly important parameter to test since there was little extraction and use of the non-renewable resource under the initial parameterization. Unsurprisingly, for higher A_u parameterizations the extraction rate u becomes substantially higher.

With a more productive non-renewable resource, production uses a greater level of u as an input (see Eq. 6.6). The direct result of this is the increased level of CO_2 emissions in the atmosphere. Figure 6.14 shows emissions falling persistently in the baseline case $A_u = 100$ (in red) until the terminal constraint pulls it up in the final periods. For $A_u = 300$ (in green) and $A_u = 500$ (in blue), emissions rise rapidly in the early phases of the model before trending down to the fixed terminal point $M(T) = 2.5$.

The six panels in Fig. 6.15 show the pathways under $A_u = 100$, 300 and 500 for public g and private K capital, consumption C and public debt b, and the allocation of public capital productivity enhancements ν_1 and climate change mitigation ν_3. In general, the large change (a 3- to 5-fold increase) in the A_u efficiency index leads to relatively small shifts in these variables' pathways. Further, in each case the shift in trajectories move the expected direction. Public and private capital accumulation rises faster and peaks at a higher level as one of the inputs to production is made cheaper (viz. more productive). Consumption is higher in all periods as A_u rises, and the allocation to productivity-enhancing capital ν_1 falls relative to the baseline as the higher carbon emissions rise. In place of ν_1, public capital is shifted toward greater CO_2 mitigation efforts, ν_3, in the third phase of the model as a response to the higher emissions generated from non-renewable resource-intensive production.

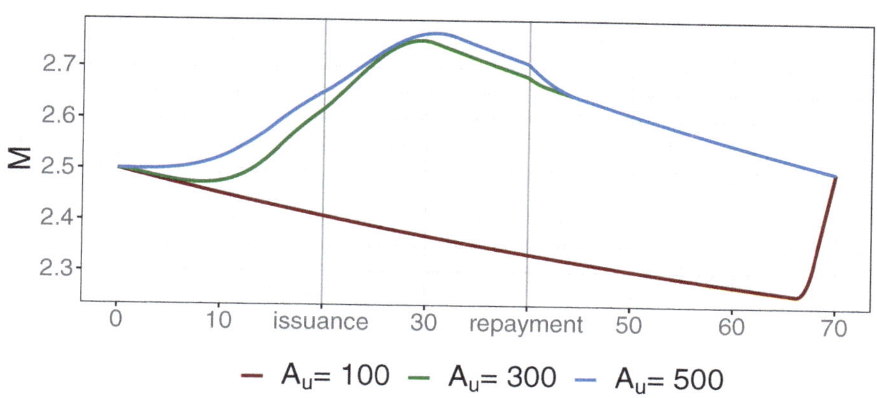

Fig. 6.14 Sensitivity test of CO_2 emissions

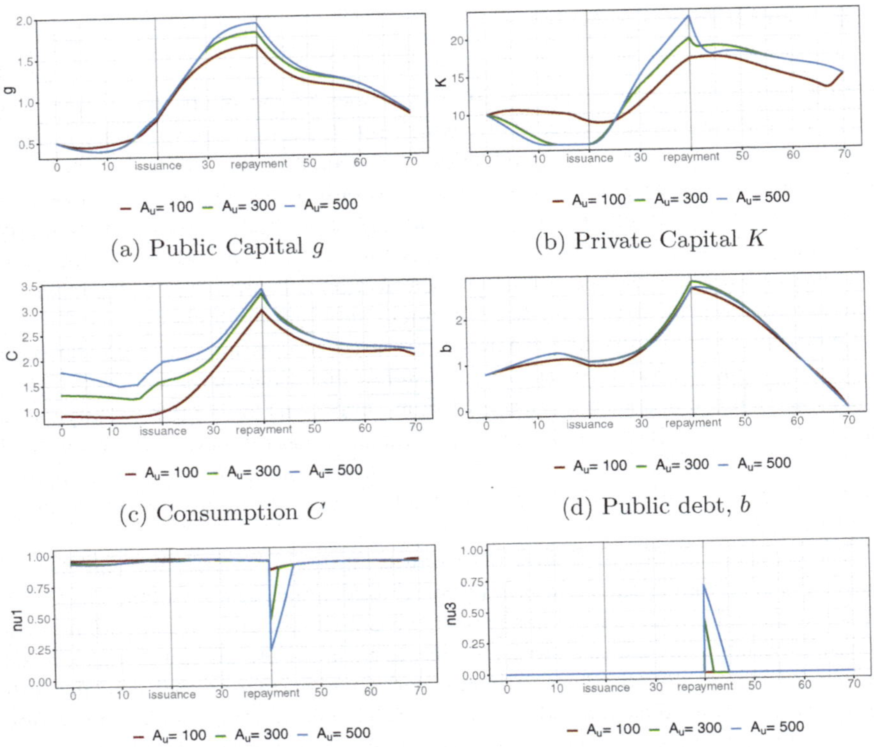

(a) Public Capital g (b) Private Capital K

(c) Consumption C (d) Public debt, b

(e) Allocation to private capital ν_1 (f) Allocation to CO_2 mitigation ν_3

Fig. 6.15 Sensitivity test of selected state and control variables

6.7 Conclusion

In Semmler et al. [15] we developed an integrated assessment model (IAM) explicitly accounting for the extraction and use in production of CO_2-emitting resources, as well as the optimal allocation of public finances to counter climate change. We have extended that IAM framework here to consider how new policies, specifically green bond financing, could be introduced to the set of available policies. To achieve this, we posited a 3-regime model in which green bonds are (i) non-existent, (ii) issued and (iii) repaid. The multi-regime model was shown to be Pareto-superior to the single-regime baseline, and enhanced intergenerational equity.[12]

Overall, the IAM developed here is an advancement both in terms of the solution algorithm employed and in its use of novel, multi-phase dynamics (namely, APM). As mentioned, the modelling of non-renewable resource extraction and detailed public

[12] In this context, a recent discussion of proposals for central banks to accept climate bonds as collateralizable securities is available in Flaherty et al. [6].

sector policies on climate change are new but important features in the IAM literature. In addition we have treated the damage function of climate change as impacting social welfare directly, as opposed to indirectly through reductions in the rate at which output is produced. While neither approach is perfect, we have employed the direct-utility impact version because we believe it is better able to capture the multitude of physical, ecological and societal losses that are likely to be induced by unabated climate change. Regardless of how the damage function is introduced, our framework allows for a multi-phase approach in which new, unforeseen policies, events and dynamics of the state equations can be introduced and responded to by the policymaker oriented toward a limited time horizon. We believe this to be a far more natural framework to address climate-economic-financial questions over a known, finite period of time.

Appendix: Multi-phase Optimal Control Problems and Their Numerical Solution

Multi-process (multi-phase) optimal control problems have been studied by Clarke and Vinter [4, 5] and later by Augustin and Maurer [1]. Suppose that a dynamic economic process on a given time interval $[0, T]$ consists of $(s+1)$ phases (regimes) that switch at the transition times $t_k \in (0, T)$, $k = 1, ..., s$. The switching times are ordered according to

$$0 = t_0 < t_1 < t_2 < \cdots < t_s < t_{s+1} = T. \tag{6.11}$$

In each interval $[t_k, t_{k+1}]$, $k = 0, 1, \ldots s$, the dynamics and objectives may be different.

Let $x \in \mathbb{R}^n$ be the state variable and $u \in \mathbb{R}^m$ the control variable. The dimensions of the state vector and control vector may be different in different phases. For simplicity, we refrain here from discussing this general case and assume the same dimensions in each subinterval. Hence, the dynamics of the economic process in the interval $[t_k, t_{t+1}]$ is given by the ordinary differential equation,

$$\dot{x}(t) = f_k\big(x(t), u(t)\big), \quad t_k \le t \le t_{k+1} \quad (k = 0, 1, \ldots, s), \tag{6.12}$$

where the right-hand side of the ODE is a C^1 function $f_k : \mathbb{R}^n \times \mathbb{R}^m \to \mathbb{R}^n$. The time t_k in (6.12) is understood from the right, while the time t_{k+1} is taken from the left. The initial condition and terminal constraints are given as

$$x(0) = x_0, \quad \psi(x(T)) = 0. \tag{6.13}$$

We further impose control constraints in each interval,

$$u_{k,\min} \leq u(t) \leq u_{k,\max}\,, \quad t_k \leq t \leq t_{k+1} \quad (k = 0, 1, \ldots, s), \tag{6.14}$$

with $-\infty \leq u_{k,\min} < u_{k,\max} \leq +\infty$.

A *continuous* state trajectory $x(t)$ on the whole interval $[0, T]$ is obtained by imposing the continuity condition

$$x(t_k) = x(t_k-), \qquad k = 1, \ldots, s. \tag{6.15}$$

Note that the continuity of the state variables in (6.15) does not automatically ensure the continuity of the control variables; in fact, these can jump, as we demonstrate below, when the system transitions between policy phases are studied. Moreover, we can prescribe interior (transition) conditions for the state variables by

$$\varphi_k(x(t_k-) = 0, \quad k = 1, \ldots, s, \tag{6.16}$$

with C^1 functions $\varphi_k : \mathbb{R}^n \to \mathbb{R}$.

In each interval one may also have different objectives which are defined by functions $L_k : \mathbb{R}^n \times \mathbb{R}^m \to \mathbb{R}^n$, $k = 0, 1, \ldots, s$. Then the optimal multi-phase control problem is defined by the following objective:

$$J(x, u) = \max_u \left\{ \sum_{k=0}^{k=s} \int_{t_k}^{t_{k+1}} e^{-r_k \cdot t} L_k(x(t), u(t))\, dt \right\}, \tag{6.17}$$

subject to the constraints (6.12)–(6.15), and $r_k > 0$ for $k = 0, 1, \ldots, s$.

To solve the optimal multi-phase control problem, we implement the arc-parametrization in Maurer et al. [12] in conjunction with discretization and non-linear programming methods. Although they apply the arc-parametrization method (APM) only to bang-bang control problems, the APM can easily be extended to continuous, multi-phase control problems as follows. Let

$$\xi_k = t_{k+1} - t_k, \quad k = 0, 1, \ldots, s, \tag{6.18}$$

denote the *arc lengths* (or, arc durations) of the multi-process. The time interval $[t_k, t_{k+1}]$ is mapped onto the fixed interval $[k/(s+1), (k+1)/(s+1)]$ by the linear transformation

$$t = a_k + b_k \tau, \quad \tau \in \left[\frac{k}{s+1}, \frac{k+1}{s+1} \right], \tag{6.19}$$

where $a_k = t_k - k\xi_k$ and $b_k = (s+1)\xi_k$. Taken together, the complete time interval $[0, T]$ is thereby mapped onto the unit interval $[0, 1]$. Identifying $x(\tau) = x(a_k + b_k\tau) = x(t)$ in the relevant intervals, we obtain the scaled ODE system

$$\frac{dx}{d\tau} = \xi_k(s+1) \cdot f_k\big(t_k + \tau \cdot \xi_k, x(\tau), u(\tau)\big) \tag{6.20}$$

for $\tau \in \left[\frac{k}{s+1}, \frac{k+1}{s+1}\right]$. Note that ξ_k are treated as optimization variables if the transition times t_k are free.

The time transformation leads us to rescale the objective function (6.17) as follows:

$$J(x,u) = \max_u \left\{ \sum_{k=0}^{k=s} \int_{\frac{k}{s+1}}^{\frac{k+1}{s+1}} \xi_k(s+1) e^{-r_k \cdot (a_k + b_k \tau)} L_k(x(\tau), u(\tau))\, d\tau \right\}, \tag{6.21}$$

and subject to (6.12)–(6.15) and rescaled according to (6.19). For the purposes of exposition, we fix the transition points t_k to reflect the exogenous (and often sub-optimally protracted) nature of introducing and implementing new policies.

References

1. Augustin, D., & Maurer, H. (2000). Second order sufficient conditions and sensitivity analysis for optimal multiprocess control problems. *Control and Cybernetics, 29*, 11–31.
2. Bonen, A., Semmler, W., & Klasen, S. (2014). *Economic damages from climate change: A review of modeling approaches*. SCEPA working paper 2014–03, Schwartz Center for Economic Policy Analysis.
3. Chiarella, C., Semmler, W., Hsiao, C.-Y., & Mateane, L. (2016). *Sustainable asset accumulation and dynamic portfolio decisions*. Springer.
4. Clarke, F. H., & Vinter, R. B. (1989). Applications of optimal multiprocesses. *SIAM Journal of Control and Optimization, 27*, 1048–1071.
5. Clarke, F. H., & Vinter, R. B. (1989). Optimal multiprocesses. *SIAM Journal of Control and Optimization, 27*, 1072–1091.
6. Flaherty, M., Gevorkyan, A., Radpour, S., & Semmler, W. (2017). Financing climate policies through climate bonds: A three stage model and empirics. *Research in International Business and Finance, 42*, 468–479.
7. Heine, D., Semmler, W., Mazzucato, M., Braga, J. P., Flaherty, M., Gevorkyan, A., et al. (2019). Financing low-carbon transitions through carbon pricing and green bonds. *Quarterly Journal of Economic Research, 88*, 29–49.
8. Herbst, E., & Schorfheide, F. (2015). *Bayesian estimation of DSGE models*. Princeton University Press.
9. Hotelling, H. (1931). The economics of exhaustible resources. *Journal of Political Economy, 39*(2), 137–175.
10. Lin, Q., Loxton, R., & Teo, K. L. (2014). The control parameterization method of optimal control problems: A survey. *Journal of Industrial and Management Optimization, 10*, 275–309.
11. Loxton, R., Lin, Q., Rehbock, V., & Teo, K. L. (2012). Control parameterization for optimal control problems with continuous inequality constraints: New convergence results. *Numerical Algebra, Control and Optimization, 2*, 571–599.
12. Maurer, H., Büskens, C., Kim, J.-H., & Kaya, C. Y. (2005). Optimization methods for the verification of second order sufficient conditions for bang-bang controls. *Optimal Control: Applications and Methods, 26*, 129–156.

13. Orlov, S., Rovenskaya, E., Puascunder, J., & Semmler, W. (2018). *Green bonds, transition to a low-carbon economy, and intergenerational fairness: Evidence from an extended dice model*. IIASA working paper.
14. Sachs, J. (2014). Climate change and intergenerational well-being. In L. Bernard & W. Semmler (Eds.), *The Oxford handbook of the macroeconomics of global warming* (pp. 248–259). Oxford University Press.
15. Semmler, W., Maurer, H., & Bonen, A. (2018). An extended integrated assessment model for mitigation & adaptation policies on climate change. In G. Feichtinger, R. Kovacevic, & G. Tragler (Eds.), *Control systems and mathematical methods in economics* (Vol. 687). Lecture notes in economics and mathematical. Springer.
16. Tobin, J. (1969). A general equilibrium approach to monetary theory. *Journal of Money, Credit and Baking*, *1*(1), 15–29.

Chapter 7
On Scientific Innovations and Constraints: A Dynamic Analysis

Yuri Yegorov and Franz Wirl

7.1 Introduction

Research is a special economic activity. It combines individual and collective efforts in order to discover new knowledge; even Sir Isaac Newton admitted that he had to stand on the shoulders of giants in order to see farther. Science consists of many fields and sub-fields of which it is often unclear at the beginning which will yield successful (not necessarily useful) knowledge. The crucial feature that we address in this paper is how a new field gets started. The obstacles are that initiating a new field provides, at least at the beginning, almost no rewards compared with work in established and respected and thus well-cited field.

In the past research was a kind of hobby pursued by few. Nowadays it is a productive sector, financed by research positions and grants paid by the government as well as research within industries, see e.g., the survey of Diamond [1]. Global expenditures are around US$ 1.7 trillion (according to UNESCO) with the leading countries spending up and around to 3% of GDP on R&D. Nevertheless, not all is well. There is high inequality in scientific productivity, discovered about 100 years ago by Lotka [2], and later described in Merton [3], and currently expressed in the good fittings of the power law (Pareto) with respect to the numbers of published papers and of citations. Another crucial feature of science is the observation of serendipity, when a researcher looks for 'a' but discovers 'b', Merton (the 'father of the economist') and Barber [4]. The focus of this paper is on modeling research activity, both at individual and collective levels, and in particular on the emergence and dynamics of a new research field.

We thank Markus Eigruber for his support (discussions and Figs. 7.1 and 7.2).

Y. Yegorov · F. Wirl (✉)
University of Vienna, Vienna, Austria
e-mail: franz.wirl@univie.ac.at

J. L. Haunschmied et al. (eds.), *Dynamic Economic Problems with Regime Switches*, Dynamic Modeling and Econometrics in Economics and Finance 25,
https://doi.org/10.1007/978-3-030-54576-5_7

Kuhn [5] describes the phenomenon of "scientific revolution". Scientific revolutions involve discontinuity in commonly accepted paradigms, cited literature, but mostly important, a rapid change in the relative number of scientists working in different fields. They happen not frequently due to some obstacles. We suggest that scale economies can be responsible for that, and model them explicitly. Increasing returns were studied by different authors from different perspective. Optimal intertemporal use subject to increasing returns to scale were first addressed in Skiba [6], Sehti [7] and Dechert and Nishimura [8] with applications in many and different fields: Brock [9] on lobbying, Brock and Dechert [10] on regulation and more recently in environmental economics, more precisely about shallow lakes, Maler in [11], and in endogenous growth models starting with Romer [12]; the analysis of Arthur [13–16]) is complementary. Summarizing, scale economies represent an important but not the only explanation for multiplicity of equilibria, see Wirl and Feichtinger [17].

The economics of science aims to understand the impact of science on the advance of technology, to explain the behavior of scientists, and to understand the efficiency or inefficiency of scientific institutions [1]. In this paper we contribute to the 2nd and 3rd issue, that is: how incentives shape the behavior of individual researchers, how they interact in expanding old fields and forming new research schools, whether this process leads to socially efficient outcomes and what should be the lessons for the policy of scientific institutions.

Stephan [18] notes that "Compensation in science is generally composed of two parts: one portion is paid regardless of the individual's success in races, the other is priority-based and reflects the value of the winner's contribution to science." She argues that while the 1st part (salary) is rather flat, the 2nd part is "much less flat as the scientist gains prestige, journalistic citations, paid speaking invitations, and other such rewards". While the second part creates incentives for individual growth, it also creates skewness in rewards, discovered by Merton [19].

This paper focuses on policies drawing on mathematical models about productivity, incentives, and social interactions in science. However, instead of aiming for a"master" model encompassing all relevant aspects, we present different models that highlight one aspect at the time. Section 7.2 addresses research productivity differentiating between an established and a new field from an individual and a collective perspective. The crucial characteristic of individual rewards is that it depends also on collective efforts (possibly cumulative, i.e., all papers published and results known in a certain field) and that it offers little or no return at all at the very beginning of a new field; rewards can start even negatively if one counts the attacks from researchers working in the at the time dominating paradigm. This initial phase is followed by increasing returns to scale (IRS) that are ultimately declining (decreasing returns to scale, DRS). Section 7.3 addresses these aspects and presents suitable functional forms for the returns if entering a new field. In Sect. 7.4 we consider the (static) decisions of individual researchers facing exogenous shifts in the productivity in the new field. Section 7.5 presents instead of the static a dynamic, two-stage set up of a researcher who can work either in research (first) or (if, then) in another research-related activity (consulting, management, etc.). Finally (Sect. 7.6), we try to integrate individual and collective actions and analyze how and under which conditions a new field gets started in a competitive setting. The result is: whether a direction of research is pursued depends not only on the initiation (i.e., the first promising results) but also on the researchers' collective expectations about the new field's prospects. Indeed, most of the presented models show a non-trivial intertemporal pattern. Section 7.7 draws some policy implications from our analysis.

7.2 Research Productivity and Individual Incentives

The goal of this section is to give microeconomic analysis of research activity of a scientist, addressing social value of research, individual incentives and human capital as the necessary input.

7.2.1 Human Research Capital

Research differs from other economic activities, and this paper focuses on those differences and tries to model them from different perspectives. At the individual level, research productivity depends on the researcher's own field-specific human capital, h. Entering a new field and then continuing requires investment into kind of human capital. It can be accumulated by investment (i) in learning, reading, participation in conferences, etc., but depreciates at the (constant) rate δ, i.e., human capital follows the dynamics of physical capital accumulation,

$$\mathrm{d}h/\mathrm{d}t = i - \delta h.$$

The difference is the shape of the marginal productivity of capital and its dependence on the efforts of other researchers, currently and maybe also in the past. Furthermore, human capital can be used in research but also alternative activities, like teaching, consulting or in management.

7.2.2 Social Value of Scientific Research

Scientific research creates social value via the implementation of scientific results in practical applications. It is often impossible to tell a priori which of the many fields will become socially beneficial and even the experts of the field can get it wrong. That is why mathematicians work in many abstract fields of which only few find an application, sometimes long after their discovery (e.g., projective or non-Euclidean geometry for the general theory of relativity). Nevertheless, any new field (or intersection of two or more fields, i.e., interdisciplinary or even transdisciplinary work) is potentially beneficial. A social planner should probably put at least a few researchers in every known field in order to maximize the number of future inventions. Indeed, Acemoglu argues in a number of papers, e.g., Acemoglu [20], for such a "directed technical change" in order to counter the positive feedback loops generated within a particular field, e.g., the automobile producers have tendency to improve the combustion engine with which they are familiar rather than the electric car. Unfortunately, public funding is conservative. For example, Azoulay, Manso and Gra Zivin [21] show that medical researchers funded by the National Institutes of Health often pursue less ambitious projects. The sociologist Kuhn [5] attacked Popper's [22] (normative theory), the logic of scientific discovery, as an appropriate description of science and characterized it instead as a rather conservative enterprise sticking to old paradigms even if facing contradictory result until scientific revolutions sweeps them away. A reason is that entering a new field is risky and may be ignored by the mainstream (at least it will receive no or only few citations). Therefore, such an innovative strategy lacks incentives for an individual scientist.

7.2.3 Productivity of a Researcher

Research is both an individual and a collective activity. Therefore, it is important to account for the interactions between researchers. Joint activity of researchers creates collective human capital in a field,[1] which (together with private human capital) influences the productivity of each researcher there.

Let us measure current productivity by the volume of additional knowledge produced, which is the quality weighted volume of currently written papers. Working in such a field can be compared with mining.[2] At the initial stage, immediately after the discovery of a mine, productivity is very low due to lack of equipment and knowledge about the size and precise location of the mine. The more is invested, the more is known and the higher is the return. Therefore, we have scale economies (IRS) in the beginning of exploiting the mine as information about the location and the size of the resource size are revealed. However, as experience is gained, the volume remaining in the mine will shrink thereby lowering productivity. The output will reach a maximum and decline thereafter, i.e., decreasing returns to scale (DRS) take over. Similarly, a new field of research has a productivity of the IRS-DRS shape, which has important consequences on the outcome, in particular, whether a field turns into a prospective one or is abandoned.

---

[1] The field can correspond, for example, to the (first) letter in the JEL classifications in economics and similar in other sciences. A subfield is then identified by the following two numbers of the JEL-code. The only crucial point is that we assume that a subfield can be invented by a researcher (but rarely a larger area which we call field) and generate several publications by one or many persons.

[2] This is only used as a metaphor. Both, mining and research, are looking for useful elements that are mixed with things useless. The analogy is in growing skills to find a mine (or research subfield) and in lowering the probability to find something useful after some time of exploitation. Both processes can be described by similar mathematical models.

We define with H_i the overall knowledge about a field i and call it aggregate (field specific) human capital. Although overall knowledge is the aggregate over time (i.e., all past findings) and space (i.e., all researchers active in the field) we ignore the cumulative effect over time for analytical reasons (its analysis added another dynamic relation) and let H_i denote the human capital of all researchers currently active in field i. This level of knowledge (H_i) determines jointly with the field-specific human capital h^i_j of a researcher j the individual output in each period t,

$$y^i_j(t) = F(H_i(t), h^i_j(t)), \tag{7.1}$$

with $\partial F/\partial h^i_j > 0$. That is, researchers with higher field-specific human capital will also be more productive in this field. However, the partial derivative w.r.t. collective human capital is positive only for small H_i (IRS) but turns negative at large values of H_i (DRS part). Along the decreasing returns to scale (DRS) part, marginal productivity will drop, but agents with sufficiently high h^i_j will still be able to achieve high output.

7.2.4 Incentives and Private Returns

A researcher selects initially one field, possibly because a professor assigns him to this field. It is impossible to start a new field alone, because of the lack of knowledge foundation, $F(0) = 0$. Only an experienced or risk friendly researcher can do this. And even if right, he may not obtain the rewards. There are many famous examples and Ignaz Semmelweis is an often quoted and tragic example. He discovered that "childbed fever" could be drastically cut by the use of hand disinfection in obstetrical clinics based on the observation that the assistance of medical doctors instead of midwives increased the death toll substantially. Of course, this insulting hypothesis faced severe opposition by his colleagues[3] for a long time and because he could offer an at that time (Louis Pasteur came years later) acceptable scientific explanation. Another historical example is that of Alfred von Wegener, a meteorologist. He proposed the idea of a super continent from the observation that Latin America fits into the Gulf of Guinea. This theory was the laughing stock of geologists until they discovered a few decades later plate tectonics themselves, which allowed them to understand what Alfred von Wegener conjectured a long time ago. A more recent example is Yoshinori Ohsumi whose research outside mainstream led to path breaking experiments (autophagy) and was rewarded with the Nobel prize 2016. Or doctors ignore very rare illnesses (due to the obvious lack of incentives, in particular financial ones) and instead people privately involved take up the research and even succeed, compare the movie Lorenzo's oil. Therefore, it is important to understand the potential costs and benefits from such an undertaking, which will be done in the next section.

---

[3] He was treacherously committed to an asylum by his colleague where died 14 days later after being beaten by the guards.

Consider the incentives for a researcher who starts to work in a well-developed field. There are several incentives to continue work in this field even in the presence of DRS. First, there are many publications in this field, and it is easier to publish (if one finds something marginally new) because many referees are familiar with the field and can easily evaluate its contribution. Indeed, from our own experience we observed that papers that make a point along a well-trodden path have a much higher acceptance rate than papers that try to make a new point. Second, it is easier to get a grant in a well-established field, because grants are largely determined by the cumulative publications of an applicant within the field, which grows monotonically even beyond the point of decreasing returns. Moreover, since the reward (salary and of course personal reputation or vanity[4]) depends on cumulative publications, it continues growing even when the flow of output is declining. In contrast, the probability to publish an original and new but possibly still imperfect result is low and its reception presumably even lower. Suppose that such a scientist has reached his point of maximal productivity, e.g., he can write 10 papers per year in this new field. If he is the only one working in this new field, nobody would understand the major, probably subtle points and therefore most of his papers will be rejected. And even if accepted, nobody (or at best very few) will take notice let alone appreciate them. Therefore, the incentives to develop a new field are very low, even given high individual productivity in this new field.

7.3 Modeling

We present a sequence of models that address different issues, partially in isolation. The first model is about the exogenous transition (Sect. 7.4.1) and the optimal timing for shifting to a new field. The model in Sect. 7.4.2 takes collective actions of scientists into account when a new research field emerges. The models of Sect. 7.5 are about individual optimization with the possibility to work just in one research field but with the additional option to use the human capital in an alternative activity (teaching, consulting, or management). Section 7.6 analyzes an intertemporal competitive equilibrium of individual researchers.

The following features seem crucial and are therefore addressed in (at least one of) our models:

1. At least one of the fields has IRS-DRS type of production function, where the marginal productivity first increases and then decreases ("when almost everything is said, but maybe not by everyone," to quote Karl Valentin). Then there exists an optimal size of the field.

[4] Scientists are among the most vain people as Carl Djerassi, the discoverer of the anti-baby pill observed. Richard Blair wrote under the pseudonym George Orwell, but the idea that Einstein had published his general theory of relativity under a pseudonym is absurd. Sir Isaac Newton is another example in his fight against Gotthold Ephraim Leibniz about priority over calculus.

2. Changes can be either exogenous (a given evolution over time as in the baseline model) or endogenous defined by the history of scientific activity in the field. The second case is more complex and only considered partially.
3. Agents (scientists) can be either homogeneous or heterogeneous. Even if they are assumed to be homogeneous, strategic interactions can lead to coordination failures. For example, the shift to a promising new sector with currently low but potentially higher future productivity can be beneficial only if many scientists join and mutually reinforce their productivity via accumulation of collective human capital in the new sector; compare Krugman [23] on the transition from agriculture (CRS) to industry (IRS).
4. The heterogeneity of agents can be of two types: age and human capital. While in the first case the dynamics of age is given, in the second it is defined by individual paths of capital accumulation. Since many different types of agents with different human capital lead to complex interactions, only simplified dynamic games can be considered in this paper.

Scientific research differs from the production technology in most other economic areas. Research means finding something new. This resembles geologic search, e.g., drilling for oil, and then mining. Qualification of a researcher is similar to qualification of a geologist who finds a new mine. Each mine can produce only final cumulative stock of output. Thus, its productivity must ultimately decline over time and eventually must vanish given the "finite" resource analogy.

Consider a mine (or a scientific field) with one unit deposit. Let x denote the skill of a geologist and miner (as one person), who has exploited the subinterval $[0, x]$, and thereby gained the skill x (learning by doing). The remaining deposit is of the size $1 - x$. Thus the (marginal) productivity of a miner will depend on the cumulative output, e.g., $mp(x) = x(1 - x)$. Using this metaphor, a scientist with specialization in a certain area has the marginal productivity proportional to both, his human capital h (growing proportionally to exploitation) and the remaining deposit.

$$mp(h) = h(1 - h).$$

Considering probabilities, suggests also an alternative formalization,

$$mp(h) = he^{-h}.$$

The probability to find something new for a unit of effort (proportional to human capital) declines exponentially, but never vanishes to zero.[5]

[5] In physics, when we move to the height h above the sea level, the density of atmosphere declines exponentially. This is also an example of a deterministic outcome for the law of large numbers, due to very large number of particles. The similar deterministic equivalence of random outcome is assumed here in "research mining," but it is less justifiable and is assumed for simplicity.

Based on the above analogy with a mine but accounting for the fact that overall individual (or marginal) productivity (mp) depends on the aggregate stock of knowledge (H ignoring sub- and superscripts in this subsection), we consider the following shapes for research output mp per unit of individual human capital,

$$mp(H) = H(1 - H), \tag{7.2}$$
$$mp(H) = He^{-H}, \tag{7.3}$$
$$mp(H) = H^2e^{-H}. \tag{7.4}$$

The first shape, the logistic one, is mathematically simpler. The possibility of negative productivity for $H > 1$ is not a problem as this domain is irrelevant for any economic consideration. The exponential shape will be used in Sect. 7.5 for individual dynamic optimization problem with infinite horizon but in this using only individual human capital as argument.

Why do we assume for the individual reward (π),

$$\pi = hmp(H)?$$

It is because scientific recognition is very low for a scientist with low human capital who operates in a new field.[6] Both functions, $mp(h) = he^{-h}$ and $\pi(h) = h^2e^{-h}$, have a maximum for finite h. mp has it at $h^* = 1$, while π has it at a higher level, $h^{**} = 2$, at already lower and declining productivity. This captures delays in rewards. Nobel Laureates usually get the prize for their accumulated reputation (life work) rather than for their production at a certain point in time and when their productivity is presumably lower (it is all downhill as Samuelson observed about the productivity of economists).

7.4 Exogenous Evolution of Productivity

This section analyzes interactions of researchers working in the same field. In the first model it is assumed that a field has inverse-U shape of productivity over time given exogenously. Different fields have different times of productivity peak. At any time moment, one researcher enters it and another exits (retires), and researchers can decide what time is optimal for shift across fields. The second model (4.2) makes the productivity endogeneous, depending on the mass of researchers there. Two

[6] Scientific reward is not directly linked to productivity due to the so-called Matthew effect ("For to everyone who has will more be given, and he will have abundance; but from him who has not, even what he has will be taken away," Matthew 25:29). This point was first addressed by the sociologist Merton [19], in order to explain why eminent scientists get disproportionately credit for their contributions, while relatively unknowns get disproportionately little. Therefore, a person with many written, maybe partially unpublished, papers gets lower reward per written paper compared to a famous scientist, even after accounting for quality.

cases, without switching cost and with it, are considered. In this section there is no accumulation of individual human capital (this will be considered in Sect. 7.5), and all researchers take sectoral human capital as given, and their productivity equals to it, $h = H$. While in 4.1 $h = H(t)$, in the model of 4.2 it depends not on time but on mass of researchers X, so that $h = H(X)$. Finally, Sect. 7.4.3 deals with age heterogeneity and switching cost.

7.4.1 Identical Researchers

Since research fields have similar to products and technologies a life cycle, there are also waves sometimes even fashions in research activity. Considering economics, applications of chaos theory (quite fashionable during the 1980-ies) are clearly in decline, while economic experiments are on the rise. Consider an exogenously given time-dependent productivities in the (only) two scientific fields A already in decline and B growing,

$$y_A(t) = (t + G/2)(G/2 - t) = G^2/4 - t^2, \qquad y_B(t) = t(G - t),$$

where $G > 0$ is a given and constant parameter that shapes the productivity in two fields A and B. The maximal productivity of $G^2/4$ is reached at $t = 0$ for sector A and at $t = G/2$ for sector B.[7] The scientific lifetime of a researcher is normalized to 1 (say around 30–40 years). The life cycle of scientific field lasts for $G > 1$, i.e., it is longer than an individual scientific lifetime.

All researchers have the same qualification, supply inelastically one unit of labor, and earn the value of their productivity. There is no saving, no investment and no discounting so they consume everything in the same period. The lifetime utility of a scientist (we use π in different places but always referring to a scientist's payoff) is simply the integral over outputs:

$$\pi = \int_0^1 [y_A(t) + y_B(t)]dt.$$

The only decision is when to shift from sector A to sector B. Given the persistence of scientific paradigms, it may happen only once in a scientist's lifetime, say at time $\tau < 1$.[8]

[7] G corresponds to both output and time. G can be calibrated as typical life cycle of a subfield, for example, 100 years. If the unit of time (t = 1) is, for example, 40 years then G = 2.5. If the maximal productivity in this subfield is G2/4, it can be measured by maximal intensity of total publications per unit of time (here 40 years). If G2/4 = 2.52/4 = 3.125 in this example, while in reality is 312 papers, then a unit of publication is 100 papers.

[8] For example, the authors of this paper have presumably missed the transition to experiments.

The scientist's lifetime utility when switching at date τ is

$$\pi(\tau) = \int_0^\tau (t + G/2)(G/2 - t)dt + \int_\tau^1 t(G - t)dt = G/2 - 1/3 + \frac{G^2}{4}\tau - \frac{G}{2}\tau^2. \tag{7.5}$$

The optimal time (τ^*) for an individual scientist starting his career at $t = 0$ follows from solving the first-order condition of maximizing $\pi(\tau)$,

$$\pi' = G^2/4 - G\tau = 0 \Rightarrow \tau^* = G/4.$$

In this baseline model it is individually optimal to work in the field A up to $t = G/4$ and then to shift to sector B. If $G > 4$, the productivity decline in A is so slow that it is not optimal to shift to B during the scientist's lifetime. Thus, we have already two regimes (work in A and then work in B) in this simple baseline case.

7.4.2 *Static Equilibria for Costless Shifting*

Now consider endogenous formation of sector sizes. Let old sector A has constant productivity. In contrast, the productivity in the new field B is endogenous, more precisely, depends on the number of researchers working in field B. The new sector B is born at $t = 0$ and productivity is of IRS-DRS shape,

$$q(X) = Q(X) = X(G - X),$$

where X is the mass of researchers already working in the field B.[9] As above, maximal productivity of $G^2/4$ is obtained at the mass $X = G/2$. If $G/2 < 1$, there is no incentive to develop the new sector, because even this mass of $G/2$ will not deliver a productivity(and reward) exceeding 1, which is earned in the old sector. In order to rule this trivial case out, we assume $G > 2$.

This problem is similar to the one considered in Mascarilla-i-Miro and Yegorov [24]: Two cities have IRS-DRS shape of net benefits for their population and the fixed population has to be split across two of them. If the total population is above the sum of optimal population values for both cities, the intersection is on the DRS part, and this equilibrium is stable. If the intersection is on IRS, such equilibrium is unstable, and all other initial conditions result in asymmetric equilibria, where all choose to live in one city.

Figure 7.1 shows that productivity in sector B is above one in sector A for $X_1 < X < X_2$ and reaches its maximum at $X = X^*$.

---

[9] Here X is the mass (quantity) of researchers. It is normalized in such a way that maximal productivity is for $X = G/2$. Both can be calibrated. For example, we know that maximal productivity, $G^2/4$, is for 10 researchers in subfield. Then $X = G/2$ corresponds to 10 researchers. A researcher lives between t and $t + 1$, so that q is measured in the units of lifetime salary, and G can be defined correspondingly.

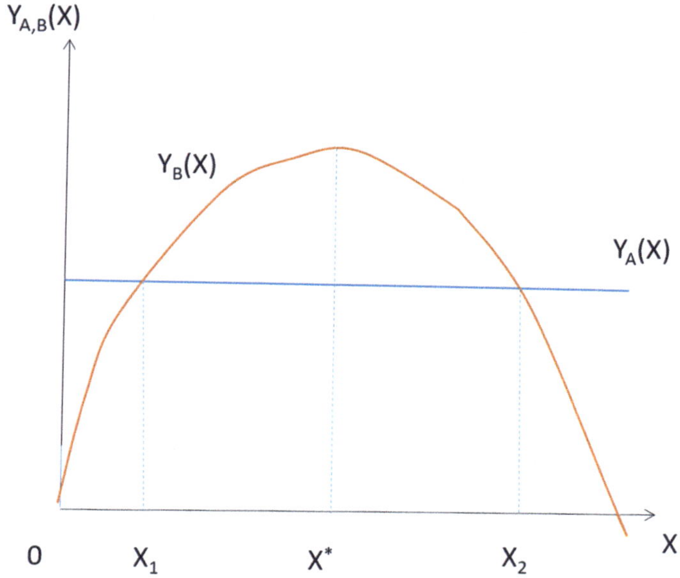

Fig. 7.1 Productivity in sector A, B as the function of sector size X

An economic equilibrium must satisfy the no arbitrage condition. Then salaries in both sectors should be equal: $X(G - X) = 1$. This equation yields the roots:

$$X_{1,2} = 0.5(G \pm \sqrt{G^2 - 4}),$$

of which X_2 is the lower one. Let μ denotes the total mass of researchers. If $X_2 < \mu$, then there is sufficient mass to create the new field with a higher productivity.

The problems are as follows: (a) how to implement these equilibria from different initial conditions, (b) whether they are socially efficient.

We have to consider individual and collective incentives for shifting from A to B in three different regions separately: (1) $X \in [0, X_1]$, (2) $X \in [X_1, X^*]$, and (3) $X \in [X^*, X_2]$. In region 1 there is neither an individual nor a collective incentive. In region 2, there is an incentive for individuals and the collective. Consider a sequential decision of rational workers. Every marginal worker with number X in this region gains from shifting personally, and at the same time the collective of researchers already in B also gains. In region 3 there is still an individual benefit for shifting to B, but the collective of those who already work in B is harmed (because their productivity drops) and also the social welfare is reduced (average productivity drops).

Note that the size of sector B at the maximum of $q(X)$ equals to $G^2/4$ and is obtained at $X = G/2$. It gives the highest productivity for sector B and the highest overall GDP (calculated as the mass of researchers in different sectors multiplied by their productivity). But this outcome is not an equilibrium, because there is an incentive for further shift to B and individual possibility of an arbitrage.

Suppose that $X_2 < \mu < X_1$. Then the mass of researchers is sufficient for a mutually beneficial shift. Shift of an optimal fraction of researchers from sector A to B yields higher productivity in B. It will be beneficial for all who shift and the society as a whole.

Finally, for $\mu \geq X_1$, the shift will stop at $X = X_1$. At this point, both sectors have the same productivity of 1. The old sector will stay with the mass $\mu - Y_1$, while Y_1 will move to the new sector.

When a new sector is created, somebody has to be the first. The first ones will have to be from the youngest cohort to be followed by older cohorts. In the absence of switching costs, all would benefit from a collective simultaneous shift (cascade) of the mass X such that $q(X) > 1$.

Will rational expectations will drive a cascade of simultaneous shifts? No, because there is still an unsolved collective action problem. The interaction between agents is as follows. The youngest can be the first mover, since he can benefit longer from a higher salary in a new sector. in the future. His move should trigger the move of the next, and so on. However, irrational (or bribed by a third party) guys can block this mutually beneficial collective shift. Free markets cannot implement coordination (but also not the above blocking).

However, such a shift can be implemented in a planned economy or by a dictator. We know examples from the USSR that offered proper incentives to young researchers to build a nuclear bomb or a space rocket. Indeed, the USSR had the first man in space but its heating systems lacked proper valves (also less interesting from an individual researcher's perspective). With the only alternative to lose freedom and the state support in financing new sector, the coordination problem disappears. The extreme case of prisoners-researchers has been described in the novel by A. Solzhenytsyn "In the first circle."

7.4.3 Heterogeneous Age and Switching Cost

Coexistence of cohorts of scientists of different age $a \in [0, 1]$ is considered in this subsection. Again, there is no individual accumulation of human capital, and everybody's productivity (and wage) $q(t)$ equals to the sectoral productivity $Q(t)$ in this moment. Scientists can switch across sectors, paying only a one time cost c at the beginning. It represents the cost of learning the tricks of the new field B.

The following argument follows Yegorov [25]. We consider cohorts entering research at different degrees of maturity of field A, which choose their intertemporal research strategy. A typical assumption is that a researcher starts in A (e.g., due to the human capital obtained during his education) but considers switching to B. Furthermore, assume that the agents have a different age, $0 < a < 1$, that each agent has a time horizon of length 1 and, like in the baseline model, that there is no time discounting, but a cost c when switching from A to B.

The lifetime income of each worker of age has two components, a for the years working in sector A (before the shift) and $(1 - a)q(X)$ for working the rest of his

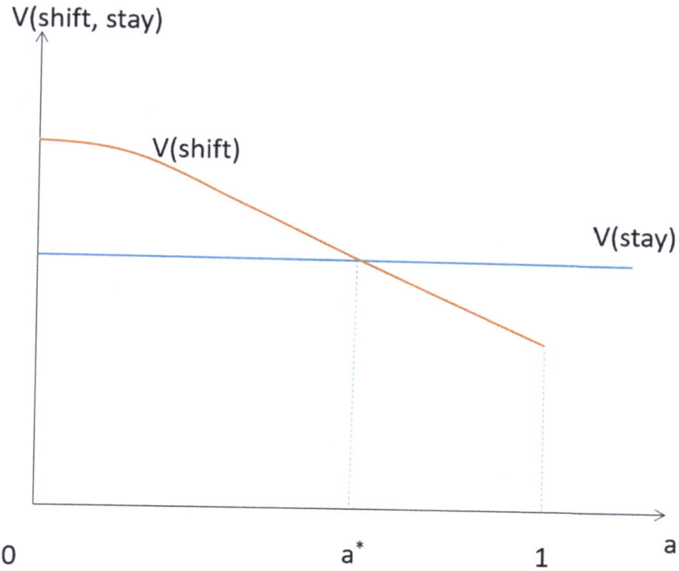

Fig. 7.2 Value from collective shift with cost ***c*** of ages below ***a*** for age ***a*** versus value of staying

life in sector B after paying cost c for shifting. The larger is age a, the lower are the benefits from this shift, because c should be compensated by the wage gain over smaller duration $(1-a)$. Hence, the incentive to shift is monotonically declining with respect to the age of the scientist. Consider the marginal worker of age a^* who is indifferent between shifting and not. Then all younger would shift, and the mass of workers will be a. Then there exists some age threshold a^*, so that all younger agents, $0 < a < a^*$, choose B, because the gains from working in the new sector will exceed the switching costs even for the oldest agent in this group, $a = a^*$. All older agents, $1 \geq a > a^*$ continue working in field A. Figure 7.2 shows the value for researchers of different age from such a collective shift. Then the benefit from shift, $V(a)$, will be a decreasing function of age, $V'(a) < 0$. The critical age, $a = a^*$, is defined by the equation

$$1 - \hat{a} = Q^*(1 - a^*) - c.$$

The scientist of critical age is indifferent between shift and stay if net benefits are equal for both options. All younger scientists will obtain larger benefits from shifting.

Furthermore, we ignore how the collective action problem of initiating research in B is solved and assume a group of researchers of mass X works already in B. Therefore, everybody working in B earns $q(X) = Q(X)$ per period. Therefore, an agent of age a will switch iff,

$$(q(X) - 1)(1 - a) \geq c,$$

i.e., the increase in earning during the remaining lifetime justifies the investment costs of entering the new field. Since benefits decrease monotonically in age (a), only the young,

$$a < 1 - \frac{c}{q(X) - 1}, \qquad q(X) > 1$$

will switch and only if X is sufficiently large.

If we assume perfect coordination among the young researchers to circumvent the collective action problem then we can determine the size of the jump from $X = 0$ to $X > 0$, i.e., the field B is born. Needless to say, this is very unlikely and the calculations serve to determine an "efficient" benchmark. The mass of shifting agents equals to $X = \mu\bar{a}$, where $\bar{a}$ denotes the limit age of shift. The oldest agents should be indifferent between shift or not. Then the following equation defines $a = a^*$:

$$(\mu a^*(G - \mu a^* - 1)(1 - a^*) \geq c.$$

In the case of switching cost the wages in both sectors can differ, but nobody will benefit from further shift.

Here it was assumed that a shift takes place once and at once. However, real world dynamics are more complex, because there will be a future entry (of new cohorts) and exits (to retirement) in both sectors.

The above models show the complexity of interactions between researchers even without accounting for human capital accumulation, which will be the focus of the following models.

7.5 Intertemporal Optimization of Individual Activities

In this section, we endogenize human capital accumulation by considering the intertemporal optimization problem of a scientist. However to do so, we make a few simplifying assumptions. First and also different from the analysis so far (and also below in Sect. 7.6) is that there is only one research field. This can be justified if entering a new field of research requires high fixed cost. Instead of two research fields, the scientist can choose between two activities either research or other activities like teaching, taking up administrative and managerial tasks (as a dean or rector), consultancy, and practising (medical doctors, psychologists, civil engineers). Second, there is only an individual scientist with an infinite planning horizon who can invest into his human, field specific, capital, $h(t)$, and who discounts (r) future benefits and costs. Third, and in contrast to the continuous shifting and the working in both fields for the high equilibrium in Sect. 7.6, the scientist must choose either or between two possible activities: research ($j = 1$) and the alternative activity ($j = 2$). At each point of time, the current level of human capital is mapped into marginal productivity (mp_j) and reward π_j for both activities $j = 1, 2$: $mp_j = f(h(t))$, $\pi_j = g(h(t))$.

Human capital (h) is subject to IRS-DRS returns in the area of research but delivers its returns linearly in the alternative activity,

$$\pi_1(t) = h(t)e^{-h(t)}, \qquad \pi_2(t) = wh(t),$$

where $w < 1$ is the relative wage in activity $j = 2$. Productivity in research is maximized at $h = 1$ and deteriorates for $h > 1$ because too little is left in the scientist's research field (again exploiting the metaphor of a mine). Since return in the second activity grows linearly with h, it becomes profitable to shift to this activity for h exceeding a threshold $(\bar{h})$.

A scientist often faces alternative options like teaching, consulting, and managing (either as a dean or in science organizations, just think of a big shot like Watkins, one of the discoverers of DNA, or in business). Normally a person has only one main employment, but it may correspond to a bundle of activities (like position of professor assumes research and teaching).

The dynamic optimization model can be formulated as follows:

$$\max_{i(t)\geq 0} \int_0^\infty e^{-rt}[\pi(t) - i(t) - \frac{c}{2}i^2]dt \tag{7.6}$$

$$s.t. \qquad \dot{h} = i - \delta h, \qquad h(0) = b > 0$$

$$\pi_1 = he^{-h}, \qquad \pi_2 = wh.$$

Given the binary choice between the two activities, the profit in each period is simply the maximum,

$$\pi(t) = \max\{\pi_1(t), \pi_2(t)\} = \begin{matrix} h(t)e^{-h(t)} \\ wh(t) \end{matrix} \quad \text{iff} \ \begin{matrix} \leq \\ > \end{matrix} \bar{h} := -\ln w,$$

because it is optimal to devote all activity to research if research delivers the higher return, i.e., $he^{-h} > wh$, which holds for not too high human capital (not a compliment for us researchers), $h < \bar{h}$, and of course, this threshold decreases with higher wages paid outside the university.

Remark: It is possible to abstract from adjustment costs, i.e., by setting $c = 0$. This eliminates interior solutions, i.e., the policy will be bang-bang following a most rapid approach path. Of course, the precise specification of the payoff in research is not crucial and we will therefore consider the case $\pi_1 = h^2e^{-h}$ too.

7.5.1 *Optimality Conditions for the Auxiliary Problems*

First, before addressing the problem of switching from activity 1 to activity 2 at some future date, we consider two problems, each forcing the agent to stick to the initial choice, either research (7.7) or making money in consulting (7.8).

$$\max_{i(t)\geq 0} \int_0^\infty e^{-rt}[he^{-h} - i(t) - ci^2(t)]dt \qquad he^{-h} > wh, \tag{7.7}$$

$$\max_{i(t)\geq 0} \int_0^\infty e^{-rt}[wh - i(t) - ci^2(t)]dt \qquad he^{-h} < wh, \tag{7.8}$$

$$s.t. \qquad \dot{h} = i - \delta h. \tag{7.9}$$

7.5.1.1 Stage 1 - Lifelong Research

The Hamiltonian for the objective (7.7) is

$$\mathcal{H}_1 = h(t)e^{-h(t)} - i - ci^2 + \lambda[i - \delta h].$$

It follows from $\mathcal{H}_j = 0$ that $i = (\lambda - 1)/(2c)$. Then the dynamic system for the case, $e^{-h} > w$, is

$$\dot{\lambda} = \lambda(r + \delta) + (h - 1)e^{-h}, \tag{7.10}$$

$$\dot{h} = \frac{\lambda - 1}{2c} - \delta h. \tag{7.11}$$

This long run outcome is only meaningful if the steady state of human capital remains below the threshold $\bar{h} = -\ln w$.

Computing the isoclines,

$$\dot{\lambda} = 0 \Longleftrightarrow \lambda = \frac{(1-h)e^{-h}}{r+\delta},$$

$$\dot{h} = 0 \Longleftrightarrow \lambda = 1 + 2c\delta h, \tag{7.12}$$

we find that the 1st is downward sloping and the 2nd is upward sloping and transgressing the entire positive quadrant of the phase plane, we have a unique and positive steady state, (h^*, λ^*) . The steady state is saddlepoint stable but the solution is only meaningful if $h^* < \bar{h}$. Of course, by choosing w sufficiently small, one can ensure that h^* is indeed the long run outcome of the original two-stage problem. The steady state of human capital is a solution of the following equation,

$$\frac{(h-1)e^{-h}}{2c\,(r+\delta)} - \frac{1}{2c} - \delta h = 0, \tag{7.13}$$

which unfortunately lacks an analytical solution.

To complete, we use a numerical example with $r = 0.05$, $\delta = 0.1$, $c = 1$, $w = 0.2$. Given the low wage for consulting, the outcome of this isolated case is even the optimal one for starting with small h_0 since the steady state is at $h = 0.668 < \bar{h} = -\ln w = -\ln 0.2 = 1.61$, see Fig. 7.3.

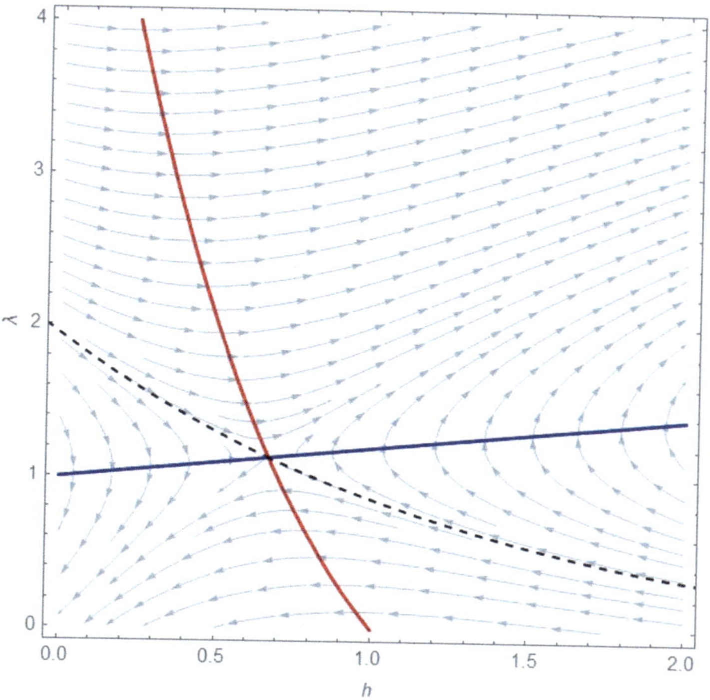

Fig. 7.3 Saddle for the case 1. r = 0.05, del = 0.1, c = 1

7.5.1.2 Stage 2 - Lifelong Consulting

For the objective (7.8) and thus $e^{-h} < w$, we denote the costate by μ and then get the canonical equations:

$$\dot{\mu} = \mu(r + \delta) - w, \tag{7.14}$$

$$\dot{h} = \frac{\mu - 1}{2c} - \delta h. \tag{7.15}$$

The corresponding canonical equations system can be explicitly integrated,[10]

$$\mu(t) = \frac{w}{r + \delta}, \qquad t \geq T,$$

$$h(t) = \frac{(w - r - \delta)\left(1 - e^{-\delta t}\right)}{2c\delta(r + \delta)} + h_0 e^{-\delta t},$$

[10]With the first isocline coinciding with the stable path.

implying the steady state,

$$h = \frac{1}{2c\delta}\left(\frac{w}{r+\delta} - 1\right).$$

Therefore, if $w > r + \delta$, stationary human capital is positive and unique, while there is no positive solution in the opposite case; i.e., if $w < r + \delta$ then it is optimal remain forever in stage 1.

We sketch also the phase diagram as in case 1 and determine the corresponding isoclines

$$\dot{\mu} = 0 \Leftrightarrow \mu = w/(r+\delta),$$
$$\dot{h} = 0 \Longleftrightarrow \mu = 1 + 2c\delta h.$$

The first isocline is horizontal, and the 2nd is upward sloping. Therefore both intersect at above computed steady state, which is a saddle. Using the same parameters but a larger wage $w = 0.5$, such that the steady state of the first case exceeds the threshold (Fig. 7.4).

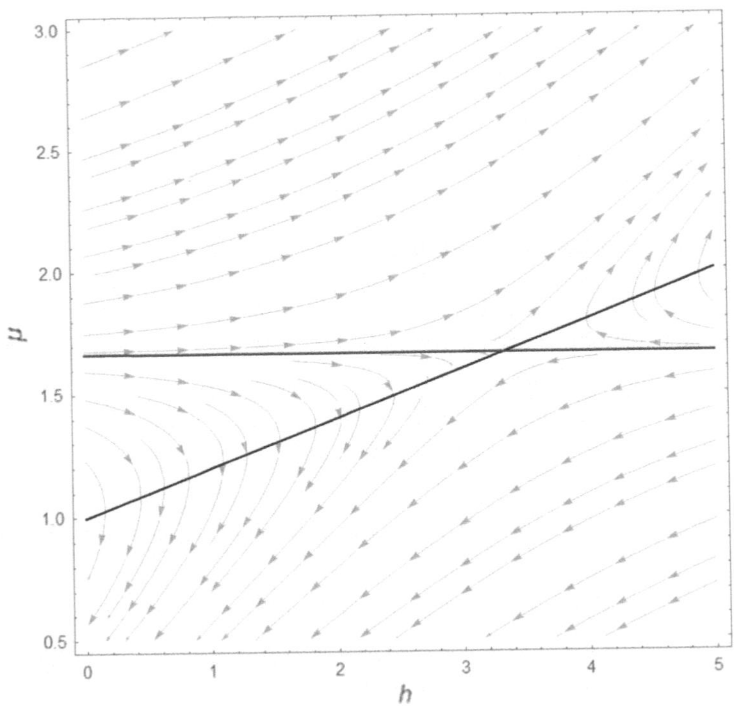

Fig. 7.4 Saddle for the case 2. r = 0.05, del = 0.1, c = 1, w = 0.25

7.5.2 Optimal Transition

Since by assumption the consulting phase is, if at all, the second one, one can rewrite the general two-stage problem as,

$$\max_{i(t)\geq 0} \int_0^T e^{-rt}[\pi_1(t) - i(t) - \frac{c}{2}i^2]dt + e^{-rT}S(h(T)), \tag{7.16}$$

in which S is the salvage value from the second phase, more precisely, the solution of the case 2 from above. Since, the corresponding canonical equation system have been explicitly integrated, the salvage value can be analytically computed,

$$S(h(T)) = \frac{(w-r-\delta)^2}{4c\delta(r+\delta)^2} + \frac{wh(T)}{r+\delta}, \tag{7.17}$$

and it is linear in $h(T)$.

The corresponding optimality conditions for the first phase become then,

$$\dot{\lambda} = \lambda(r+\delta) + (h-1)e^{-h}, \lambda(T) = S' = \frac{w}{r+\delta}, \tag{7.18}$$

$$\dot{h} = \frac{\lambda - 1}{2c} - \delta h, h(0) = h_0, \tag{7.19}$$

and the optimal stopping condition is

$$\mathcal{H}_1(T) = rS(h(T))$$

$$\mathcal{H}_1(T) = h(T)e^{-h(T)} - \frac{\frac{w}{r+\delta}-1}{2c} - c\left(\frac{\frac{w}{r+\delta}-1}{2c}\right)^2 + \frac{w}{r+\delta}\left(\frac{\frac{w}{r+\delta}-1}{2c} - \delta h(T)\right).$$

Therefore,

$$h(T)e^{-h(T)} - \frac{\frac{w}{r+\delta}-1}{2c} - c\left(\frac{\frac{w}{r+\delta}-1}{2c}\right)^2 + \frac{w}{r+\delta}\left(\frac{\frac{w}{r+\delta}-1}{2c} - \delta h(T)\right) = \frac{r(w-r-\delta)^2}{4c\delta(r+\delta)^2} + \frac{rwh(T)}{r+\delta}$$

and expressed in terms of the arbitrage between research and alternative activity,

$$h(T)e^{-h(T)} - wh(T) = \frac{r(w-r-\delta)^2}{4c\delta(r+\delta)^2} + \frac{\frac{w}{r+\delta}-1}{2c} + c\left(\frac{\frac{w}{r+\delta}-1}{2c}\right)^2 - \frac{w}{r+\delta}\frac{\frac{w}{r+\delta}-1}{2c}$$

we get finally the condition,

$$h(T)e^{-h(T)} - wh(T) = \frac{r(w-r-\delta)^2}{4c\delta(r+\delta)^2} + \frac{w-r-\delta}{2c(r+\delta)}\left(1 + \frac{w-r-\delta}{2(r+\delta)} - \frac{w}{r+\delta}\right), \tag{7.20}$$

which cannot be solved analytically with respect to $h(T)$. The right hand side is a constant, and equating the left hand side to zero yields $\bar{h} := -\ln w$, which is the point of identical marginal productivity in both fields. However, we conjecture that the transition takes place before the return on the first activity (research) falls below the return on the second activity, at least if the right hand side is positive.

7.5.3 Delayed Rewards

A kind of Matthew effect in science as noted in Merton [3] is also generated if the individual reward is delayed. There are several difficulties for a scientist starting new field (especially when he is not famous). Although he may be able to formulate many new problems and solve some of them, there are several disadvantages. First, the papers in a new field have a lower probability to be published, because reviewers may not fully understand its contribution and thus reject it.[11] Second, even if a paper is published, it might receive fewer citations, because people with human capital in this new field are lacking.[12] As a consequence, this endeavor results in a lower record in publications and citations even for a successful researcher in a new field. And it will be more difficult to win grants and to attract new researcher in this field, even when it has promising future.

In order to model this case, it is possible to use the idea about reward being proportional not only to net productivity he^{-h}, but also to human capital of a researcher, h. Thus, the returns to research are

$$\pi_3(h) = h^2 e^{-h}.$$

Then we can consider two subproblems, like in the previous case:

$$\max_{i(t)\geq 0} \int_0^\infty e^{-rt}[h^2e^{-h} - i(t) - ci^2(t)]dt \qquad h^2e^{-h} > wh, \tag{7.21}$$

$$\max_{i(t)\geq 0} \int_0^\infty e^{-rt}[wh - i(t) - ci^2(t)]dt \qquad h^2e^{-h} < wh, \tag{7.22}$$

$$s.t. \qquad \dot{h} = i - \delta h. \tag{7.23}$$

But the difference is that there are normally 2 positive roots (and one zero) for the transcendental equation

$$h^2e^{-h} = wh.$$

Denote them as $h_1, h_2, h_1 < h_2$. Then the production function will stimulate research only in the interval $h_1 < h < h_2$. If $h < h_1$, it is optimal to alternative job (think about

---

[11] This also often happens with interdisciplinary papers.

[12] And they also have low incentives to do that.

teaching assistantship for Ph.D. students). There is also a possibility to do science in a classical field with DRS and accumulate human capital to start a new field (but we do not model this formally).

The Hamiltonian for the 1st case is

$$\mathcal{H}_1 = h^2(t)e^{-h(t)} - i - ci^2 + \gamma[i - \delta h].$$

Again, from $\mathcal{H}_j = 0$ it follows, that $i = (\lambda - 1)/(2c)$. Then the dynamic system for the 1st case, $he^{-h} > w$, is:

$$\dot{\gamma} = \gamma(r+\delta) + (h^2 - 2h)e^{-h},$$
$$\dot{h} = \frac{\gamma - 1}{2c} - \delta h. \tag{7.24}$$

The 2nd case, with wh, is identical to the one considered above.

The isocline $\dot{\gamma} = 0$ generates the equation

$$(2h - h^2)e^{-h} = \gamma(r+\delta).$$

Since the l.h.s. is positive for $0 < h < 2$ and negative afterward, there are typically 2 solutions to it.

Consider the parameters $r = 0.05$, $\delta = 0.1$, $c = 1$, $w = 0.2$, the same as in the previous model. Then we have two equilibria, $h_1 = 0.087$ and $h_2 = 1.435$. The left is an unstable focus, while the right is a saddle. The saddle for the 2nd system (when wh dominates) is the same as before; $h_w = 1.667$. Again we have $h_2 = 1.435 < 1.667 = h_w$, but two saddles are now closer (comparing to the previous example).

The general behavior of the problem may result in convergence to the 1st or the 2nd saddle. If parameter c becomes smaller, the saddle h_w grows, and this may stimulate lock in the lower saddle, with never leaving research field. The full investigation of this problem is not provided here. It is similar in complexity to the previous one, and analytical solution cannot be obtained.

Policy issues are important for this model. In the domain $0 < h < h_1$ it is not optimal to do research, because the research productivity is too low. It might be also impossible to survive on a research salary (without incurring personal debt), because the returns are quadratic in h (for low h), while the replacement cost of depreciated capital contains linear term. The previous model did not contain that interval. It is one more hurdle in the development of new sector (here for human capital accumulation on individual level). Young researchers often get their income from alternative activity (like teaching assistantship), before they reach some threshold in research productivity.

7.6 An Intertemporal Competitive Equilibrium

One way to analyze the collective actions of individual researchers, which was ignored in the previous sections, is to depart from Krugman's [20] labor market model. Assume that by chance a jump occurs, $X \to X_0$, X_0 possibly small. There are various reasons for such a jump. For example, some stubborn persons (recall the few examples from above) pursue for idiosyncratic reasons a research so far ignored or considered as either not promising or as unsolvable. Or alternatively, the research in field B is initiated by spillovers from a completely different field. Recent developments in economics offer two examples: Chaos theory in economics was created as a spillover from natural science (economists visiting the Santa Fe Institute). Economic experiments are to a large extent a follow up of the work of psychologists as the Nobel prize awarded to Daniel Kahneman documents. At the beginning it was by many ("neoclassical" economists) considered as no economics at all. However, it offered opportunities for different (and with hindsight) some fresh views and all this with little technical hurdles (knowledge of more or less elementary statistics was sufficient, and the games were not only off the shelves but the most simple ones like the ultimatum game) compared with traditional economics. In the meantime it has received its fair (presumably more than that) share of Nobel prize winners.

Let us consider how such an event affects the decision of an individual scientist. The scientist's temporal endowment at each point in time is normalized to 1, and he can spend y hours in field A delivering the gain y (due to constant and unitary marginal productivity) and x hours in the new activity B delivering the individual output $xQ(X)$ in which the capital letter (X) refers to the time spent in the activity B by all researchers (having the measure 1). Yet entering and then expanding his activity in the new field (u) is costly (K), in terms of time, $K = \frac{c}{2}u^2$. This leads to an individual scientist's optimization problem

$$\max_{u(t)\geq 0} \int_0^{\infty} e^{-rt}[xQ(X) + \left(1 - x - \frac{c}{2}u^2\right)]dt, \tag{7.25}$$

$$s.t. \qquad \dot{x} = u, \qquad x(0) = x_0 > 0. \tag{7.26}$$

Although we assume an infinite planning horizon, we consider the appearance of the single new scientific paradigm B during the scientists "infinite" lifetime.

The specification of the productivity in B is $Q = Q(X)$, i.e., with respect to the total time (X) of all researchers spent actively in B. It is important to note that the individual has no control over what the other researchers are doing and must therefore treat $X(t)$ as exogenously given path when solving the individual intertemporal optimization problem (7.25) and (7.26). However, assuming a symmetric and competitive equilibrium we can set,

$$X(t) = x(t)\, \forall t \geq 0. \tag{7.27}$$

Therefore, (7.25)–(7.27) does not constitute a standard optimal control and the implied first-order optimality conditions have important different consequences (as the possibility of a non-unique outcome shown below). Allowing for the hump-shaped quadratic relation of Q as argued above allows in contrast to Krugman's model for an interior steady state.

Remark 1 Clearly, this is the most simple setup that can be extended in various dimensions. Starting with costs, one could imagine that learning as well spillover effects from a larger research community lower the costs, $K(u, X)$ and $K_X \leq 0$. Another alternative is to relate the productivity in the field to the cumulative research output, e.g.,

$$\dot{\Xi} = X - \omega \Xi$$

if allowing for depreciation.

Remark 2 Wirl [26] indicates that similar results hold in a setting of strategically acting researchers, not necessarily many, let alone infinitely many. Wirl [27] considers general intertemporal competitive equilibria and the corresponding conditions for thresholds and indeterminacy.

Setting up the Hamiltonian for an individual researcher's optimization problem,

$$\mathcal{H} = xQ(X) + \left(1 - x - \frac{c}{2}u^2\right) + \lambda u,$$

implies the first-order conditions,

$$u = \frac{\lambda}{c},$$
$$\dot{\lambda} = r\lambda + 1 - Q,$$

and thus the steady-state condition,

$$\lambda = 0 \wedge Q = 1. \tag{7.28}$$

Assuming a hump-shaped relation, e.g., a variation of one of the above suggested specifications,

$$Q(X) = X(G - X),$$

in which $G > 2$ (otherwise the new field can never reach the productivity of the field) scales the potential of this new field and $X = G/2$ is the point of maximal productivity (=maximum sustainable yield in resource economic terms) implies the following characterization of a competitive equilibrium (including the isoclines for the analysis below),

$$\dot{x} = \frac{\lambda}{c},\ x(0) = x_0,\ \dot{x} = 0 \Leftrightarrow \lambda = 0$$

$$\dot{\lambda} = r\lambda + 1 - x\,(G - x)\,,\ \dot{\lambda} = 0 \Leftrightarrow \lambda = \frac{(G - x)\,x - 1}{r},$$

which results from thus substituting (7.27). The system has two interior steady states, the two roots,

$$x_{12} = \frac{G}{2} \pm \sqrt{\frac{G^2}{4} - 1},$$

one above and one below arg max Q and a boundary solution,

$$x \to 0,$$

all three combined with $\lambda \to 0$.

Computing the Jacobian,

$$J = \begin{pmatrix} 0 & 1/c \\ 2x - G & r \end{pmatrix}$$

implies that the lower steady state, x_1, is unstable, since the corresponding eigenvalues have positive real parts,

$$e_{12} = \frac{1}{2}\left(r \pm \frac{\sqrt{r^2c - d}}{\sqrt{c}}\right),\ d := 4\sqrt{G^2 - 4}$$

and are complex unless the product r^2c exceeds d which is unlikely. The second steady state implies the eigenvalues

$$e_{34} = \frac{1}{2}\left(r \pm \frac{\sqrt{r^2c + d}}{\sqrt{c}}\right)$$

and is thus a saddlepoint. As a consequence, indeterminacy, both eigenvalues being negative (or have negative real parts) is impossible, (compare Wirl [28]) for corresponding conditions that are not satisfied here.

Hence the long run outcome for research in the field B can depend on the initial stimulus X_0 as well as on how the agents coordinate. To see this, consider the example,

$$G = 3, r = 0.10, c = 1.$$

Then $x_1 = 0.38$ is the unstable and $x_2 = 2.61$ the (saddlepoint) stable steady state. Figure 7.5 shows the phase diagram and the potential intertemporal and long run outcomes. It highlights that the large and stable steady state is globally reachable. This means that even a spark can be sufficient to trigger this beneficial outcome.

However, a warning is here in place because for X_0 almost as large as 0.75, the termination of research in B, i.e., $x \to 0$, is also possible and thus for quite substantial initiations, i.e., very prospective first results (maybe the role chaos theory played in economics for some time fits this scenario). Hence, an initiation of research of the size ($X_0 \in [0, .75)$ can result in the termination or large expansion of this particular area of research., i.e., indeterminacy.

However, the outcome in Fig. 7.5 that the "good" equilibrium can be reached even from initial "sparks" depends crucially on the specifics. More precisely, that example assumes a fairly farsighted research community (10% is a low discount rate considering the uncertainties of a scientific career, a lottery according to Max Weber), low costs, and a substantial potential gain, 50% over continuing along the trodden path. Less optimistic assumptions, e.g., $G = 2.5$, $r = 0.50$, $c = 4$, result in the scenario shown in Fig. 7.6. The lower steady state at $x = 1/2$ is an unstable

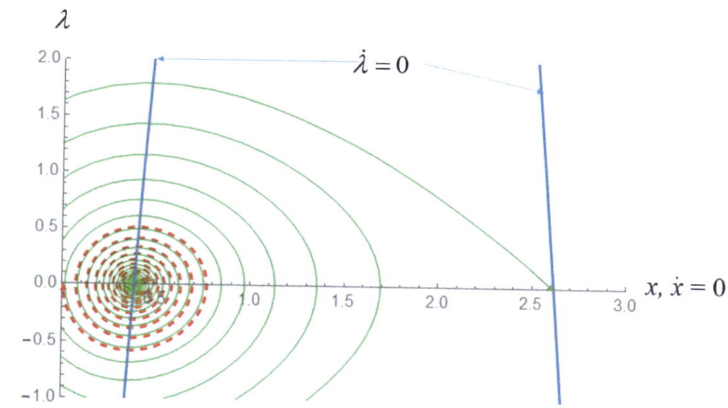

Fig. 7.5 Phase diagram for $r = 0.10$, $G = 3$, $c = 1$

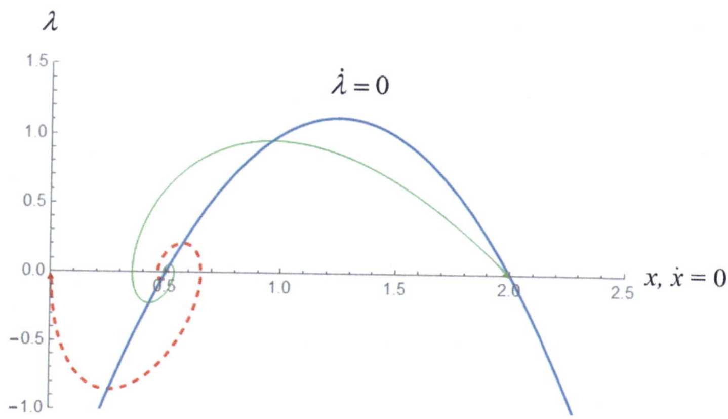

Fig. 7.6 Phase diagram for $r = 0.50$, $G = 2.5$, $c = 4$

spiral and the larger one at $x = 2$ is again a saddlepoint. However now the overlap between the saddlepoint branches is rather small. It requires an initial seeding of at least $X_0 > 0.35$ to get the research in B started and this outcome cannot be granted unless $X_0 > 0.65$. In this case, we have three regions (see Fig. 7.6). For $X_0 < 0.35$, the new research activity converges to zero, for $X_0 > 0.65$ it converges to the high steady state $X = 2$ and the outcome is indeterminate, i.e., depends on the researchers joint expectations, for $X_0 \in [0.35, 0.65]$. Both cases but in particular the second and less attractive case supports Acemoglu's argument of directed technical change, i.e., the necessity of a public push into unexplored research and development areas, e.g., in Acemoglu [20].

7.7 Policy Implications and Concluding Remarks

We have formulated a few models and solved some of them, about socio-economic interactions of scientists focusing on individual and collective choices and in particular about the birth of a new field (and to trigger scientific revolutions, which are rare according to Kuhn [5]). The crucial feature of a new field is that the individual returns for the first published papers are very low at the time of publishing them. Furthermore note that the more or less immediate recognition and appreciation of such an endeavor is crucial for the career of a young scientist. In contrast, working in the dominating research paradigm yields more predictable results (more readers, more citations). However, the productivity in an established field will inevitably decline while a new research field offers the possibility of higher future productivity.

A crucial question is how a group of researchers enters a new scientific field. It is always very difficult to be the first, because of scale economies. Therefore, only altruistic or stubborn researchers can develop new fields. Perfect coordination could make such a collective shift to a new field possible, but this is very unlikely in an unregulated, i.e., market economy. However, government incentives, to pay for research outside mainstream could foster such coordinations even in market economies, which is similar to please for directed technical change, e.g., Acemoglu [20].

In the case of computing an intertemporal competitive equilibrium, we derived that there are three areas of initial (aggregate) human capital. For very low values there is inefficient outcome of the elimination of this research field. At an intermediate level there is indeterminacy, the research in the new field gets going or is phased out. Only high initiation levels of a new research paradigm guarantee that it will be pursued.

We also model delayed returns. This indicates even larger obstacles for the development of a new field. In this case, the difficulties are not linked to the build-up of the human capital stock, specific for the new field, but to the management and finance of research that award grants based on past research records. Summarizing this but also the above observation lends critique to the current research practice. More precisely, the current policy (especially for grants but also for journal publications) provides insufficient incentives for researchers to encourage them to take the risk of working in and exploring new fields.

Of course, our paper attempts to address just a few of the critical issues to understand the enterprise of scientific research better. Therefore, there are many conceivable extensions and variations. A natural one is to account for the uncertainty associated with any research undertaking (recall the quote from Max Weber) and in particular in new areas. It is also important to differentiate between vertical (complementary skills between scientists) and horizontal (professor-student) cooperation. Here the first case tends to foster new ideas, while the second one adds to the conservatism of science.

References

1. Diamond Jr, A. M. (2008). Economics of science. *The New Palgrave Dictionary of Economics*, *7*, 328.
2. Lotka, A. (1926). The frequency distribution of scientific productivity. *Journal of the Washington Academy of Sciences*, *16*(12), 317–323.
3. Merton, R. K. (1988). The Matthew effect in science, II: Cumulative advantage and the symbolism of intellectual property. *isis*, *79*(4), 606–623.
4. Merton, R. K., Barber, E. (2004). *The travels and adventures of serendipity: A study in sociological semantics and the sociology of science*. Princeton: Princeton University Press.
5. Kuhn, T. (1962). *The structure of scientific revolutions*. University of Chicago Press.
6. Skiba, A. K. (1978). Optimal growth with a convex-concave production function. *Econometrica*, *46*, 527–539.
7. Sethi, S. P. (1979). Optimal advertising policy with the contagion model. *Journal of Optimization Theory and Applications*, *29*(4), 615–627.
8. Dechert, D. W., Nishimura, K. (1983). Complete characterization of optimal growth paths in an aggregative model with a non-concave production function, *Journal of Economic Theory*, *31*, 332–354.
9. Brock, W. A. (1983). Pricing, predation and entry barriers in regulated industries. In D. S. Evans (Ed.), *Breaking Up Bell* (pp. 191–229). Amsterdam, North Holland.
10. Brock, W. A., Davis, W. D. (1985). Dynamic ramsey pricing. *International Economic Review*, *26*, 569–591.
11. Mäler Karl Göran. (2000). Development, ecological resources and their management: A study of complex dynamic systems, Joseph Schumpeter lecture. *European Economic Review*, *44*, 645–665.
12. Romer, P. (1986). Increasing returns and long-run growth. *Journal of Political Economy*, *94*(5), 1002–1037.
13. Arthur, W. B. (1994). *Increasing returns and path dependence in the economy*, p. 201. University of Michigan Press.
14. Arthur, W. (1989). Brian, competing technologies, increasing returns, and lock-in by historical events. *Economic Journal*, *99*, 116–131.
15. Arthur, W. B. (1990). Positive feedbacks in the economy. *Scientific American*.
16. Arthur, W. B. (1996). Increasing returns and the new world of business, *Harvard Business Review*, 100.
17. Franz, Wirl, & Feichtinger, G. (2005). History dependence in concave economies. *Journal of Economic Behavior and Organization*, *57*, 390–407.
18. Stephan, P. E. (1996). The economics of science. *Journal of Economic literature*, *34*(3), 1199–1235.
19. Merton, R. K. (1968). The matthew effect in science: The reward and communication systems of science are considered. *Science*, *159*(3810), 56–63.

20. Daron, Acemoglu, Aghion, Philippe, Bursztyn, Leonardo, & Hemous, David. (2012). The environment and directed technical change. *The American Economic Review*, *102*, 131–166.
21. Pierre, Azoulay. (2011). Gustavo manso and Joshua Graff Zivin, incentives and creativity: Evidence from the academic life sciences. *RAND Journal of Economics*, *42*(3), 527–554.
22. Popper K. R. (1935). *Logik der Forschung*, Springer-Verlag Wien.
23. Krugman, P. (1991). History versus expectations. *The Quarterly Journal of Economics*, *106*(2), 651–667.
24. Mascarilla-i-Miro, O., Yegorov, Y. (2005). Modelling functional area and commuting flows. *Cuadernos de economia*, *28*, 39-56.
25. Yegorov, Y. (2009). The dynamics of research schools and the demand for teaching. In *Proceedings of INTED 2009* (p. 12). Valencia, Spain; ISBN 978-84-612-7578-6.
26. Franz, Wirl. (2016). Indeterminacy and history dependence of strategically interacting players. *Economics Letters*, *145*, 19–24.
27. Franz, Wirl. (2011). Conditions for indeterminacy and thresholds in neoclassical growth models. *Journal of Economics*, *102*, 193–215.
28. Franz, Wirl, & Feichtinger, G. (2007). Social interactions within a dynamic competitive economy. *Journal of Optimization Theory and Applications*, *133*, 385–400.

Chapter 8
On the Structure and Regularity of Optimal Solutions in a Differential Game with Regime Switching and Spillovers

Anton Bondarev and Dmitry Gromov

8.1 Introduction

Examples of non-smooth dynamics in Economics are numerous, but papers dealing with full formal complexity of dynamics are scarce, see Brito et al. [5], Brito et al. [6], for example. Recently, some interest emerged in regime-switching differential games, e.g., Dawid et al. [8], Gromov and Gromova [19], Long et al. [22], Bondarev, and Greiner [4]. In there, different types of switching conditions and different types of solutions are proposed. Note that a lot of economic phenomena naturally conform to switching dynamics: consider, e.g., a transition to new production technology, changing leadership in oligopolistic markets with imitation, resource extraction games, or advertising games.

On the theory side, there is an increasing number of papers dealing with switching systems or sliding dynamics, see Di Bernardo et al. [10] for an overview. However, this strand of literature does not consider optimal dynamics and is focused on a qualitative behavior of non-smooth dynamical systems in the vicinity of the switching manifold.

On the other hand, the optimal dynamics of switched systems is extensively studied in hybrid optimal control theory, where a number of important results were obtained, see, e.g., Boltyansky [2], Azhmyakov et al. [1], Shaikh and Caines [27] among many others. A particularly important result consists in the formulation of Hybrid Maximum Principle [27] that extends the classical maximum principle to the class of hybrid control systems, that is systems experiencing structural changes due

A. Bondarev
International Business School Suzhou, Xi'an Jiaotong-Liverpool University, Ren'ai Road 111, Suzhou 215123, People's Republic of China
e-mail: anton.bondarev@xjtlu.edu.cn

D. Gromov (✉)
Faculty of Applied Mathematics and Control Processes, St. Petersburg State University, 7/9 Universitetskaya Emb., 199034 Saint Petersburg, Russia
e-mail: dv.gromov@gmail.com

Krasovskii Institute of Mathematics and Mechanics, Ural Branch of the Russian Academy of Sciences, 620990 Yekaterinburg, Russia

J. L. Haunschmied et al. (eds.), *Dynamic Economic Problems with Regime Switches*, Dynamic Modeling and Econometrics in Economics and Finance 25,
https://doi.org/10.1007/978-3-030-54576-5_8

to some exogenous or endogenous events. We refer the interested reader to Lunze and Lamnabhi-Lagarrigue [23] for a detailed account on hybrid systems and related fields.

While the hybrid maximum principle is capable of addressing a wide class of multi-modal control problems its application is pretty much restricted due to the rapidly increasing difficulty of obtaining an analytical solution as the model complexity increases. To overcome this difficulty, an approach based on the extension of the classical maximum principle was proposed in Gromov and Gromova [19]. This approach, albeit less general, allows one to solve a pretty wide range of multi-modal optimal control problems in a rather intuitive way.

In this paper, we continue this line of research and apply the previously described methodology to the analysis of a particular class of switched differential games that has been studied recently in Bondarev [3], Bondarev and Greiner [4]. The goal of this study is to provide a detailed analysis and thorough understanding of the consequences of non-smooth dynamics for economic models. We wish to note that after this paper was submitted for publication, the authors came across a recent preprint by Reddy et al. [24] and the paper [25] that consider similar problems, albeit from a somewhat different perspective.

We intentionally consider the simplest possible model with a single common state which has linear dynamics. It has been demonstrated (see Dockner and Nishimura [12], Wirl and Feichtinger [31]) that even this class of models can have surprisingly rich dynamics, including thresholds, history dependence and multiple equilibria.

We consider only two specific cases: the open-loop Nash equilibrium with two players and the social cooperation case. These two suffice to demonstrate the main qualitative findings, whereas the method itself is no way limited to these situations.

The contribution of this paper is twofold. First, we explicitly derive optimal trajectories both for the cooperative and non-cooperative cases for a regime-switching system through the application of a modified version of the classical maximum principle. This increases the tractability of results and makes explicit solution feasible. Second, we demonstrate that our results are in the agreement with the intuition given by the hybrid maximum principle while being more tractable and intuitive. We show in which cases the problem admits regular cases of finite-time switches and study the conditions for the emergence of new complicated types of dynamics such as the sliding motion along the switching manifold.

The paper is organized as follows: in Sect. 8.2, we describe the multi-modal differential game with spillovers that forms the subject of our study. In Sects. 8.3 and 8.4, a detailed analysis of optimal solutions both for the cooperative and the Nash equilibrium cases is presented in detail. Section 8.5 contains a discussion of possible extensions associated with more complex dynamic patterns that can occur in the considered game. Finally, Sect. 8.6 presents brief conclusions.

8.2 A Multi-modal Dynamic Game

As a starting point of our analysis, consider a differential game with two players and one state variable such that the objectives of players are *interdependent* (referred to simply as a game further on):

$$\forall i \in \{1, 2\} : \ J_i = \int_0^\infty e^{-\rho t}(a_i x + c_i x u_{-i} - \frac{1}{2}u_i^2)dt \tag{8.1}$$

where x is the common state variable and u_i are the controls (strategies) of both players with the subscript $-i$ denoting the complement to i: $u_{-1} = u_2$ and so on. The objective (8.1) contains the cross-term $c_i x u_{-i}$ which measures the indirect benefit of a given player from efforts of the other player. We interpret this in terms of advertising models, whereas the goodwill accumulation is affecting both agents. Furthermore, both the controls and the state are required to be (almost everywhere) differentiable. Note that the payoff functional (8.1) has a linear-quadratic form and hence enjoys a number of important properties. In particular, it is known that Hamilton–Jacobi–Bellman approach and Maximum Principle yield the same controls if they are restricted to linear-feedback forms, see Dockner et al. [11].

We additionally impose the following non-negativity constraints on controls and the state:

$$x \in \mathbb{R}_+, \ u_{\{1,2\}} \in U \subset \mathbb{R}_+ \tag{8.2}$$

where U is the set of admissible controls and $\mathbb{R}_+ = [0, \infty)$. These are standard for economic applications, where controls $u_{\{1,2\}}$ are interpreted as investments. Thus, by (8.2), we just require investments to be non-negative and the resulting stock to be bounded by zero.

This game has a bilinear-quadratic structure, similar to advertising and marketing models (see the seminal example by Deal et al. [9] and more recent [20] for a review) and includes a *spillover effect* modeled by the term $c_i x u_{-i}$. This term is novel and rarely appears in economic applications. It can represent a positive or negative impact on the value of firm i by state and investments product of firm j, hence the term spillover effect. Such effects are typically present in advertising and goodwill models, where the value of advertising for one firm positively depends on advertising efforts of the other firm provided they have similar products.

The dynamic constraint is given by

$$\dot{x} = b_1 u_1 + b_2 u_2 - \delta x, \tag{8.3}$$

i.e., the stock of (advertising, technology, resource, capital) is changing due to the common investments/extractions of both players and depreciates over time. In this equation, coefficients b_i represent the *efficiency* of investments of the firm i.

It can be transformed into a multi-modal game, once we let efficiency coefficients b_i to vary across regimes, $b_i^+ \neq b_i^-$ with either *time-dependent* or *state-dependent (autonomous) switching*. For instance, the situation with $b_i^+ > b_i^-$ would refer to joint learning while both firms become more efficient after reaching the threshold, whereas the case $b_i^+ > b_i^-$, $b_{-i}^+ < b_{-i}^-$ refers to the changing leadership situation, while the firm which becomes the leader is more efficient in investments (learns more).

Let $f(x, t)$ be a smooth map, $f : \mathbb{R}_+ \times \mathbb{R}_+ \to \mathbb{R}$, such that the rank of Df is equal to 1 for all $(x, t) \in \mathbb{R}_+ \times \mathbb{R}_+$. The level set $f(x, t) = 0$ is the *switching manifold*. Define

$$b_1 = \begin{cases} b_1^+, & f(x(t), t) \geq 0, \\ b_1^-, & f(x(t), t) < 0 \end{cases} \qquad b_2 = \begin{cases} b_2^+, & f(x(t), t) \geq 0, \\ b_2^-, & f(x(t), t) < 0 \end{cases} \tag{8.4}$$

We limit ourselves to the two basic cases (using the terminology from [19] and [27]):

1. Time-driven (controlled) switch: $f(x(t), t) = t - \tau^*$ with τ^* fixed
2. State-driven (autonomous) switch: $f(x(t), t) = x(t) - x^*$ with x^* fixed, but t left free.

In Long et al. [22], a somewhat similar problem is considered but with τ being subject to decision of one of the players. Here, we mainly focus on the derivation of optimal solutions with the help of the standard Maximum Principle, whereas in the aforementioned paper an alternative piece-wise Nash solution concept is developed.

Many economic problems can be put into this simple framework. As examples, consider resource extraction problems with regime switches (e.g., Long et al. [22]), technological transitions where efficiency of investments change after initial transition time (e.g., Dawid et al. [8]), pollution control (e.g., Gonzalez [18]), and patent races (e.g., Fudenberg et al. [15]).

In the following, we will use τ and $x(\tau)$ to denote the switching time and the switching state. Furthermore, we put the asterisk ($*$) to denote which component of the solution is fixed at the switching. That is to say, we use $x^*(\tau)$ or simply x^* when referring to a state-driven switch and $x(\tau^*)$, resp., τ^* when referring to a time-driven switch.

Whichever type of switching is considered, we assume that the system has a *fixed initial mode*, that is, $0 < \tau^*$, resp., $x(0) < x^*$ and thus refer to $T^- = [0, \tau^*)$, resp., $T^- = \{t \in \mathbb{R}_+ | x(t) < x^*\}$ as the first interval and consequently, $T^+ = \mathbb{R}_+ \setminus T^-$ as the second interval (note that T^+ can be empty).

In the following, we will use the minus and the plus superscripts to refer to the first, resp. the second interval. We make some remarks further on what changes in our results if the switching sequence is reversed. Finally, we assume that all parameters in the model are non-negative:

$$\{a_{1,2}, c_{1,2}, b_{1,2}^{\pm}, \delta, \rho\} \in \mathbb{R}_+. \tag{8.5}$$

It is worth noting that the distinction between the two types of switches is of substantial rather than notational character. For instance, for the time-driven switch, the switching time τ^* is always eventually reached as the system evolves. In contrast to it, the switching state is not always reached as the following definition suggests.

Definition 8.1 The game (8.1)–(8.3) is said to be in the *normal mode*, if the switching threshold $f(x(\tau), \tau)$ is never reached by the optimal trajectory and no switching occurs. Otherwise, the game is said to be in the *switching mode*.

In other words, if the game is in the normal mode and under the fixed switching sequence condition, Definition 8.1 implies that $T^- = [0, \infty)$ and the second interval is never reached.

We observe that under the assumption of fixed sequence, the only case when the game can be in the normal mode is when the switching condition is defined to be dependent on x and the equilibrium of the dynamics in the first interval is such that $x_{eq}^- < x^*(\tau)$. This agrees with the assumption that both equilibria are *regular* (terminology follows Di Bernardo et al. [10]) in the case of a state-driven switch,[1] that is:

Assumption 8.1 If equilibria of the dynamic system (8.3) exist, they are regular:

$$x_{eq}^- < x^*(\tau) < x_{eq}^+ \tag{8.6}$$

with $x_{eq}^{\pm}$ denoting the (potential) equilibria of the system below and above the threshold value x^*.

By employing this assumption, we restrict the attention to the case of at most one switching event. Indeed, if the optimal trajectory contacts the switching manifold, the optimal jump in the co-state trajectory will immediately select the extension after the threshold lying on the stable manifold of the associated equilibrium (since this is, by construction of the model, of a saddle type). However, in more complex settings, multiple (or even infinite number of) switches are potentially possible. We will further discuss the relevance and importance of this assumption in Sect. 8.5.

Below, we explore the solution technique for this game in more details. We consider separately the social cooperative and the (open-loop) Nash equilibrium solutions to the differential game (8.1)–(8.4).

8.3 Cooperative Game

8.3.1 Second Interval

The solution for the cooperative game is obtained by solving the optimal control problem given by dynamic constraint (8.3) and an objective being the sum of individual ones:

$$W = J_1 + J_2 = \int_0^{\infty} e^{-rt} \left[(a_1 + a_2)x + (c_1 u_2 + c_2 u_1)x - \frac{1}{2}(u_1^2 + u_2^2) \right] dt \tag{8.7}$$

[1] Since in the case of a time-driven switch this notion does not bear any meaning.

The Hamiltonian is

$$H(x,\psi,u) = e^{-\rho t}\left((a_1+a_2)x + (c_1u_2+c_2u_1)x - \frac{1}{2}(u_1^2+u_2^2)\right) + \psi(b_1u_1+b_2u_2-\delta x). \tag{8.8}$$

The optimal controls are obtained from the first order optimality condition to be

$$u_i^C = c_{-i}x + e^{\rho t} b_i \psi. \tag{8.9}$$

Plugging (8.9) into (8.3) and introducing a new variable $\lambda = e^{\rho t}\psi$ representing the current value of the adjoint ψ we get a system of two autonomous DEs:

$$\begin{aligned} \dot{x} &= (b_1^2+b_2^2)\lambda + (c_2b_1+c_1b_2-\delta)x, \\ \dot{\lambda} &= \rho\lambda - \left((c_1^2+c_2^2)x + a_1 + a_2\right) - (b_1c_2+b_2c_1-\delta)\lambda. \end{aligned} \tag{8.10}$$

Using vector-matrix notation and introducing some abbreviations the system (8.10) can be rewritten as

$$\dot{z} = Cz + g, \tag{8.11}$$

where $z = \begin{bmatrix} x \\ \lambda \end{bmatrix}$, $C = \begin{bmatrix} r & b \\ -c & \rho - r \end{bmatrix}$, and $g = \begin{bmatrix} 0 \\ -a \end{bmatrix}$. In (8.11), we also used the following short notation: $a = a_1 + a_2$, $b = b_1^2 + b_2^2$, $c = c_1^2 + c_2^2$, $q = b_1c_2 + b_2c_1$, and $r = q - \delta$.

Note that the problem (8.11) constitutes a system of two linear ODEs. Thus, there is at most only one equilibrium in each interval and the resulting piece-wise system has at most two equilibria (or exactly two if the equilibria are regular), each one associated with the corresponding regime.

The equilibrium state of the system (8.11) is

$$\begin{bmatrix} x_{eq}^C \\ \lambda_{eq}^C \end{bmatrix} = -C^{-1}g = -\frac{1}{\det(C)}\begin{bmatrix} \delta - q + \rho & -b \\ c & q - \delta \end{bmatrix}\begin{bmatrix} 0 \\ -a \end{bmatrix} = \frac{a}{\det(C)}\begin{bmatrix} -b \\ q - \delta \end{bmatrix}. \tag{8.12}$$

Since x_{eq}^C must be non-negative,[2] we impose the following regularity assumption.

Assumption 8.2 The coefficients of the matrix C have to satisfy $\det(C) < 0$ or, equivalently,

$$(\delta - q)^2 + (\delta - q)\rho - bc \geq 0.$$

The matrix C has two eigenvalues:

$$\sigma_{1,2} = \frac{\rho}{2} \pm \frac{\sqrt{[2(\delta - q) + \rho]^2 - 4bc}}{2}.$$

[2]Also note that if Assumption 8.2 doesn't hold, we have $\mathrm{trace}(C) = \rho > 0$ and hence, the equilibrium is a source (either an unstable focus or an unstable node). This implies that there does not exist an initial state (x_0, λ_0) such that the solution to (8.11) converges to the equilibrium.

We will enumerate the eigenvalues in the way that $\sigma_2 > \sigma_1$. Assumption 8.2 implies, in particular, that $\sigma_1\sigma_2 < 0$ and hence, the equilibrium point of (8.11) is a saddle.

To simplify further analysis, we rewrite C in terms of its eigenvalues. So, we get

$$C = \begin{bmatrix} r & \frac{(r-\sigma_1)(r-\sigma_2)}{c} \\ -c & \sigma_1+\sigma_2-r \end{bmatrix}.$$

The two-point boundary value problem associated with the optimal control problem (8.3), (8.7) consists in solving the system of DEs (8.11) while satisfying the following boundary conditions:

$$\begin{aligned} x(\tau^*) &= x^* \\ \lim_{t\to\infty} e^{-\rho t}\lambda(t) &= 0. \end{aligned} \tag{8.13}$$

The conditions (8.13) allow us to determine the initial condition on the adjoint variable as stated below. We define $\lambda^{*+} = \lim_{t\to\tau+0}\lambda(t)$.

Proposition 8.1 *The initial value* λ^{*+} *guaranteeing the fulfilment of* (8.13) *is uniquely defined as*

$$\lambda^{*+} = \frac{a}{\sigma_2} - \frac{c}{r-\sigma_2}x^*.$$

Proof The solution of (8.11) is dominated by its largest positive eigenvalue σ_2, which is larger than ρ due to Assumption 8.2. This implies that the initial conditions $(x^*(\tau^*), \lambda(\tau^*))$ must be chosen in the way that the solution's component at $e^{\sigma_2 t}$ is equal to 0. This is equivalent to saying that the initial condition must lie on the stable manifold of the saddle. As time elapses, the state and the adjoint variable will approach their equilibrium values (8.12).

Let $v_{1,x}$ and $v_{1,\lambda}$ be the respective components of the eigenvector corresponding to σ_1. The linear subspace corresponding to v_1 can be written as $V_1 = \{(x^C_{eq} + \alpha v_{1,x}, \lambda^C_{eq} + \alpha v_{1,\lambda}) \mid \alpha \in \mathbb{R}\}$. Setting the first component to x^*, we recover α and the respective λ-component of the vector of initial conditions. □

Now we can compute explicit expressions for the optimal solution $x(t)$ and $\lambda(t)$ to get

$$\begin{aligned} x(t) &= a\frac{(r-\sigma_1)(r-\sigma_2)(e^{\sigma_1(t-\tau)}-1)}{c\sigma_1\sigma_2} + x^* e^{\sigma_1(t-\tau)} \\ \lambda(t) &= a\frac{\left(r-(r-\sigma_1)e^{\sigma_1(t-\tau)}\right)}{\sigma_1\sigma_2} - \frac{ce^{\sigma_1(t-\tau)}}{r-\sigma_2}x^* \end{aligned} \tag{8.14}$$

One can easily check that $x(\tau) = x^*$ and $\lim_{t\to\infty} x(t) = x^C_{eq}$. Furthermore, we observe that $x(t)$ changes monotonously and the sign of $\dot{x}(t)$ depends on whether

$x(t)$ is smaller or greater than x^C_{eq}: if $x(t) < x^C_{eq}$, then $\dot{x}(t) \geq 0$ and vice versa. The optimal control is

$$u_i^C(t) = \frac{ab_i e^{\sigma_1(t-\tau)}}{\sigma_2} + \frac{a\left(e^{\sigma_1(t-\tau)} - 1\right)}{\sigma_1\sigma_2}(bc_{-i} - b_i r) + \left(c_{-i} - \frac{b_i c}{r - \sigma_2}\right) x^* e^{\sigma_1(t-\tau)} \tag{8.15}$$

We know that all the previous results stay valid only for the case that the optimal control (8.15) is non-negative. We have the following result.

Proposition 8.2 *The optimal controls u_i^C are non-negative if the following condition holds:*

$$\min\left(\frac{ab_i}{\sigma_2} + \left(c_{-i} - \frac{b_i c}{r - \sigma_2}\right)x^*, \frac{-a}{\sigma_1\sigma_2}(bc_{-i} - b_i r)\right) \geq 0, \quad i = 1, 2.$$

Proof To check if this is the case, we compute the derivative of $u_i^C(t)$ w.r.t. time:

$$\frac{d}{dt}u_i^C = -\frac{(b_i c - c_{-i}(r - \sigma_2))(ab + \sigma_1\sigma_2 x^*)}{\sigma_2(r - \sigma_2)} e^{\sigma_1(t-\tau)}$$

We see that the sign of the derivative is constant for all $t \in T^+$ and is determined only by the parameters and the value of the switching state x^*.

Depending on the sign of the derivative, the minimal value of the control is attained either at $t = \tau$ or at $t = \infty$. Thus, both of the following two expressions must be non-negative to ensure that the control falls in with the bounds:

$$\lim_{t\to\tau+0} u_i^C(\tau^*) = \frac{ab_i}{\sigma_2} + \left(c_{-i} - \frac{b_i c}{r - \sigma_2}\right)x^*, \quad \lim_{t\to\infty} u_i^C(t) = \frac{-a}{\sigma_1\sigma_2}(bc_{-i} - b_i r).$$

This yields the required result. □

Note that $\sigma_1\sigma_2 < 0$, while, say, $r - \sigma_2$ can be of either sign depending on parameters.

Finally, substituting (8.14) into the expression for the payoff function, integrating and performing some algebraic simplifications, we get

$$J(\tau^*, x^*) = \frac{1}{2}\left[\frac{2a}{\sigma_2}x^* - \frac{r - \sigma_1}{b}(x^*)^2 + \frac{a^2 b}{\sigma_2^2\rho}\right] e^{-\tau^*\rho}. \tag{8.16}$$

Note that the value function depends both on τ^* and x^*. Relaxing one of the arguments, we recover either a state-dependent or time-dependent switch.

8.3.2 First Interval

8.3.2.1 Normal Mode

We start by considering the normal mode as introduced in Def. 8.1. The social cooperative game is given by the optimal control problem with joint maximization. Under given assumptions, there exists only one Skiba-threshold in a system (8.3):

Lemma 8.1 *The indifference point (DNSS-point)* x_S^C *exists for cooperative game* (8.1)–(8.3) *and* $x_S^C < x^*(\tau)$.

Proof Any bi-stable system without heteroclinic connections possesses such a point, see Wagener [30], Bondarev and Greiner [4]. The state–co-state system associated with (8.3) in cooperative case is given by (8.10) is linear in each interval and as such has a unique equilibrium. The overall piece-wise canonical system is thus bi-stable (it has two regular equilibria by Assumption 8.1) and it does not have heteroclinic connections since we assume fixed switching sequence and there is no unstable focus in between. By definition of the DNSS-point, it has to be $x_S^C < x^*(\tau)$. □

Denote the associated equilibrium of the first interval flow cooperative game by $x_{eq}^{C,-}$, and assume $0 \leq x_{eq}^{C,-} < x^*(\tau)$. We use the minus superscript to denote quantities associated with the first interval (so C^- denotes the equivalent of the matrix C for the first interval).

Proposition 8.3 *Let* $x(0) < x^*(\tau)$ *and Assumption 2 holds for matrix* C^-. *Then:*

- *If* $x(0) < x_S^C$, *the cooperative game is in the normal mode and* $x_{eq}^{C,-}$ *is realized as the long-run equilibrium of the cooperative game.*
- *If* $x(0) > x_S^C$, *the cooperative game is in the switching mode and* $x_{eq}^{C,-}$ *realizes as the long-run equilibrium with a unique switching event at* $x^*(\tau)$.
- *Outcome is indeterminate only at* $x_{eq}^{C,-} = x_S^C$ *which has zero measure.*

Proof If Assumption 8.2 holds, the $x_{eq}^{C,-}$ is positive. We also assumed that $x_{eq}^{C,-} < x^*(\tau)$ so it can be reached by the $x(t)$ process without crossing the threshold. Now if $x(0) < x_S^C$ holds, it means that by definition of the Skiba point it is optimal to converge to this value $x_{eq}^{C,-}$. Once it is a saddle, no arcs entering the $x > x(\tau)$ region can be part of the optimal trajectory converging to this (lower) equilibrium. Game is in the normal mode.

On the contrary, once $x(0) > x_S^C$, it is optimal to converge to the equilibrium $x_{eq}^{C,+} > x^*(\tau)$ and crossing occurs. It is the unique one since the equilibrium is a saddle (so there are no arcs re-entering the first region).

At last, once $x(0) = x_S^C$ the dynamics is indeterminate, as is standard for this types of models (see discussion in Krugman [21]), but this is a non-generic point. □

We further on assume away the indeterminacy by letting $x_{eq}^{C,-} \neq x_S^C$.

8.3.2.2 State-Driven Switch

In the first interval, we optimize the objective function given by

$$J^-(0, x_0) = \int_0^\tau e^{-rt}\left[ax + (c_1u_2 + c_2u_1)x - \frac{1}{2}(u_1^2 + u_2^2)\right]dt + J^+(\tau, x).$$

Here, the optimal value of the objective function enters as the terminal cost for the respective optimization problem. Note that we put the superscripts $-$ and $+$ to distinguish between the parameters and variables that take different values in the first, resp., second phases.

There are two possible cases: when the switching state x^* is fixed and the switching time τ is free and the opposite. The former corresponds to the state-driven switch, while the latter to the time-driven one.

We start be considering the state-driven switch. Here, the terminal time τ is free and hence, we can use the result presented in the Appendix. According to (8.31), we evaluate $H(t, x, \psi)$ at $t = \tau$ and equate the resulting expression to $-\frac{d}{d\tau}J^+(\tau)$, where $J^+(\tau)$ is given in (8.16). Replacing ψ with $e^{-\rho\tau}\lambda$, we get the following quadratic equation in λ:

$$b^-\lambda^2 + 2r^-\lambda x^* = \frac{a^2b^+}{(\sigma_2^+)^2} + \frac{2a(\rho - \sigma_2^+)}{\sigma_2^+}x^* - \frac{\rho\left(r^+ - \sigma_1^+\right) + b^+c}{b^+}(x^*)^2. \quad (8.17)$$

Solving (8.17), we get two candidates for the end-point values of the adjoint variable $\lambda^{*-} = \lim_{t\to\tau-0}\lambda(t)$. We require that $\lambda^{*-}(\tau)$ yields $\dot{x}(\tau) > 0$, which is equivalent to (cf. 8.10):

$$((b_1^-)^2 + (b_2^-)^2)\lambda^{*-} + (c_2b_1^- + c_1b_2^- - \delta)x^* > 0.$$

When λ^{*-} is determined, the final step consists in solving the system (8.10) with conditions $x(0) = x_0$, $x(\tau) = x^*$ and $\lambda_i(\tau) = \lambda_i^{*-}$. This allows us to determine τ, for instance, by integrating (8.10) backward in time with initial conditions (x^*, λ_i^{*-}) and determining τ from $x(\tau) = x_0$.

8.3.2.3 Time-Driven Switch

If the switching time τ^* is fixed, the endpoint condition on λ^{*-} is uniquely determined by

$$\lambda^{*-} = \frac{d}{dx}J(\tau^*, x(\tau^*))\bigg|_{t=\tau^*} = \frac{a}{\sigma_2} - \frac{r - \sigma_1}{b}x(\tau^*),$$

which equals λ^{*+} (see Proposition 8.1). Again, this agrees with the Hybrid Maximum Principle in that the adjoint variable is continuous at an autonomous switch.

Finally, the switching state is computed by solving the system (8.10) over the interval $[0, \tau^*]$ with boundary conditions $x(0) = x_0$, and $\lambda_i(\tau^*) = \lambda_i^{*-}$. The obtained solution is used to determine the switching state $x(\tau^*)$.

8.4 Nash Equilibrium Solution

8.4.1 Second Interval

We start by determining the Nash equilibrium solution to the differential game (8.1)–(8.5) in the second interval T^+. In doing so, we will initially suppose that both the initial time and the state are fixed to (τ^*, x^*). Relaxing respective terms, we will recover either the state- or the time-driven switch.

Since the presented below results can be of interest on their own, we first drop the $+$ superscript indicating the value of the parameter b_i in the second interval and restore it later when making a connection to the first interval.

When determining the Nash equilibrium solution one has to solve simultaneously as many optimization problems as many players there are. That is to say, for $i \in \{1, 2\}$, we maximize J_i w.r.t. u_i while assuming that u_{-i} is chosen to satisfy $u_{-i} = u_{-i}^{NE}$. Following the standard procedure, we write the individual Hamiltonian for each optimization problem:

$$\begin{aligned} H_1(x, \psi_1, u) &= e^{-\rho t}\left(a_1 x + c_1 u_2 x - \frac{1}{2}u_1^2\right) + \psi_1(b_1 u_1 + b_2 u_2 - \delta x), \\ H_2(x, \psi_2, u) &= e^{-\rho t}\left(a_2 x + c_2 u_1 x - \frac{1}{2}u_2^2\right) + \psi_2(b_1 u_1 + b_2 u_2 - \delta x). \end{aligned} \tag{8.18}$$

The respective optimal controls are found to be $u_i^{NE} = e^{\rho t} b_i \psi_i$. Plugging u^{NE} into (8.3), and going over to the current values of the adjoint states $\lambda_i = e^{\rho t}\psi_i$, we recover a set of autonomous DEs corresponding to Hamiltonians (8.18):

$$\begin{aligned} \dot{\lambda}_1 &= \delta\lambda_1 - a_1 + \lambda_1\rho - b_2 c_1 \lambda_2, \\ \dot{\lambda}_2 &= \delta\lambda_2 - a_2 + \lambda_2\rho - b_1 c_2 \lambda_1. \end{aligned} \tag{8.19}$$

The resulting system of differential equations for x and λ_i has the following form:

$$\frac{d}{dt}\begin{bmatrix} x \\ \lambda_1 \\ \lambda_2 \end{bmatrix} = \begin{bmatrix} -\delta & b_1^2 & b_2^2 \\ 0 & \delta + \rho & -b_2 c_1 \\ 0 & -b_1 c_2 & \delta + \rho \end{bmatrix}\begin{bmatrix} x \\ \lambda_1 \\ \lambda_2 \end{bmatrix} + \begin{bmatrix} 0 \\ -a_1 \\ -a_2 \end{bmatrix} = A\begin{bmatrix} x \\ \lambda_1 \\ \lambda_2 \end{bmatrix} + f. \tag{8.20}$$

The matrix A is block-diagonal with the first eigenvalue $\sigma_1 = -\delta < 0$. Thus, the character of the system's behavior in long run is determined by the eigenvalues of the second block submatrix. The respective eigenvalues are $\sigma_{\{2,3\}} = \delta + \rho \pm \sqrt{b_1 b_2 c_1 c_2}$.

Assumption 8.3 The coefficients of the matrix A satisfy

$$(\delta + \rho)^2 - b_1 b_2 c_1 c_2 \geq 0. \tag{8.21}$$

We have the following result.

Proposition 8.4 *The system* (8.20) *has the following long-run solution*

$$\begin{bmatrix} x_{eq} \\ \lambda_{1,eq} \\ \lambda_{2,eq} \end{bmatrix} = \begin{bmatrix} \dfrac{(a_1 b_1^2 + a_2 b_2^2)(\delta + \rho) + (a_1 b_2 c_2 + a_2 b_1 c_1) b_1 b_2}{\delta\left((\delta + \rho)^2 - b_1 b_2 c_1 c_2\right)} \\ \dfrac{a_1(\delta + \rho) + a_2 b_2 c_1}{(\delta + \rho)^2 - b_1 b_2 c_1 c_2} \\ \dfrac{a_2(\delta + \rho) + a_1 b_1 c_2}{(\delta + \rho)^2 - b_1 b_2 c_1 c_2} \end{bmatrix} \in \mathbb{R}^3_+ . \tag{8.22}$$

with $x_{eq} \geq 0$ *iff Assumption 8.3 holds. Furthermore, it holds that* $\lambda_i(t) = \lambda_{i,eq}$ $\forall t \in T^+$.

Proof The equilibrium solution to (8.20) is obtained as $-A^{-1} f$. The non-negativity of the equilibrium state x_{eq} follows from Assumption 8.3.

On the other hand, the same Assumption implies that at the equilibrium point the system (8.20) has two positive and one negative eigenvalues. This means that the initial values are to be located along the respective eigenvector $v_1 = [1, 0, 0]^\top$, which, in turn, implies that $\lambda_i(\tau^*) = \lambda_{i,eq}$. Since the r.h.s. of (8.19) do not depend on x, we have that $\lambda_i(t) = \lambda_{i,eq}$ $\forall t \in T^+$. □

Solving the equation for $x(t)$ with initial condition $x(\tau^*) = x^*$, we get

$$x(t) = \\ x^* e^{-\delta(t-\tau^*)} + \frac{\left((a_1 b_1^2 + a_2 b_2^2)(\delta + \rho) + b_1 b_2 (a_1 b_2 c_2 + a_2 b_1 c_1)\right)}{\delta\left((\delta + \rho)^2 - b_1 b_2 c_1 c_2\right)} (1 - e^{-\delta(t-\tau^*)}).$$

At this point, we note that the system state changes monotonically (given (8.5)); furthermore, at the switching time τ^* the phase vector $\dot{x}(\tau^*)$ points toward the region $x > x^*$ if $x^* < x_{eq}$ (that is, Assumption 8.1 holds).

Finally, the value functions are computed to be

$$J_1^+(\tau^*, x^*) = \frac{\left(a_1(\delta+\rho) + a_2 b_2^+ c_1\right)}{c_1\rho\left((\delta+\rho)^2 - b_1^+ b_2^+ c_1 c_2\right)} \times$$
$$e^{-\rho\tau^*}\left[\frac{\left(a_1(\delta+\rho) + a_2 b_2^+ c_1\right)\left(c_1(b_1^+)^2 + 2b_2^+\delta + 2b_2^+\rho\right)}{2\left((\delta+\rho)^2 - b_1^+ b_2^+ c_1 c_2\right)} - \left(a_1 b_2^+ - c_1\rho x^*\right)\right]$$
$$J_2^+(\tau^*, x^*) = \frac{a_2(\delta+\rho) + a_1 b_1^+ c_2}{c_2\rho\left((\delta+\rho)^2 - b_1^+ b_2^+ c_1 c_2\right)} \times$$
$$e^{-\rho\tau^*}\left[\frac{\left(a_2(\delta+\rho) + a_1 b_1^+ c_2\right)\left(c_2(b_2^+)^2 + 2b_1^+\delta + 2b_1^+\rho\right)}{2\left((\delta+\rho)^2 - b_1^+ b_2^+ c_1 c_2\right)} - \left(a_2 b_1^+ - c_2\rho x^*\right)\right],$$

where we restored the + superscript. Note that the value functions depend on both τ^* and x^*. Letting one of them be free, we recover a state-driven switch or a time-driven switch, respectively.

8.4.2 *First Interval*

In this subsection, we compute the optimal controls for the first interval in three cases (including the normal mode) and discuss the specific aspects of each case.

8.4.2.1 Normal Mode

We start by considering the conditions under which the normal mode is realized in the system. We assume that an equivalent of Assumption 8.3 holds for the first interval, i.e., the x_{eq}^- value is positive and the equilibrium is of the saddle type $(1, 0, 0)$.

Recall first also that in piece-wise smooth system the so-called Skiba (DNSS) points (thresholds) may exist even if dynamics is linear in each of the intervals (see Skiba [28], Sethi [26], Caulkins et al. [7] for original definition and Bondarev and Greiner [4] for (pseudo) DNSS-points in piece-wise smooth systems).

Definition 8.2 The value x_S^i is called a (pseudo) DNSS-threshold for player i in a differential game given by a piece-wise smooth dynamical system if converging from this threshold to the x_{eq}^- or x_{eq}^+ yields the same value for player i:

$$x_S^i : J_i^+(0, x_S^i) = J_i^-(0, x_S^i) \tag{8.23}$$

with $J_i^\pm(x(0) = x_S^i)$ being value functions of player i with initial condition set at the threshold while converging to $x_{eq}^\pm$.

Let A^- be the system matrix for the dynamical system associated with the first interval[3] analogous to A defined above.

[3] It may easily be derived by substituting for $b_{1,2}^-$ in (8.20).

Then we get the following:

Proposition 8.5 *Let $x(0) < x^*(\tau)$ and Assumption 8.3 holds for the matrix A^-. Then:*

1. *Once $x(0) < \min\{x_S^1, x_S^2\}$ the normal mode realizes with x_{eq}^- being the long-run equilibrium of the game*
2. *Once $x(0) > \max\{x_S^1, x_S^2\}$ the switching mode realizes and the optimal trajectory reaches the switching manifold in finite time*
3. *Once $x(0) \in [x_S^i, x_S^{-i}]$, the outcome is indeterminate*

If Assumption 8.3 does not hold for the matrix A^-, only the switching mode realizes as the outcome of the game.

Proof If Assumption 8.3 holds, x_{eq}^- is positive. We also assumed that $x_{eq}^- < x^*(\tau)$ so it can be reached by the $x(t)$ process without crossing the threshold. Now if $x(0) < \min\{x_S^1, x_S^2\}$ holds, it means that by definition of the Skiba point it is optimal to converge to this value x_{eq}^-. Once it is a saddle type $(1, 0, 0)$, no arcs entering the $x > x(\tau)$ region can be part of the optimal trajectory converging to this (lower) equilibrium. Game is in the normal mode.

On the contrary, once $x(0) > \max\{x_S^1, x_S^2\}$, it is optimal to converge to the equilibrium $x_{eq} > x^*(\tau)$ and crossing occurs. It is a unique one, since the equilibrium is a saddle type $(1, 0, 0)$ (so there are no arcs re-entering the first region).

At last, once $x(0) \in [x_S^i, x_S^{-i}]$ the dynamics is indeterminate, as is standard for this types of models (see discussion in Krugman [21]), provided $x_S^i \neq x_S^{-i}$. □

To avoid further complications, we will assume for the rest of this subsection that $x(0) \notin [x_S^i, x_S^{-i}]$.

This condition implies that there are exactly two branches of the optimal trajectory: once $x(0) < \min\{x_S^1, x_S^2\}$, the optimal trajectory cannot cross the indeterminacy region $[x_S^i, x_S^{-i}]$ and since the switching manifold is located to the right of this region, the switching mode cannot realize as the optimal one. We thus consider only the normal mode, that is a game with the only *one* regular equilibrium which is the lower one. On the other hand, once $x(0) > \max\{x_S^1, x_S^2\}$, the lower equilibrium cannot be reached by the optimal trajectory since it is located to the left of the indeterminacy region and we could consider the switching trajectory only, whereas there is only one feasible equilibrium located to the right of the switching manifold, x_{eq}^+. The fact that the optimal trajectory cannot cross the region $[x_S^i, x_S^{-i}]$ follows from the definition of the Skiba-point: once the game starts to the left from this region, it is not profitable for both players to move to the upper equilibrium and vice versa. So we can observe that the existence of these (pseudo) Skiba-points actually simplifies the analysis by helping us to select the proper mode for the game.

8.4.2.2 State-Driven Switch

Similar to the cooperative case we assume that the switching state is fixed, i.e., $x(\tau) = x^*$, while the switching instant is left free. This means, in particular, that the adjoint variables corresponding to the state x are not fixed at the final time and that the functions J_i^+ should now be considered as functions of τ. In this case, we have to employ a slightly modified version of the Pontryagin's maximum principle as described in Appendix.

To use the condition (8.31), we rewrite the Hamiltonians (8.18) replacing u with respective optimal controls u^{NE}. Thus, we get

$$\begin{aligned} H_1^*(t,x,\psi) &= e^{-\rho t}a_1x + b_2^- c_1\psi_2 x + \psi_1\left(\frac{1}{2}(b_1^-)^2 e^{\rho t}\psi_1 + (b_2^-)^2 e^{\rho t}\psi_2 - \delta x\right)\\ H_2^*(t,x,\psi) &= e^{-\rho t}a_2x + b_1^- c_2\psi_1 x + \psi_2\left(\frac{1}{2}(b_1^-)^2 e^{\rho t}\psi_1 + (b_2^-)^2 e^{\rho t}\psi_2 - \delta x\right). \end{aligned} \tag{8.24}$$

Finally, we write (8.31) while replacing ψ_i with $e^{-\rho t}\lambda_i$. This results in the following set of equations:

$$(b_1^-)^2{\lambda_1}^2 + 2(b_2^-)^2\lambda_1\lambda_2 + 2b_2^- c_1\lambda_2 x^* - 2\delta\lambda_1 x^* + 2a_1x^* = \frac{a_1(\delta+\rho) + a_2b_2^+c_1}{(\delta+\rho)^2 - b_1^+b_2^+c_1c_2}$$
$$\times\left[\frac{\left(a_1(\delta+\rho)+a_2b_2^+c_1\right)\left(c_1(b_1^+)^2 + 2b_2^+\delta + 2b_2^+\rho\right)}{c_1\left((\delta+\rho)^2 - b_1^+b_2^+c_1c_2\right)} - \frac{2\left(a_1b_2^+ - c_1\rho x^*\right)}{c_1}\right] \tag{8.25}$$

$$2(b_1^-)^2\lambda_1\lambda_2 + 2b_1^- c_2\lambda_1 x^* + (b_2^-)^2{\lambda_2}^2 - 2\delta\lambda_2 x^* + 2a_2x^* = \frac{a_2(\delta+\rho) + a_1b_1^+c_2}{(\delta+\rho)^2 - b_1^+b_2^+c_1c_2}$$
$$\times\left[\frac{\left(a_2(\delta+\rho)+a_1b_1^+c_2\right)\left(c_2(b_2^+)^2 + 2b_1^+\delta + 2b_1^+\rho\right)}{c_2\left((\delta+\rho)^2 - b_1^+b_2^+c_1c_2\right)} - \frac{2\left(a_2b_1^+ - c_2\rho x^*\right)}{c_2}\right] \tag{8.26}$$

Solving the system (8.25), (8.26) with respect to λ_i, we obtain a number of candidates for the end-point values of the respective adjoint variables at $t=\tau$: $\lambda_i^{*-} = \lim_{t\to\tau-0}\lambda_i^-(t)$. Note that according to Bézout's theorem [16], there cannot be more than four such candidates. Choosing an appropriate solution to (8.25), (8.26) is a separate problem that requires some extra analysis. One obvious test consists in checking the direction of the state phase vector at τ: it should hold that $\dot{x}(\tau) > 0$. This is equivalent to requiring that $(b_1^-)^2\lambda_1^{*-} + (b_2^-)^2\lambda_2^{*-} > \delta x^*$. We conjecture that there will be at most two feasible solutions out of four. A rigorous proof of this fact is yet to follow.

When λ_i^{*-} are determined, the final step consists in solving the system (8.20) with conditions $x(0) = x_0, x(\tau) = x^*$ and $\lambda_i(\tau) = \lambda_i^{*-}$. This will allow us to determine τ, for instance, by integrating (8.20) backward in time with initial conditions (x^*, λ_i^{*-}) and determining τ from $x(\tau) = x_0$. If there are more than one feasible solution to (8.25)–(8.26), the optimal solution is obtained by comparing the values of the objective functions.

8.4.2.3 Time-Driven Switch

If τ^* is fixed, while the switching state $x(\tau^*)$ is unrestricted one can make use of the standard transversality condition from optimal control theory and compute the end-point values of adjoint variables as

$$\psi_i = \frac{d}{dx} J_i^+(x)\big|_{x=x^*}.$$

Expressed in terms of current values of adjoint variables, this yields $\lambda_i^{*-} = \lambda_{eq}$ (cf. (8.22)). This agrees with the Hybrid Maximum Principle in that the adjoint variables do not undergo a discontinuity at a time-driven, i.e., controllable switch.

Finally, the switching state is computed following the same procedure as in Sect. 8.3.2.3. We note that since the DEs for λ do not depend on x, the problem can be conveniently solved in two runs: first, the DEs for λ are solved backward in time to recover $\lambda_i(0)$; next, the whole system (8.20) is solved forward to yield the required value of the state.

8.5 Extensions

It is of interest what would be the global dynamics of the system (8.3) for the state-driven switch in case Assumption 8.1 does not hold. In particular, since x^* is an arbitrary value, it could be the case that one of $x_{eq}^+ > x^*$ and $x_{eq}^- < x^*$ do not hold or both do not hold. These are cases of *virtual* (again following terminology of Di Bernardo et al. [10]) equilibria. We discuss these two cases separately.

One virtual equilibrium. If only one of the equilibria is virtual the dynamics of both cooperative and non-cooperative cases becomes somewhat simple. In the case of the fixed switching sequence, we observe that:

Proposition 8.6 *Once either $x_{eq}^+ > x^*$ or $x_{eq}^- < x^*$ but not both, the equilibrium which remains regular is reached by the optimal trajectory in finite time.*

Once $x_{eq}^+ > x^*$ but $x_{eq}^- > x^*$ and $x(0) < x^*$, there is a unique switching trajectory with a unique stitching event at some $x^*(\tau)$ which reaches x_{eq}^+.

Once $x_{eq}^+ < x^*$ but $x_{eq}^- < x^*$ and $x(0) > x^*$, there is a unique switching trajectory with a unique stitching event at some $x^*(\tau)$ which reaches x_{eq}^-.

Once $x_{eq}^+ > x^*$ but $x_{eq}^- > x^*$ and $x(0) > x^*$, there is a unique smooth trajectory in the second interval which reaches x_{eq}^+.

Once $x_{eq}^+ < x^*$ but $x_{eq}^- < x^*$ and $x(0) < x^*$, there is a unique smooth trajectory in the second interval which reaches x_{eq}^-.

Proof It has been shown in numerous literature (see e.g. Feichtinger and Wirl [13]) that in (optimally controlled) systems with a threshold there exist optimal solution candidates crossing the threshold in finite time. Once we get only one (thus unique) regular equilibrium, it is reached by the optimal trajectory either by crossing the threshold or not. □

Now the case of one virtual equilibrium is not exhausting the full list of global configurations. The other one[4] is the case of *two* virtual equilibria.

Two virtual equilibria. If both equilibria are virtual, that is, they are infeasible, there are no optimal control candidates leading to any equilibrium. This is the case where we have to apply the alternative solution concept to find a suitable solution. We thus may resort to the sliding mode control (see, e.g., Gamkrelidze [17] and further works). We abstain here from the formal proof of the optimality of such control (leaving this for future extension) and limit ourselves to the following observation:

Proposition 8.7 *For a system* (8.3) *with a state-driven switch and once both* $x_{eq}^+ < x^*$ *and* $x_{eq}^- > x^*$*, the only optimal control candidates are those leading to the sliding mode dynamics such that*

$$\forall t > \tau : \dot{x} = 0, \; x(t) = x^*. \tag{8.27}$$

Proof If both equilibria of regular flows are virtual, there are no candidate trajectories crossing the threshold, Thus, the only sufficiently long trajectory is the one leading to the threshold x^* and this is the only optimal control candidate. □

This type of dynamics will not come up from the maximum principle (as is noted already in Gamkrelidze [17]) and in general is obtained as a (linear) combination of controls, leading the trajectory of the state to the switching manifold. Once the trajectory reaches the switching manifold, it stays there the rest of the game, converging to the *pseudoequilibrium* (see Di Bernardo et al. [10]) which is defined as the equilibrium of the (in our case one-dimensional) sliding flow $\dot{\lambda}_S := conv\{\dot{\lambda}^+, \dot{\lambda}^-\}|x = x^*$ while state remains fixed at the threshold value. There are several methods of defining this flow (see Filippov [14], Utkin [29]) which in general lead to equivalent results.

We also note that even in the case the pseudoequilibrium is repelling; it still can be reached from the outside of the switching manifold by the suitably designed sequence of controls. We stop our discussion of the sliding mode here since it is not easy to prove the optimality of such a candidate and this is an entirely different problem requiring much more complicated analysis left for further research.

The other question is whether the study undertaken here is applicable to a wider variety of problems besides those linear in the state. We claim that the core method is valid for many piecewise systems which allow for derivation of steady states. This, of course, includes a lot of non-linear problems. We do not claim however that *any* piece-wise system may be treated this way, since non-linear systems may exhibit rather rich additional dynamics. Checking the derivations above we observe that the

[4] We neglect the case of boundary equilibria as having zero measure in the space of parameters.

explicit solution for the underlying canonical system is not actually necessary to derive the global dynamics. We still need to know where both equilibria are located (are they regular or virtual), which requires explicit derivation of those equilibria. The method for obtaining boundary conditions at the switching time is quite general and does not rely on the linear-quadratic structure of the problem. This has been taken only for the sake of simplicity of exposition and by no way limits the applicability of our results to a larger class of piece-wise smooth problems.

8.6 Conclusions

In this paper, we considered an example of a multi-modal differential game with two players and one state variable. It is demonstrated that this differential game can possess a rich variety of types of dynamics, including the normal mode (with smooth solution trajectory), the switching mode (with optimal trajectory consisting of two parts) and even the sliding mode, which requires some further analysis.

We applied a modified version of a standard Maximum Principle and obtained full analytic results for the switching trajectories including the conditions on adjoint states at the threshold. We studied both time-driven and state-driven switching conditions. In so doing, we explore the main difficulties arising in these two formulations (leading, in general, to different optimal control problems).

Moreover, we also find out that the game, although linear-quadratic in each of the intervals, is overall non-linear and as such possesses so-called indifference (DNSS) points for each of the players. It is remarkable that this effect arises solely because of the piece-wise structure of the state equation (8.3), leading to the multiplicity of equilibria in the overall game, both for cooperative and non-cooperative solutions. To see this, we observe that every sub-system (lower and upper ones) possesses a unique equilibrium despite the presence of the $c_i x u_{-i}$ term in objective functionals. As such, the smooth system itself cannot exhibit Skiba-points (since the multiplicity of equilibria is a necessary condition for this, see Wagener [30]). However, the *combined* piece-wise smooth system has two equilibria and thus can exhibit Skiba-points even for linear-quadratic systems.

The main insight from our analysis so far is the following: The relevant method for obtaining an optimal solution depends on the global configuration of the respective dynamic system's equilibria. While in the normal mode conventional optimal control methods can be used, in the case of switching additional boundary constraints on adjoint variables have to be taken into account. In the even more special case of sliding, conventional tools are inapplicable at all. Thus, the general algorithm to solve this kind of problems would be as follows: we have to define the configuration of equilibria of the game first and then select an appropriate solution technique. This issue is frequently neglected in economic applications of regime-switching systems, where usually only the switching mode is studied.

Acknowledgements The work of D. Gromov was supported by the Russian Science Foundation (project no. 17-11-01093).

Appendix

Consider the following optimal control problem:

$$\begin{cases} J = \int_0^T f_0(t, x, u)dt + \phi(T) \to \max, \\ \dot{x} = f(x, u), \quad x(0) = x_0, \end{cases} \tag{8.28}$$

where the final time T is assumed to be free. A particular feature of this problem statement is that the terminal payoff function depends on the final time, rather the final state. To accommodate the known results to this case, we reformulate the system and include an auxiliary variable θ that evolves according to $\dot{\theta} = 1$ with initial condition $\theta(0) = 0$. Following [Pontryagin], we write the Hamiltonian function as

$$\mathcal{H}(x, \theta, \psi, \psi_\theta, u) = H(\theta, x, \psi, u) + \psi_\theta,$$

where $H(t, x, \psi, u) = \langle \psi, f(x, u) \rangle + f_0(t, x, u)$ is the "original" Hamiltonian function. Following the standard procedure, we obtain the optimal control u^o by maximizing $H(\theta, x, \psi, u)$, i.e., $H^*(\theta, x, \psi) = \max_u H(\theta, x, \psi, u)$ and write the differential equations for the adjoint variables as

$$\begin{cases} \dot{\psi} = -\frac{\partial H}{\partial x} \\ \dot{\psi}_\theta = -\frac{\partial f_0}{\partial t}. \end{cases} \tag{8.29}$$

While the end-point conditions on the adjoints ψ are determined according to the standard procedure, for ψ_θ, we have $\psi_\theta(T) = \frac{d}{dt}\phi(t)\big|_{t=T}$. Finally, we recall that, for an optimal control problem with free final time, we have

$$\mathcal{H}^*(x, \theta, \psi, \psi_\theta) = H^*(\theta, x, \psi) + \psi_\theta = 0 \tag{8.30}$$

along the optimal trajectory. Evaluating (8.30) at $t = T$ and noting that $\theta(T) = T$, we obtain

$$H^*(T, x(T), \psi(T)) = -\frac{d}{dt}\phi(t)\big|_{t=T}. \tag{8.31}$$

That is to say, in contrast to the case when the terminal payoff expressed in terms of $x(T)$ defines the final values of the adjoint variables, the terminal payoff expressed in terms of the final time imposes an additional restriction on the value of the Hamiltonian function at $t = T$.

Note that an alternative approach would be to use the jump condition from Boltyansky [2]. However, the above-described approach seems to be more appropriate as it is tailored to a particular class of multi-modal optimal control problems as contrasted to the general formulation proposed by Boltyansky.

References

1. Azhmyakov, V., Attia, S., Gromov, D., & Raisch, J. (2007). Necessary optimality conditions for a class of hybrid optimal control problems. In A. Bemporad, A. Bicchi, & G. Buttazzo (Eds.), *Hybrid systems: Computation and control. HSCC 2007* (Vol. 4416, pp. 637–640). Lecture notes in computer science. Springer.
2. Boltyansky, V. G. (2004). The maximum principle for variable structure systems. *International Journal of Control, 77*(17), 1445–1451.
3. Bondarev, A. (2018). *Games without winners: Catching-up with asymmetric spillovers*. WWZ working paper 2018/12, Wirtschaftswissenschaftliches Zentrum (WWZ), Universität Basel, Basel.
4. Bondarev, A., & Greiner, A. (2018). Catching-up and falling behind: Effects of learning in an R&D differential game with spillovers. *Journal of Economic Dynamics and Control, 91*, 134–156.
5. Brito, P. B., Costa, L. F., & Dixon, H. (2013). Non-smooth dynamics and multiple equilibria in a Cournot-Ramsey model with endogenous markups. *Journal of Economic Dynamics and Control, 37*(11), 2287–2306.
6. Brito, P. B., Costa, L. F., & Dixon, H. D. (2017). From sunspots to black holes: Singular dynamics in macroeconomic models. In K. Nishimura, A. Venditti, & N. C. Yannelis (Eds.), *Sunspots and Non-linear dynamics: Essays in honor of Jean-Michel Grandmont* (pp. 41–70). Cham: Springer International Publishing.
7. Caulkins, J. P., Feichtinger, G., Grass, D., Hartl, R. F., Kort, P. M., & Seidl, A. (2015). Skiba points in free end-time problems. *Journal of Economic Dynamics and Control, 51*, 404–419.
8. Dawid, H., Kopel, M., & Kort, P. (2013). R&D competition versus R&D cooperation in oligopolistic markets with evolving structure. *International Journal of Industrial Organization, 31*(5), 527–537.
9. Deal, K. R., Sethi, S. P., & Thompson, G. L. (1979). A bilinear-quadratic game in advertising. In P. T. Liu & J. G. Sutinen (Eds.), *Control theory in mathematical economics* (pp. 91–109). New York: Marcel Dekker.
10. Di Bernardo, M., Budd, C., Champneys, A., Kowalczyk, P., Nordmark, A., Olivar, G., et al. (2008). Bifurcations in nonsmooth dynamical systems. *SIAM Review, 50*(4), 629–701.
11. Dockner, E., Jorgensen, S., Long, N., & Sorger, G. (2000). *Differential games in economics and management sciences*. Cambridge: Cambridge University Press.
12. Dockner, E. J., & Nishimura, K. (2005). Capital accumulation games with a non-concave production function. *Journal of Economic Behavior & Organization, 57*(4), 408–420. Multiple Equilibria, Thresholds and Policy Choices.
13. Feichtinger, G., & Wirl, F. (2000). Instabilities in concave, dynamic, economic optimization. *Journal of Optimization Theory and Applications, 107*(2), 275–286.
14. Filippov, A. F. (1988). *Differential equations with discontinuous righthand sides*. Dordrecht: Kluwer Academic.
15. Fudenberg, D., Gilbert, R., Stiglitz, J., & Tirole, J. (1983). Preemption, leapfrogging and competition in patent races. *European Economic Review, 22*(1), 3–31. Market Competition, Conflict and Collusion.
16. Fulton, W. (1989). *Algebraic curves: An introduction to algebraic geometry*. Addison-Wesley.
17. Gamkrelidze, R. V. (1986). Sliding modes in optimal control theory. *Proceedings of Steklov Institute for Mathematics, 169*, 180–193.

18. Gonzalez, F. (2018). Pollution control with time-varying model mistrust of the stock dynamics. *Computational Economics*, *51*(3), 541–569.
19. Gromov, D., & Gromova, E. (2017). On a class of hybrid differential games. *Dynamic Games and Applications*, *7*(2), 266–288.
20. He, X., Prasad, A., Sethi, S. P., & Gutierrez, G. J. (2007). A survey of Stackelberg differential game models in supply and marketing channels. *Journal of Systems Science and Systems Engineering*, *16*(4), 385–413.
21. Krugman, P. (1991). History versus expectations. *The Quarterly Journal of Economics*, *106*(2), 651–667.
22. Long, N. V., Prieur, F., Tidball, M., & Puzon, K. (2017). Piecewise closed-loop equilibria in differential games with regime switching strategies. *Journal of Economic Dynamics and Control*, *76*, 264–284.
23. Lunze, J., & Lamnabhi-Lagarrigue, F. (Eds.). (2009). *Handbook of hybrid systems control: Theory, tools, applications*. Cambridge University Press.
24. Reddy, P. V., Schumacher, J. M., & Engwerda, J. (2019). *Analysis of optimal control problems for hybrid systems with one state variable*. Technical report.
25. Seidl, A. (2019). Zeno points in optimal control models with endogenous regime switching. *Journal of Economic Dynamics and Control*, *100*, 353–368.
26. Sethi, S. P. (1977). Nearest feasible paths in optimal control problems: Theory, examples, and counterexamples. *Journal of Optimization Theory and Applications*, *23*(4), 563–579.
27. Shaikh, M. S., & Caines, P. E. (2007). On the hybrid optimal control problem: Theory and algorithms. *IEEE Transactions on Automatic Control*, *52*(9), 1587–1603.
28. Skiba, A. K. (1978). Optimal growth with a convex-concave production function. *Econometrica*, *46*(3), 527–539.
29. Utkin, V. I. (1992). *Sliding Modes in Control and Optimization*. Berlin: Springer.
30. Wagener, F. O. O. (2003). Skiba points and heteroclinic bifurcations, with applications to the shallow lake system. *Journal of Economic Dynamics and Control*, *27*(9), 1533–1561.
31. Wirl, F., & Feichtinger, G.(2005). History dependence in concave economies. *Journal of Economic Behavior & Organization*, *57*(4), 390–407. Multiple Equilibria, Thresholds and Policy Choices.

Chapter 9
Optimal Switching from Competition to Cooperation: A Preliminary Exploration

Raouf Boucekkine, Carmen Camacho, and Benteng Zou

JEL Classification: C61 · D71

9.1 Introduction

The recent years have noticed the emergence of numerous debates on the opportunity and timing of very different types of transitions, each associated with a bunch of academic works in the economic and operational research literatures: environmental transitions (in particular, the energy transition, see the early work of Tsur and Zemel [17]), political transitions (among others, transition to democracy following Acemoglu and Robinson [1]), organizational transitions (either in markets or in the workplace, see Vallée and Moreno Galbis [18]), etc. In most of these works, the optimal timing of transitions (if any) are only implicitly tackled though the vast majority of the models developed are dynamic.[1]

There exists however an increasing number of papers interested in optimal regime transition and the inherent timing. As a common feature, all these papers use multistage optimal control techniques, first developed in economics by Tomiyama [16].

[1] This is especially true in the political transitions literature, with the notable exception of Boucekkine et al. [5].

R. Boucekkine (✉)
Aix-Marseille University (Aix-Marseille School of Economics), CNRS, and EHESS, Marseille, France
e-mail: Raouf.Boucekkine@univ-amu.fr

C. Camacho
Paris Sciences et Lettres, Paris School of Economics and CNRS, Paris, France
e-mail: carmen.camacho@psemail.eu

B. Zou
University of Luxembourg, Luxembourg City, Luxembourg
e-mail: benteng.zou@uni.lu

J. L. Haunschmied et al. (eds.), *Dynamic Economic Problems with Regime Switches*, Dynamic Modeling and Econometrics in Economics and Finance 25,
https://doi.org/10.1007/978-3-030-54576-5_9

This notably covers technological regime switching ([2], or [13]) and environmental transitions (see [4], or [12]).[2] Not surprisingly, given the nature of problems tackled in the latter literature, the decisions are taken by a single player, say the central planner. However, in many cases, from one stage to another, the decision makers may differ. For example, from dictatorship to democracy, we have to move from a regime in which initially almost all the decisions are taken by a dictator, to a regime in which at the very least, the decisions are taken in a more collective way (see [1], in a frame without multistage optimal control, and [5], with multistage optimal control). Similar considerations arise when analyzing international climate agreement processes where we typically switch from a regime with country-level decision-making and no cooperation, for example, prior to the 2015 Paris agreement, to a regime of institutional cooperation with joint decision-making. On a more technical ground, the economic literature is rather very poor in papers merging multistage optimal control and dynamic game ingredients. The corresponding operational research literature is less poor (see, for example, Boucekkine et al. [3]).

In this paper, we shall deal with a generic regime switching problem where the decision-making process is not the same from one regime to another. Precisely, we consider a simple model of optimal switching from competition to cooperation. We believe this problem is generic enough to cover a large set of problems. In addition to the dawn of multinational agreements, it can be also applied to political party or company mergers within a country.[3] Needless to say, the reverse can happen, and one can notice situations in which long time cooperation comes to an end and a further competition regime sets in: Canada pulled out of the Kyoto Protocol in 2011, USA withdrew from the Paris Agreement in 2017, and the UK left the EU on the January 31, 2020. In this kind of situation, the decision-making process changes as well.

A quite rich set of questions arises from the examples given above: what are the trade-offs involved in the decision to move from competition to cooperation (and vice versa), and what is the optimal timing for the institutional regime change if any? Since addressing these questions involves dealing with strategic trade-offs in dynamic settings, it implies embedding dynamic game ingredients in multistage optimal control problems. As outlined above, the economic literature is rather thin in this respect. There is a substantial applied game theory literature of endogenous coalition formation (see, for example, Di Bartolomeo et al. [8], who study how coalitions among fiscal and monetary authorities are formed and what their effects are on the stabilization of output and prices). However, it generally pays no attention to the regime switching problem described above.[4]

---

[2] Applications to macroeconomic policy switching, e.g., Zampolli et al., or to workplace organization as in the above-cited paper by Vallée and Moreno Galbis [18], can be also found.

[3] For example, in December 2003, the Progressive-Conservative Party and the Reform/Canadian Alliance parties merged and created a new right-wing political formation, the Conservative Party of Canada. In October 2007, the two most important Italian left-wing parties merged into a single political entity, the Democratic Party.

[4] Also, and even more clearly, our research questions and inherent settings are quite different from the classical literature on cooperative and noncooperative R&D [7, 15], research joint ventures

In this paper, we propose a preliminary exploration of the latter switching problem. To this end, we solve a two-stage optimal control problem. In the first stage, two players "compete" on a common state variable which could be public good or public bad. They engage in a dynamic game in this first stage, and we solve for open-loop strategies. Arguably, under the current setting, the interesting equilibrium is the *path strategies*, i.e., open-loop strategies, given the fact that usually the commitment extends over the entire future time, and negotiation results depend on the initial condition. Thus, the strategy may not be subgame perfect by definition, which gives possibility for changing in some future time. The *decision rule strategies*, i.e., the Markovian strategies, are subgame perfect Nash equilibria, but no commitment at all is possible, thus they do not fit our examples above. As the state equation and the individual payoffs are linear-quadratic, the resulting dynamic game would be a conceptually trivial problem if no perspective of switching to a cooperative regime emerges. We do introduce such a possibility with the associated joint optimization of the sum of the individual payoffs.

The paper is organized as follows. Section 9.2 briefly presents the differential game setting. Section 9.3 provides the open-loop strategies during the noncooperative and cooperative periods, thus the optimal switching conditions can be obtained. In Sect. 9.4, we turn to the numerical analysis that illustrates and deepens the theoretical findings. Section 9.5 provides extra discussion of the other direction of the game, that is, from cooperative to noncooperative, and Sect. 9.6 concludes.

9.2 A Simple Model of Optimal Switching

There are two players: players 1 and 2, who share a common variable, $y \in [0, Y]$, which could be public good or public bad. Each player chooses the level of a variable $x_i \in [0, X] \subset [0, +\infty)$, which provides her with utility. At the same time, their choice increases the level of y, which induces a loss in utility. Let us assume that at time 0, players play a noncooperative dynamic game, choosing their optimal trajectory for x_i. Then, since their individual choices affect equally the common variable, they could decide to switch strategies at a time T and continue playing strategically. Indeed, in this paper players can optimally choose a time T to start playing cooperatively.

Let us provide an economic example that will accompany us throughout the paper. We assume that there exists a unique final good, which requires only a polluting resource as input. With a quantity x_i of pollution, the firm produces an amount $a_i x_i$ of the final good. Consumption provides the player with utility, but at the same time it increases the level of CO_2 emissions, y. Obviously, the level of CO_2 affects both players. In the end, the player can obtain utility directly from the consumption of x_i, but she also suffers from pollution, so she will receive disutility from y.

---

[10], or the empirical work of Cassiman and Veugelers [6]. Essentially, these models are static and cannot show time switching conditions from noncooperative to cooperative games and vice versa.

The objective of player $i = 1, 2$ is to maximize overall welfare, defined as

$$\max_{x_i} W_i = \int_0^{+\infty} e^{-rt}\left[a_i x_i - \frac{x_i^2}{2} - \frac{b_i y^2}{2}\right] dt,$$

which depends on her individual choice, x_i and which is subject to the dynamic constraint:

$$\dot{y}(t) = x_1 + x_2 - \delta y(t), \quad y(0) = y_0 \text{ given.} \tag{9.1}$$

Welfare is the sum of instantaneous utility from time zero. Instantaneous utility at a time $t \geq 0$ is given by $a_i x_i - \frac{x_i^2}{2} - \frac{b_i y^2}{2}$. As mentioned, x_i is the choice of player i. The cost of choosing x_i is $\frac{x_i^2}{2}$, thus the net gain from x_i is $a_i x_i - \frac{x_i^2}{2}$, with $a_i (\geq X)$ positive constant measuring the unit gain.

An increase in the state variable y has a negative effect on utility. Indeed, as in our example where y is the stock of CO_2 emissions, any increase in y induces a damage $\frac{b_i y^2}{2}$, with scaling parameter b_i, a positive constant.

Suppose that at some future date T, the two players decide to play cooperatively. Then, the join objective is

$$\max_{x_1, x_2} W_{II}(T) = \int_T^{+\infty} e^{-rt}\left[a_1 x_1 + a_2 x_2 - \frac{x_1^2 + x_2^2}{2} - \frac{(b_1 + b_2) y^2}{2}\right] dt,$$

subject to the same state equation (9.1).

For simplicity reasons, we only consider here the symmetric case where $a_i = a_j = a$ and $b_i = b_j = b$.

Remark 9.1 The difficulty of considering the asymmetric case is to provide the sharing principal in the second period. Here, in the symmetric case, the two share equally the gain.

Indeed although switching from asymmetric noncooperative game to a cooperative one is not uninteresting at all, it is a very difficult task which is beyond the scope of the current study. It is important to grasp the difficulty of the asymmetric problem. Under the current affine-quadratic setting, if we only study the differential games without switching time, it is certainly possible to obtain explicit equilibria in the asymmetric case. However, with switching, the problem becomes much trickier as the outcomes will essentially depend on the sharing strategy under cooperation when it turns to be optimal, that is at time T. Depending on the specific problem, the sharing strategy may differ. The sharing problem when joining the Kyoto Protocol, the Copenhagen Accord, Paris Climate Agreements,...etc., is different from the one involved in joining the United Nations, the EU, or the NATO, which is in turn different from the counterpart when adhering to the Universal Postal Union, the World Bank, and so on. The sharing strategy itself is the object of a stream of the literature named "conflict between groups". Hence, to tackle the asymmetric problem one needs to impose a specific sharing strategy, leading to very specific results. Instead, we aim here at providing general results even if this limits our scope to symmetric games.

In the following, we compute the optimal switching time T. Denote by $W_I(T)$ welfare of a player playing noncooperatively, from time 0 to the switching time T, when she starts playing cooperatively. Then, the player solves the following problem:

$$\max_x W_I(T) = \int_0^T e^{-rt}\left[ax - \frac{x^2}{2} - \frac{by^2}{2}\right]dt,$$

subject to the law of motion (9.1).

The optimal choice of T is given by the solution of

$$\max_T \left[W_I(T) + e^{-rT}\frac{W_{II}(T)}{2}\right]. \tag{9.2}$$

If $T = 0$, then it is optimal for the two players to play cooperatively immediately, whereas if $T = +\infty$, the noncooperative game should continue forever.

9.3 The Optimal Switching Strategy

Following Boucekkine et al. [2–5], we solve first the second-stage problem. Optimal trajectories for x and y are obtained in case the players play cooperatively from T onwards. Second, taking T as given, we solve the first noncooperative game. We obtain optimal trajectories in both phases that depend on y_0 and T. Finally, the optimal switching time T is obtained substituting the resulting W_I and W_{II} into (9.2), and solving the resulting problem. As a result, the optimal solution will be made of an optimal switching time T, and the optimal trajectories for y and x before and after the switch.

9.3.1 The Cooperative Regime

The joint symmetric optimization problem can be restated as

$$\max_{x_{II}} W_{II}(T) = \int_T^{+\infty} e^{-rt}\left[2ax_{II} - x_{II}^2 - by_{II}^2\right]dt,$$

subject to the state equation

$$\dot{y}(t) = 2x_{II} - \delta y_{II}(t), \quad t \geq T.$$

Define the associated Hamiltonian as

$$\mathcal{H}_{II}(x, y, \lambda_{II}) = 2ax_{II} - x_{II}^2 - by_{II}^2 + \lambda_{II}(2x - \delta y_{II}(t)), \tag{9.3}$$

with λ_{II} as the costate variable. The first-order conditions yield the following set of optimal conditions:

$$\begin{cases} x_{II} = a + \lambda_{II}, \\ \dot{y}_{II}(t) = -\delta y_{II} + 2\lambda_{II} + 2a, \\ \dot{\lambda}_{II}(t) = 2by_{II} + (r + \delta)\lambda_{II}, \end{cases} \tag{9.4}$$

with transversality condition $\lim_{t \to +\infty} e^{-rt} y(t)\lambda_{II}(t) = 0$.

The next proposition provides the analytical solution to (9.4). We do not reproduce here the proof since it is a standard exercise.

Proposition 9.1 *Suppose that the two players play the cooperative game from time T onwards, then the optimal effort is given by (9.4). The corresponding state and costate variables are given by*

$$\begin{cases} y_{II}(t) = 2C_1 e^{\mu t} + \overline{y}, \\ \lambda_{II}(t) = (\mu + \delta)C_1 e^{\mu t} + \overline{\lambda}, \end{cases} \tag{9.5}$$

where $t \geq T$ and $\mu = \frac{r - \sqrt{r^2 + 4[\delta(r+\delta)+4b]}}{2} (< 0)$. Constant C_1 will be determined later by the transversality condition $y(T^-) = y(T^+)$.

Moreover, there exists a unique steady state $(\overline{y}, \overline{\lambda})$ given by

$$\overline{y} = \frac{2a(r + \delta)}{\delta(r + \delta) + 4b}, \quad \overline{\lambda} = -\frac{4ab}{\delta(r + \delta) + 4b}.$$

Associated with this steady state, we can compute the steady state of x_{II},

$$\bar{x} = a + \bar{\lambda} = a\left(1 - \frac{4b}{\delta(r + \delta) + 4b}\right).$$

Corollary 9.1 *Both $\bar{y}$ and $\bar{x}$ increase with a, they decrease with b.*

Obviously if $T = 0$, that is, if the cooperation starts from the beginning, then $y(T) = y(0) = y_0$ and $C_1 = \frac{y_0 - \overline{y}}{2}$. It is easy to check that the optimal joint welfare is in this case

$$W_{II}^*(0) = \frac{2a(a + \overline{\lambda}) - (a + \overline{\lambda})^2 - b\overline{y}}{r} + \frac{(\mu + \delta)^2 C_1^2}{2\mu - r} + \frac{2C_1(b + (\mu + \delta)\overline{\lambda})}{\mu - r}. \tag{9.6}$$

9.3.2 The Noncooperative Regime

Let us solve the first period noncooperative game for a switching time T given. The first period optimization problem for player i is

$$\max_{x_i} W_I(T) = \int_0^T e^{-rt}\left[a_i x_i - \frac{x_i^2}{2} - \frac{b_i y^2}{2}\right] dt,$$

subject to

$$\dot{y}(t) = x_i + x_j - \delta y(t),$$

for $i, j = 1, 2, i \neq j$, and where $y(0) = y_0$ is given. The associated Hamiltonian is

$$\mathcal{H}_I(x_i, y, \lambda_I) = a_i x_i - \frac{x_i^2}{2} - \frac{b_i y^2}{2} + \lambda_I(x_i + x_j - \delta y(t)), \tag{9.7}$$

with λ_I as costate variable in the first period. In the symmetric setting, the first-order conditions yield the following set of optimal conditions:

$$\begin{cases} x_I = a + \lambda_I, \\ \dot{y}_I(t) = -\delta y_I + 2\lambda_I + 2a, \\ \dot{\lambda}_I(t) = by + (r + \delta)\lambda_I. \end{cases} \tag{9.8}$$

Proposition 9.2 below provides the analytical solution to the first-order conditions in (9.8). Its proof can be found in the appendix.

Proposition 9.2 *Suppose that at time T, the two players start playing the cooperative game. Then, for $t \in [0, T]$, the unique symmetric strategic Nash equilibrium is $x_I(t) = a + \lambda_I$, with state, y_I, and costate variables, λ_I, given by*

$$\begin{cases} y_I(t) = \hat{a}\, e^{\nu_1 t} + \hat{b}\, e^{\nu_2 t} + y_P(t), \\ \lambda_I(t) = \hat{c}\, e^{\nu_1 t} + \hat{d}\, e^{\nu_2 t} + \lambda_P(t), \end{cases} \tag{9.9}$$

in which ν_1 and ν_2 are eigenvalues given by

$$\nu_1 = \frac{r - \sqrt{r^2 + 4[\delta(r + \delta) + 2b]}}{2}, \quad \nu_2 = \frac{r + \sqrt{r^2 + 4[\delta(r + \delta) + 2b]}}{2}.$$

$(y_P(t), \lambda_P(t))$ is a special solution:

$$y_P(t) = \hat{y}_P\,(1 - e^{\nu_1 t}) + \overline{y}_P\,(1 - e^{\nu_2 t}),$$
$$\lambda_P(t) = \hat{\lambda}_P\,(1 - e^{\nu_1 t}) + \overline{\lambda}_P\,(1 - e^{\nu_2 t}),$$

with

$$\hat{y}_P = \frac{2a(\nu_2+\delta)}{\nu_1(\nu_1-\nu_2)}, \quad \overline{y}_P = -\frac{2a(\nu_1+\delta)}{\nu_2(\nu_1-\nu_2)},$$
$$\hat{\lambda}_P = \frac{a(\nu_1+\delta)(\nu_2+\delta)}{\nu_1(\nu_1-\nu_2)}, \quad \overline{\lambda}_P = -\frac{a(\nu_1+\delta)(\nu_2+\delta)}{\nu_2(\nu_1-\nu_2)}.$$

Constants $\widehat{a}, \widehat{b}, \widehat{c}, \widehat{d}$ *are given by*

$$\widehat{a} = \frac{-(\nu_2+\delta)y_0 + 2\lambda_0}{\nu_1-\nu_2}, \quad \widehat{b} = \frac{(\nu_1+\delta)y_0 - 2\lambda_0}{\nu_1-\nu_2}$$

and

$$\widehat{c} = \frac{(\nu_1+\delta)[-(\nu_2+\delta)y_0 + 2\lambda_0]}{2(\nu_1-\nu_2)}, \quad \widehat{d} = \frac{(\nu_2+\delta)[(\nu_1+\delta)y_0 - 2\lambda_0]}{2(\nu_1-\nu_2)},$$

where y_0 *is given initial condition, while* λ_0 *will be determined by the switching condition at time* T.

A special case arises when $T = +\infty$, that is, when no cooperation is possible. Here, the solution of (9.8) with transversality condition $\lim_{t\to+\infty} e^{-rt} y(t)\lambda_I(t) = 0$ is

$$\begin{cases} y_I(t) = 2C_{1I}\, e^{\nu_1 t} + \overline{y}_I, \\ \lambda_I(t) = C_{1I}\, e^{\nu_1 t} + \overline{\lambda}_I, \end{cases} \tag{9.10}$$

where $C_{1I} = \frac{y_0 - \overline{y}_I}{2}$ and the steady state $(\overline{y}_I, \overline{\lambda}_I) = \left(\frac{2a(r+\delta)}{\delta(r+\delta)+2b}, -\frac{2ab}{\delta(r+\delta)+2b}\right)$. In this special case, each player's social welfare is

$$W_I^*(+\infty) = \frac{a(a+\overline{\lambda}_I) - (a+\overline{\lambda}_I)^2/2 - b\overline{y}_I/2}{r} + \frac{(\nu_1+\delta)^2 C_{1I}^2}{2(2\nu_1 - r)} + \frac{C_{1I}(b + (\nu_1+\delta)\overline{\lambda}_I)}{\nu_1 - r}. \tag{9.11}$$

From (9.6) and (9.11), it is easy to check that

$$\frac{W_{II}^*}{2}(+\infty) - W_I^*(+\infty) \lesseqgtr 0,$$

depending on the combination of parameters. In other words, it is possible that the switching happens at $T = 0$, or $T = +\infty$, or $T \in (0, +\infty)$, depending on the situation under study, which can also be seen in Sect. 9.4 with numerical illustration.

9.3.3 Optimal Switching Time

At the switching time, T, and by continuity, the state variable y must verify that

$$y_I(T) = y_{II}(T). \tag{9.12}$$

Similarly, the costate variable must verify that

$$\lambda_I(T) = \lambda_{II}(T), \tag{9.13}$$

which is the standard matching condition for continuity identified by Tomiyama [16] and used in Boucekkine et al. [2, 4].

Usually in the optimal switching problems with a unique decision maker, as in Boucekkine et al. [2, 4, 5], a transversality condition is imposed on the maximized Hamiltonian and shadow values at the switching time T. In particular, within a more general setting with different types of switching, Boucekkine et al. [4] demonstrate that the present value of the Hamiltonian must be also continuous at the switching time. In other words, in the one decision maker's optimal control problem, the maximized Hamiltonian is exactly the same immediately before and immediately after the switch. However, under the current dynamic game setting, it is different. Before and after the switch, there are different decision makers as stated in Introduction. Before the switch, each individual player takes her own optimal decisions, while after the switch, it is a joint choice. Thus, it is improper to equalize the maximized Hamiltonian before and after the switch, even with identical players. Therefore, the shadow value continuity condition is the only natural and valid transversality condition in our setting. In our example where y measures the stock of CO_2, this transversality condition implies that the shadow value of pollution does not change immediately due to the signature of some agreement or protocol. Instead it takes some time for the shadow value to change, while immediate changes come from the choice variables.

Combining (9.12) and (9.13), we obtain that the values of $\lambda_0(T)$ and $C_1(T)$:

$$\begin{cases} \lambda_0(T) = \dfrac{2E_1(T) - (\mu+\delta)A_1(T)}{(\mu+\delta)A_2(T) - 2E_2(T)}, \\ C_1(T) = \dfrac{E_1(T) + E_2(T)\lambda_0}{\mu+\delta} e^{-\mu T}, \end{cases} \tag{9.14}$$

where

$$A_1(t) = \frac{(\nu_1+\delta)e^{\nu_2 t} - (\nu_2+\delta)e^{\nu_1 t}}{\nu_1 - \nu_2} y_0 + (y_P(t) - \overline{y}), \quad A_2(t) = \frac{2(e^{\nu_1 t} - e^{\nu_2 t})}{\nu_1 - \nu_2},$$

and

$$E_1(t) = \frac{(\nu_1+\delta)(\nu_2+\delta)(e^{\nu_2 t} - e^{\nu_1 t})}{2(\nu_1 - \nu_2)} y_0 + (\lambda_P(t) - \overline{\lambda}), \quad E_2(t) = \frac{(\nu_1+\delta)e^{\nu_1 t} - (\nu_2+\delta)e^{\nu_2 t}}{\nu_1 - \nu_2}.$$

Thus, substituting (9.14) into (9.9) and (9.5), we could obtain the explicit optimal trajectories in both periods.

The optimal switching time T is given by the optimization problem (9.2). With the above explicit solutions, it only remains to calculate the social welfare in the two periods, i.e., $W_I(T)$ and $W_{II}(T)$. It is easy to check that

$$\begin{aligned}\frac{e^{-rT}}{2}W_{II} = &\left(\frac{a^2-\bar{\lambda}^2-b\bar{y}^2}{2}\right)\frac{e^{-2rT}}{r} - \left[(\mu+\delta)\bar{\lambda}+2b\bar{y}\right]C_1\,\frac{e^{(\mu-2r)T}}{r-\mu}\\ &-\left[\frac{(\mu+\delta)^2+4b}{2}\right]C_1^2\,\frac{e^{(2\mu-2r)T}}{r-2\mu},\end{aligned} \tag{9.15}$$

in which $C_1 = C_1(T)$ is given by (9.14).

Similarly, the first period social welfare W_I is given by

$$\begin{aligned}W_I = &\left[\frac{a^2-(\widehat{\lambda}_P+\bar{\lambda}_P)^2-b(\widehat{y}_P+\bar{y}_P)^2}{2}\right]\frac{(e^{-rT}-1)}{-r}\\ &-\left[\frac{(\widehat{c}-\bar{\lambda}_P)^2}{2}+\frac{b(\widehat{a}-\bar{y}_P)^2}{2}\right]\frac{(e^{(2\nu_1-r)T}-1)}{2\nu_1-r}\\ &-\left[\frac{(\widehat{d}-\bar{\lambda}_P)^2}{2}+\frac{b\,(\widehat{b}-\bar{y}_P)^2}{2}\right]\frac{(e^{(2\nu_2-r)T}-1)}{2\nu_2-r}\\ &-\left[(\widehat{c}-\bar{\lambda}_P)(\widehat{d}-\bar{\lambda}_P)+b\,(\widehat{a}-\bar{y}_P)(\widehat{b}-\bar{y}_P)\right]\frac{(e^{(\nu_1+\nu_2-r)T}-1)}{\nu_1+\nu_2-r}\\ &-\left[(\widehat{c}-\bar{\lambda}_P)(\widehat{\lambda}_P+\bar{\lambda}_P)+b\,(\widehat{a}-\bar{y}_P)(\widehat{y}_P+\bar{y}_P)\right]\frac{(e^{(\nu_1-r)T}-1)}{\nu_1-r}\\ &-\left[(\widehat{d}-\bar{\lambda}_P)(\widehat{\lambda}_P+\bar{\lambda}_P)+b\,(\widehat{b}-\bar{y}_P)(\widehat{y}_P+\bar{y}_P)\right]\frac{(e^{(\nu_2-r)T}-1)}{\nu_2-r},\end{aligned} \tag{9.16}$$

in which $\widehat{a}, \widehat{b}, \widehat{c}, \widehat{d}$ depend on the switching time T as well, via $\lambda_0(T)$.

The first-order condition from (9.2) yields

$$\frac{dW_I(T)}{dT}+\frac{e^{-rT}}{2}\left[-rW_{II}(T)+\frac{dW_{II}(T)}{dT}\right]=0. \tag{9.17}$$

Intuitively, the first term should be nonnegative, i.e., the longer the time period $[0, T]$, the higher the welfare it yields (otherwise $T = 0$ already). The second term should be nonpositive (otherwise $T = +\infty$). Nevertheless, due to the complexity of expressions (9.15) and (9.16), the study of an explicit form becomes cumbersome, and the study of the properties and impacts becomes impossible. Thus, it is not wise to continue working on the search of explicit solutions. Instead, we focus in the next section on numerical simulations to illustrate the impacts of important parameters, such as the efficient parameter, a, and disutility parameter, b, as well as the initial condition, y_0. Although the model is not realistically calibrated, the qualitative pattern is illustrative.

9.4 Numerical Illustration

This section illustrates numerically the theoretical results obtained in the previous sections. First, we compare the optimal switching time as a function of the initial endowment of the state variable, y_0. In a second set of exercises, we compute the optimal dynamic trajectories of the state variable y and of the optimal choice x.

9.4.1 *Optimal Switching Time*

To seize the importance of all the elements of the model, we have computed the optimal switching time under various scenarios for both a and b. In all exercises, $r = 0.0015$ and $\delta = 0.01$. The time discount is in line with the most recent literature (see the literature following Stern [14]).

We begin by exploring the role of a. Setting $b = 0.000095$, we choose an economy which loses little utility with y. Two economies are compared: a performant economy, which can extract high utility from the same amount of the choice variable x, for which $a = 0.115$; and a second less-efficient economy, in which $a = 0.109$. Figure 9.1 shows the optimal switching time and the associated overall welfare when the initial endowment of the state variable ranges from $y_0 = 50$ to $y_0 = 100$.[5]

Interpreting y as the CO_2 stock and x as consumption, a is a measure of the efficiency of the economy. Indeed, with the same amount of the resource, the higher the a, the more of the final consumption good we obtain. Low levels of y_0 correspond to economies that are not too polluted initially. By the same token, high y_0 correspond to economies that are already very polluted at the beginning of the game.

On the left panel of Fig. 9.1, we see that players always decide to play cooperatively, that is $0 \leq T < \infty$. They cooperate from $T = 0$ when y_0 is relatively low. When the environmental quality is high, players join as soon as possible to preserve the environment and avoid the loss of utility induced by pollution. Besides, switching times are increasing functions of the initial stock of y_0. If y is the stock of CO_2 emissions, then clean economies start cooperating earlier. Indeed, earlier cooperation reduces global emissions and ensures a higher production in the future. Conversely, when economies are relatively dirtier, y_0 is relatively high, then they start cooperating later as if the damage was already too large to struggle against. Finally, note that T is always smaller for $a = 0.115$. Efficient economies cooperate sooner since their advantage in production allows them to obtain more of the final good producing the same amount of pollution. Note that better performing economies can cope better with possible losses in consumption arising upon cooperation.

Let us study next the negative effect of the state variable in utility, as captured by parameter b. The right panel in Fig. 9.1 shows optimal switching times when $a = 0.09$ for two values of b: a high level of $b = 0.00015$ and a low level $b = 0.000095$.

[5] The switching time trajectories displayed in Figs. 9.1 and 9.2 are maxima of (9.2), since they verify the second-order condition as shown in Appendix 9.2.

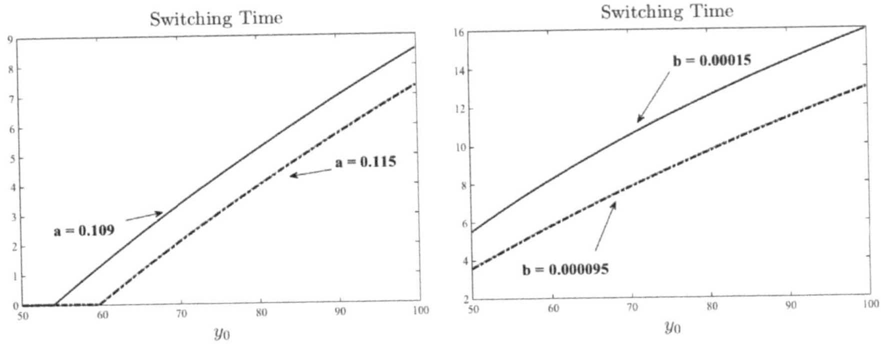

Fig. 9.1 Optimal T. Left: role of a when $b = 0.000095$. Right: role of b when $a = 0.09$

Again, understanding y as CO_2 emissions, an economy with a low b corresponds to an economy less sensitive to pollution, such as a very small open economy or an economy embodied with sufficient forest or water source to absorb important amounts of CO_2. These economies start playing cooperatively earlier in order to maintain y at a moderate level.

9.4.2 Optimal Dynamic Trajectories

Next, let us illustrate the dynamic trajectories for the economy underlining the roles of a and b in the dynamics of the optimal control and state variables. In the first exercise, $a = 0.5$ and we compare two economies which differ in their sensitivity to pollution: in the less-sensitive $b = 0.000095$ and in the more-sensitive $b = 0.00015$. In both cases the economy is initially endowed with $y_0 = 75$. Figure 9.2 shows our results for the state variable, y, and for the control variable per capita, x.

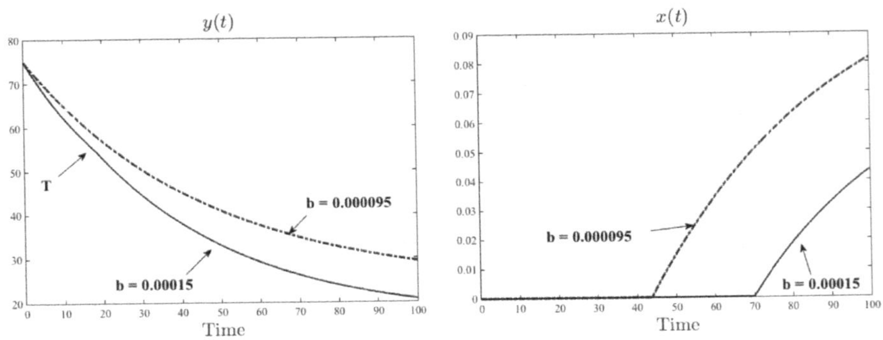

Fig. 9.2 Dynamics of $y(\cdot)$ and $x(\cdot)$. $a = 0.5$

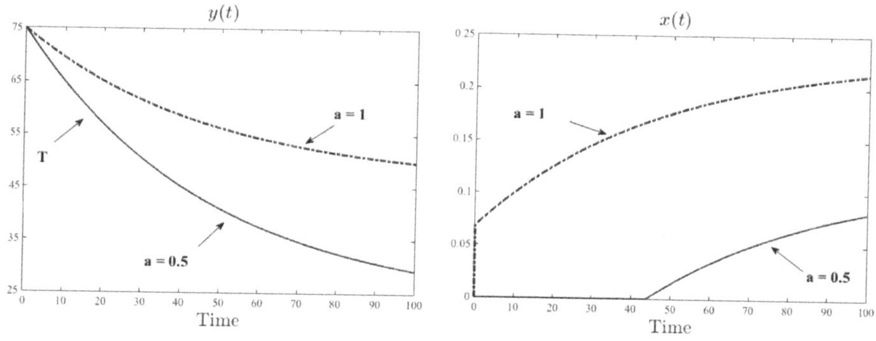

Fig. 9.3 Dynamics of $y(\cdot)$ and $x(\cdot)$. $b = 0.000095$

The results are in line with the precedent theoretical analysis. Indeed, we knew that the steady state of y, y_{ss}, decreases with b. Hence the long-term of y for $b = 0.000095$ is higher than for $b = 0.00015$. The less-sensitive economy can maintain a higher production level and afford higher consumption because it loses less welfare from CO_2 emissions. When b is low, players start playing cooperatively from the beginning and the economy accumulates more pollution.

The right panel of Fig. 9.2 reveals the thinking of the policy maker. Compared to their steady states, both economies start with a highly polluted environment, which makes them lose a significant amount of welfare. Hence, the policy maker decides not to consume during a certain period of time. This is true when $b = 0.000095$ and those players play together from $t = 0$, as well as in the more-sensitive case, when $b = 0.00015$, and that there are two policy makers from $t = 0$ till $T = 18.4$. Note that the less-sensitive economy starts consuming before and it consumes more (Fig. 9.3).

Our last exercise compares two economies which differ in their efficiency level letting $b = 0.000095$. In the first, $a = 0.5$, and in the second, the efficiency level doubles the first, $a = 1$. As shown in Proposition 9.1, the steady state of y keeps the same proportion than efficiency. Indeed, $y_{ss} = 23.23$ when $a = 0.5$ and $y_{ss} = 46.46$ when $a = 1$. Our results are displayed in Fig. 9.4. The most salient feature is that the most efficient economy does not sacrifice initial consumption in order to preserve y at a low level. Its technological advantage allows the policy maker to choose positive levels of consumption from $t = 0$.

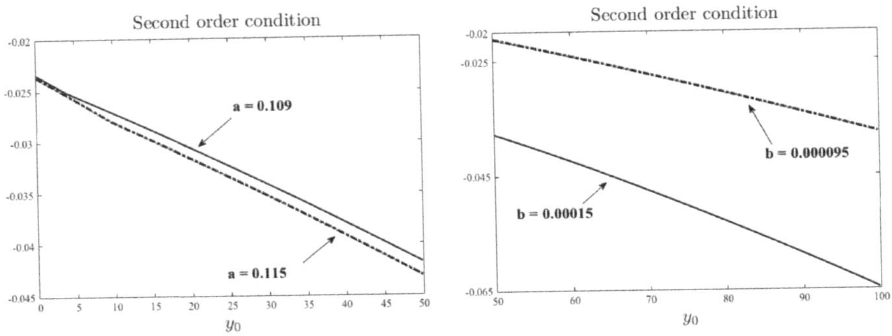

Fig. 9.4 Second-order conditions

9.5 From the Cooperative to the Noncooperative Game

In the previous sections, if players decide to join at time T, they will play cooperatively forever after. In other words, their commitment to some protocol, or union, or agreement is irreversible. However, if it were possible to exit in some future time, choices and trajectories would be more complicated. In this section, we extend the previous study to include the possibility of a future exit. The steps of the game are the following:

Step 1. The two players play the noncooperative differential game until time T_1. Under the symmetric assumption, the revenue of the identical players is $V_I(T_1, T_2)$.

Step 2. From time T_1 until T_2, the two players play cooperatively. Denote the join revenue as $V_{II}(T_1, T_2)$.

Step 3. One of the two players exits the cooperative game at T_2, thus from T_2 onwards, the two players play the noncooperative differential games again. Denote the revenue in this period as $V_{III}(T_1, T_2)$.

Then the optimal choice of switching times should be given by

$$\max_{T_1, T_2} \left[V_I(T_1, T_2) + e^{-rT_1} V_{II}(T_1, T_2) + e^{-rT_2} V_{III}(T_1, T_2) \right]. \tag{9.18}$$

Mathematically, this more complicated differential game can be solved by backward induction by reversing the calculation order of Sect. 9.3. Similarly, numerical investigation can be performed the same way as Sect. 9.4. Therefore, we do not need to do this exercise here in detail.

9.6 Conclusion

We investigate both theoretically and numerically some dynamic game settings where players choose to switch from noncooperative competition to cooperate (or vice versa) and which automatically change the decision makers as well. The explicit conditions of open-loop strategic switching conditions are presented, though it cannot be fully analytical. We rely on numerical simulation and demonstrate the importance of parameters. Not surprisingly, given the intrinsic nature of open-loop strategies, both switching time and the social welfare depend essentially on the initial condition of the game, but neither were monotonic. The efficient parameter plays a very important role in not only the decision of switching from noncooperative to cooperative, but also the choice of consumption. High efficient parameter may cover the side effects, such as leading to high pollution as a by-product. Of course, different damage parameters may be assigned to the noncooperative versus cooperative games. Furthermore, it is easy to imagine the situation where the two players are asymmetric or there are more than two players. Last but not least, the present work is silent as to Markovian strategies, which, of course, are sometimes a more adequate choice. This said, we do believe that the current study paves the way to handle a much wider class of problems, beyond the examples we presented in Introduction.

Appendix 9.1: Proof of Proposition 9.2

Recall the dynamic system as

$$\begin{cases} \dot{y}_I(t) = -\delta y_I + 2\lambda_I + 2a, \\ \dot{\lambda}_I(t) = by + (r+\delta)\lambda_I. \end{cases}$$

It is easy to obtain the associated eigenvalues

$$\nu_{1,2} = \frac{r \pm \sqrt{r^2 + 4(\delta(r+\delta)+2b)}}{2}$$

with $\nu_1 < 0$ and $\nu_2 > 0$, and associated eigenvectors

$$\overrightarrow{v_i} = \begin{pmatrix} 2 \\ \nu_i + \delta \end{pmatrix}, \quad i = 1, 2.$$

Define matrix

$$M(t) = \begin{pmatrix} \overrightarrow{v_1} & \overrightarrow{v_2} \end{pmatrix} \begin{pmatrix} e^{\nu_1 t} & 0 \\ 0 & e^{\nu_2 t} \end{pmatrix} = \begin{pmatrix} 2e^{\nu_1 t} & 2e^{\nu_2 t} \\ (\nu_1+\delta)e^{\nu_1 t} & (\nu_2+\delta)e^{\nu_2 t} \end{pmatrix}.$$

Thus, the inverse matrix of $M(t)$ is

$$M^{-1}(t) = \begin{pmatrix} -\frac{\nu_2+\delta}{2(\nu_1-2)}e^{-\nu_1 t} & \frac{e^{-\nu_1 t}}{\nu_1-\nu_2} \\ \frac{\nu_1+\delta}{2(\nu_1-\nu_2)}e^{-\nu_2 t} & -\frac{e^{-\nu_2 t}}{\nu_1-\nu_2} \end{pmatrix}.$$

The unique solution of the system is then given by

$$\begin{pmatrix} y(t) \\ \lambda(t) \end{pmatrix} = M(t)M^{-1}(0)\begin{pmatrix} y_0 \\ \lambda(0) \end{pmatrix} + M(t)\int_0^t M^{-1}(s)\begin{pmatrix} 2a \\ 0 \end{pmatrix} ds,$$

in which $\lambda(0)$ is undetermined. Substituting $M(t)$, $M^{-1}(t)$, and $M^{-1}(0)$ into the above matrix algebra and taking integrals, we obtain the explicit solution in Proposition 9.2.

That completes the proof.

Appendix 9.2: Second-Order Condition

Figure 9.4 shows the value of the second-order derivative of (9.2) associated with the examples in Sect. 9.4. The graphs show that the second derivative is negative for all values of y_0 in the four examples. Hence, the switching times T in Fig. 9.1 is always a maximum.

References

1. Acemoglu, D., & Robinson, J. (2006). *Economic origins of dictatorship and democracy*. Cambridge University Press.
2. Boucekkine, R., Saglam, C., & Vallée, T. (2004). Technology adoption under embodiment: A two-stage optimal control approach. *Macroeconomic Dynamics*, *8*(2), 250–271.
3. Boucekkine, R., Krawczyk, J., & Vallée, T. (2011). Environmental quality versus economic performance: A dynamic game approach. *Optimal Control Applications and Methods*, *32*, 29–46.
4. Boucekkine, R., Pommeret, A., & Prieur, F. (2013). Optimal regime switching and threshold effects. *Journal of Economic Dynamics and Control*, *37*, 2979–2997.
5. Boucekkine, R., Prieur, F., & Puzon, K. (2016). On the timing of political regime changes in resource-dependent economies. *European Economic Review*, *85*, 188–207.
6. Cassiman, B., & Veugelers, R. (2002). R&D cooperation and cpillovers: Some empirical evidence from Belgium. *The American Economic Review*, *92*, 1169–1184.
7. D'Aspremont, C., & Jacquemin, A. (1988). Cooperative and noncooperative R & D in duopoly with spillovers. *The American Economic Review*, *78*, 1133–1137.
8. Di Bartolomeo, G., Engwerda, J., Plasmans, J., & van Aarle, B. (2006). Staying together or breaking apart: Policy-makers' endogenous coalitions formation in the European Economic and Monetary Union. *Computers & Operations Research*, *33*, 438–463.
9. Dockner, E., Jorgensen, S., Van Long, N., & Sorger, G. (2000). *Differential games in economics and management*. Cambridge University Press.
10. Kamien, M., Muller, E., & Zang, I. (1992). Research joint ventures and R&D cartels. *The American Economic Review*, *82*, 1293–1306.
11. Mair, P. (1990). The electoral payoffs of fission and fusion. *British Journal of Political Science*, *20*, 131–142.
12. Moser, E., Seidel, A., & Feichtinger, G. (2014). History-dependence in production-pollution-trade-off models: A multi-stage approach. *Annals of Operations Research*, *222*, 455–481.
13. Saglam, C. (2011). Optimal pattern of technology adoptions under embodiment: A multi stage optimal control approach. *Optimal Control Applications and Methods*, *32*, 574–586.

14. Stern, N. (2006). *The economics of climate change: The Stern review*. Cambridge University Press.
15. Suzumura, K. (1992). Cooperative and noncooperative R&D in an oligopoly with spillover. *The American Economic Review*, *82*, 1307–1320.
16. Tomiyama, K. (1985). Two-stage optimal control problems and optimality conditions. *Journal of Economic Dynamics and Control*, *9*, 317–337.
17. Tsur, Y., & Zemel, A. (2003). Optimal transition to backstop substitutes for nonre-newable resources. *Journal of Economic Dynamics and Control*, *27*, 551–572.
18. Vallée, T., & Moreno Galbis, E. (2011). Optimal time switching from tayloristic to holistic workplace organization. *Structural Change and Economic Dynamics*, *22*, 238–246.
19. Zampolli, F. (2006). Optimal monetary policy in a regime switching economy: The response to abrupt shifts in exchange rate dynamics. *Journal of Economic Dynamics and Control*, *30*(1527), 1567.

Chapter 10
Delaying Product Introduction in a Duopoly: A Strategic Dynamic Analysis

Herbert Dawid and Serhat Gezer

JEL Classification: D43 · L13 · O31

10.1 Introduction

Technological change is a crucial driver of industrial dynamics. Improved versions of products appear regularly. Furthermore, product innovations lead to differentiated products and new submarkets arise. According to an empirical investigation by Chandy and Tellis [5], most of the product innovations has been achieved by established incumbents. Typical examples include Asus which has been active on the notebook market and has introduced netbooks in 2007 or Apple's introduction of the iPad in 2010 which generated a huge submarket for tablet computers. For a firm competing with others on a homogeneous market, a product innovation can be very valuable. Given that a product innovation has been made, the innovator has to decide whether to introduce a new product immediately, to delay the product introduction strategically, or not to introduce at all.[1] Wang and Hui [34] provide examples where the market introduction of products has been delayed, e.g. DVD players and MP3-related products which could have been introduced earlier.

[1]Several studies [1, 2, 28] have found out that a large fraction of product innovations is not brought to the market.

H. Dawid (✉) · S. Gezer
Department of Business Administration and Economics, Institute of Mathematical Economics, Bielefeld University, Bielefeld, Germany
e-mail: hdawid@wiwi.uni-bielefeld.de

S. Gezer
e-mail: sgezer@uni-bielefeld.de

J. L. Haunschmied et al. (eds.), *Dynamic Economic Problems with Regime Switches*, Dynamic Modeling and Econometrics in Economics and Finance 25,
https://doi.org/10.1007/978-3-030-54576-5_10

To analyze the question how an incumbent should optimally choose its market introduction strategy we consider two firms competing on an established homogeneous market, and assume that one of the firms has the option to introduce a new product, whereas the rival has to stick with producing the established product. Moreover, we assume that the new product is horizontally and vertically differentiated, in particular that it has a higher quality than the established product. Both firms are restricted by production capacities which they adjust over time. The setting after the introduction of the new product has been analyzed in [10]. They find that not only the innovator benefits but the non-innovator is better off as well in most cases, in particular if the products are not too differentiated. The innovator strongly reduces capacities on the established market in order to increase demand for the established product.

Adjustments of capacities of established products *prior* to a product innovation has been studied in a stochastic setting in [8] who consider a duopoly where both firms can also invest in R&D in order to increase the probability of product innovation (see [11] for an exogenous hazard rate). In contrast to those approaches, we assume that the innovation has been made already and the time of product introduction is an additional choice variable and hence is not directly linked to the time of the successful completion of an R&D project. The separation of innovation and introduction has been employed by Dawid et al. [9], however only in a three-stage model where continuous capacity adjustments are not taken into account and the timing of product introduction could not be addressed.

The game we are considering is a multi-mode differential game where one of the firms can induce a regime switch (in our context adding a second differentiated product to its product range) at any time. This is in contrast to models where a regime switch occurs when the state variable hits some critical threshold (see e.g. [29, 31]) or is governed by a stochastic process as in [8, 11].

In [16] a related setting to that in this paper has been analyzed, however abstracting from competition. An incumbent monopolist has the option to introduce an new (substitute) product in addition to the one already offered. It is shown that the firm might delay product introduction if it incurs adoption costs. By delaying the product introduction, the monopolist benefits from discounted adoption costs, which has to be paid as a lump sum at the time of product introduction. Furthermore, the monopolist can increase the marginal value of the new product by decreasing established capacities. Similar effects are also present in the duopoly considered in this paper, however strategic interaction adds substantial new effects.

Optimal timing of innovation has been analyzed extensively in the optimal stopping and real options literature (see e.g. [12, 14, 22]). Recent contributions consider for stochastic demand, both, optimal timing and capacity choice simultaneously (see e.g. [23, 24]). The latter find in a setting with two firms who have the option to enter a new market that firms invest earlier compared to the monopoly setting. In particular, the first investor overinvests in order to delay market entry of the second investor. The innovation of the present paper relative to this literature is that it considers the dynamic adjustment of capacities before and after the innovation, whereas mostly one-time investments have been treated in the real options literature.

Optimal timing has been considered only in a few differential game models. Yeung [35] derives feedback Nash equilibria for games with endogenous time horizon by restricting terminal values for state variables. Recently, Gromov and Gromova [18] formalize the class of hybrid differential games and characterize a switching manifold in the time-state space which is determined by a switching condition. They argue that deriving feedback Nash equilibria for state-dependent switching is complicated and resort to open-loop Nash equilibria, which in certain games, parametrized by initial conditions yields feedback Nash equilibria.

In terms of timing, the most related contribution is Long et al. [27] where in a differential game model with multiple regimes, the concept of piecewise-closed loop Nash equilibria (PCNE) is introduced. They derive necessary conditions for the optimal switching time in a two player setting, where both players can induce a change of the regime of the game. The timing decision is given implicitly by the state variable arriving at a certain state which is derived by optimality conditions. However, in their setting, it is assumed that firms commit to their switching time in the sense, that they would not alter that time even if the other firm would deviate from its equilibrium control path. Hence, the considered equilibrium is not fully Markov perfect with respect to the timing decision.

In our approach, we consider a case where the innovator can fully commit to its product introduction time. Hence, the competitor cannot influence the timing of the product introduction. An equilibrium is given if the choice of the product introduction time, T, maximizes the value of the game for the innovator while given this T, the investment strategies played by both players constitute a Markov-perfect Nash equilibrium in the classical sense. Note that the timing decision is made in the beginning of the game for given initial capacities and hence it is an open-loop strategy whereas the continuous control variables constitute a Markov perfect equilibrium using closed-loop strategies. Characterizing a fully closed-loop equilibrium in which the introduction of the new product is triggered if the state variable hits a switching manifold (to be optimally determined by the innovator) is technically challenging and might lead to non-existence of equilibria (see [27] for details).

From an economic perspective, the commitment to the product introduction time might be due to a preannouncement. There is a huge literature on preannouncements considering its effects on various interest groups such as consumers, competitors and others. Preannouncements are made for various purposes (cf. [26]). They are used e.g. for building interest for the new product before the market launch [3], in order to stimulate consumers to delay purchases, in particular to wait for a better product [33] or to deter entry of potential entrants or to induce a competitor to adjust capacities or to reposition (see [15, 21]).

We use dynamic programming for solving for the optimal capacity investment strategies and derive an optimality condition for the optimal timing which depends on the time-derivative of the corresponding value function at the outset of the game. In that respect our game might be interpreted as a two stage game, where in the first stage only the innovator decides on the introduction time and in the second stage both firms simultaneously choose their Markovian capacity investment strategies and apply them either starting with only the established product or with both products in case that the innovator introduces immediately.

We find that whenever it is optimal to delay the product introduction, the optimal introduction time is increasing in adoption costs. Furthermore, we find that the optimal introduction time increases in both initial capacities, i.e. the stronger the innovator or the non-innovator on the established market, the later the product introduction. The latter is in accordance with results of Dawid et al. [8] where R&D investments are negatively affected by both firms' capacities.

Additionally, we find that in a duopoly, the innovator introduces the product less often compared to a monopoly scenario and, in case of product introduction she introduces earlier compared to the monopoly. Thus, this paper contributes to the debate initiated by Schumpeter and Arrow in the sense that we show that market concentration facilitates product innovation but slows down the actual introduction of the new product.

In Sect. 10.2, we provide the model and in Sect. 10.3, we derive a general sufficient condition for delaying the product introduction. Furthermore, we derive general necessary conditions for optimal timing which has to hold at the outset of the game. In Sect. 10.4 we discuss the different dynamic patterns that can arise in equilibrium using numerical methods. In Sect. 10.5 we give some concluding remarks.

10.2 Model

We consider a duopoly where both firms, denoted by firm A and B, produce a homogeneous established product, denoted as product 1. Due to product innovation, firm A has the option to introduce a horizontally and vertically differentiated substitute product with higher quality, denoted as product 2. We call this firm the innovator whereas the other firm, firm B is called the non-innovator. The innovator incurs lumpy costs F at the time of introduction.

Both firms need to build and maintain production capacities, denoted by $K_{if}, i = 1, 2, f = A, B$, for every product they are offering. For simplicity, we assume that the innovator can only start to invest in the capacity of the new product after introduction, i.e. there are no capacities at the time of introduction for the new product, yet. In line with large parts of the literature (see e.g. [13, 24]), it is assumed that capacities are always fully used. Production costs for given capacities are normalized to zero. There is no inventory, i.e. production equals sales.

Before product introduction, i.e. for all $t \leq T$, the linear inverse demand function for the established product is given by

$$p_1^{m_1}(K_{1A}(t), K_{1B}(t)) = 1 - K_{1A}(t) - K_{1B}(t), \tag{10.1}$$

whereas after product introduction, i.e. for all $t \geq T$, the inverse demand system is given by

$$p_1^{m_2}(K_{1A}(t), K_{1B}(t), K_{2A}(t)) = 1 - \big(K_{1A}(t) + K_{1B}(t)\big) - \eta K_{2A}(t), \tag{10.2}$$

and

$$p_2^{m_2}(K_{1A}(t), K_{1B}(t), K_{2A}(t)) = 1 + \theta - K_{2A}(t) - \eta\big(K_{1A}(t) + K_{1B}(t)\big), \tag{10.3}$$

where η with $0 < \eta < 1$ measures the degree of horizontal and $\theta > 0$, the degree of vertical differentiation of the strategic substitutes.

There are two modes in the game:

- mode 1 (m_1): The new product has been developed by the innovator and is ready for market introduction which is common knowledge. However, only the established product is sold.
- mode 2 (m_2): The new product has been introduced to the market. Both products are sold.

Capacity investment is costly with quadratic costs

$$C_1(I_{1f}(t)) = \frac{\gamma_1}{2} I_{1f}^2(t), \qquad f = A, B, \tag{10.4}$$

and

$$C_2(I_{2A}(t)) = \frac{\gamma_2}{2} I_{2A}^2(t). \tag{10.5}$$

The capacity dynamics in m_1 are

$$\dot{K}_{1f} = I_{1f} - \delta K_{1f}, \qquad f = A, B, \tag{10.6}$$

for initial capacities

$$K_{1f}(0) = K_{1f}^{ini}, \; f = A, B, \tag{10.7}$$

where $\delta > 0$ measures the depreciation rate of the capacities. In m_2, there is an additional state for the capacity of the new product which evolves in the same way according to

$$\dot{K}_{2A} = I_{2A} - \delta K_{2A}, \tag{10.8}$$

$$K_{2A}(t) = 0 \quad \forall t \le T. \tag{10.9}$$

As in [10], we allow the firms to intentionally scrap capacities (i.e. investments might be negative) while capacities have to remain non-negative, i.e. $K_{1f} \ge 0 \; \forall \, t$, $f = A, B$, and $K_{2A} \ge 0 \; \forall \, t$. It should be noted that, due to the adjustment costs, in our setting scrapping capacity is associated with costs.[2]

[2]Since we use a quadratic, rather than a linear quadratic, cost function we have implicitly normalized the price of capital, both when buying and selling, to zero. Having a positive price of capital would not qualitatively change our results, as long as the price is not too large.

The innovator wants to determine the optimal time of product introduction T, i.e. the time of transition from m_1 to m_2, and the optimal strategies for investment in capacities before and after product introduction, whereas the non-innovator only determines the optimal strategy for investing in her capacity for the established product. The discounted stream of profits of the innovator is given by

$$
\begin{aligned}
J_A = &\int_0^T e^{-rt}\big(p_1^{m_1}(\cdot)K_{1A} - C_1(I_{1A})\big)dt \\
&+ \int_T^\infty e^{-rt}\big(p_1^{m_2}(\cdot)K_{1A} + p_2K_{2A} - C_1(I_{1A}) - C_2(I_{2A})\big)dt - e^{-rT}F,
\end{aligned}
\tag{10.10}
$$

which is maximized with respect to T, I_{1A} and I_{2A}. For the non-innovator, it is given by

$$
J_B = \int_0^T e^{-rt}\big(p_1^{m_1}(\cdot)K_{1B} - C_1(I_{1B})\big)dt + \int_T^\infty e^{-rt}\big(p_1^{m_2}(\cdot)K_{1B} - C_1(I_{1B})\big)dt,
\tag{10.11}
$$

where the control variable of firm B is I_{1B}.

10.3 Equilibrium Strategies

In this section, we will derive some sufficient and necessary conditions for the optimal timing of the product introduction. It should be noted that those conditions hold generally for models where two firms' controls affect the dynamics of a continuously evolving state variable and one of the firms can induce a regime switch.

For the sake of brevity, denote the capacity pair (K_{1A}, K_{1B}) by K. Let

$$\phi_{if}(K, K_{2A}, t, m),\ \ f = A, i = 1, 2 \text{ and } f = B, i = 1$$

be the Markovian investment strategies of both firms in mode m and $T = \tau$ the timing strategy of the innovator. Then, a strategy vector of the innovator is a pair $\psi_A = ((\phi_{1A}, \phi_{2A}), \tau)$ whereas the strategy of the non-innovator is given by $\psi_B = \phi_{1B}$. A strategy profile (ψ_A, ψ_B) is an equilibrium if given τ, (ϕ_{1A}, ϕ_{1B}) constitutes a Markov perfect equilibrium and τ maximizes the objective functional of the innovator.

In the case that the innovator introduces the improved product at some finite time T, there will be a structural change of the model. Denote by $V_f^{opt}(K, K_{2A}, t, m)$ the value function of firm f in mode m where the switching time from m_1 to m_2 is selected *optimally* by the innovator. Furthermore, denote by $V_f^{m_1}(K)$ and $V_f^{m_2}(K, K_{2A})$, $f = A, B$, the value functions of the corresponding infinite horizon games where the mode is fixed and hence does not change. This immediately gives $V_f^{opt}(K(t), K_{2A}(t), t, m_2) = V_f^{m_2}(K(t), K_{2A}(t))$ for all t and $V_f^{opt}(K(T), K_{2A}(T), T, m_1) = V_f^{m_2}(K(T), K_{2A}(T)) - F$, $f = A, B$ for the switching time T since in m_2,

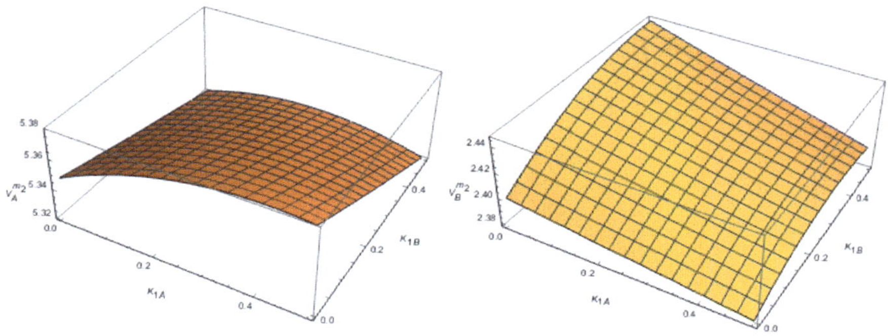

(a) (b)

Fig. 10.1 Value functions of m_2 for $K_{2A} = 0$. Parameters: $r = 0.04,\ \delta = 0.2,\ \eta = 0.5,\ \theta = 0.1,\ \gamma = 0.1$

the mode does not change anymore. The infinite horizon games are time-autonomous, and therefore we consider stationary strategies. Hence the value functions of the infinite horizon games with fixed mode do not explicitly depend on time. The subproblem of m_2 is of linear-quadratic type which can be solved easily by the dynamic programming approach.[3] Due to the linear quadratic structure of the game, the value functions have the following form:

$$\begin{aligned} V_f^{m_2} =& C_f^{m_2} + D_f^{m_2} K_{1A} + E_f^{m_2} K_{1A}^2 + F_f^{m_2} K_{1B} + G_f^{m_2} K_{1B}^2 + H_f^{m_2} K_{2A} + J_f^{m_2} K_{2A}^2 \\ &+ L_f^{m_2} K_{1A} K_{1B} + M_f^{m_2} K_{1A} K_{2A} + N_f^{m_2} K_{1B} K_{2A},\ f = A, B. \end{aligned} \tag{10.12}$$

Using this functional form, the HJB-equations can be reduced to a set of algebraic equations which has to be satisfied by the coefficients of the quadratic value functions. Coefficients can be found by standard numerical methods for a given parameter setting (cf. [10] for a similar model with slightly different inverse demand functions). Figure 10.1 illustrates the shape of the value functions in m_2. By regarding the value of the subproblem (minus adoption costs) as the salvage value of the finite time horizon problem in mode m_1, i.e.

$$S(K_{1A}(T), K_{1B}(T)) = V_A^{m_2}(K_{1A}(T), K_{1B}(T), 0) - F\ , \tag{10.13}$$

we can write the optimization problems of both firms in m_1 as

[3] When solving the investment problem in mode m_2 the non-negativity constraint for the capacities is not explicitly included in the optimization problem. Rather, we have verified ex-post that the constraint holds under the optimal investment strategy. A similar statement also applies to our solution of the problem in mode m_1.

$$\max_{T, I_{1A}} \int_0^T e^{-rt}\big(p_1 K_{1A} - C_1(I_{1A})\big)dt \\ + e^{-rT}\Big(V_A^{m_2}\big(K_{1A}(T), K_{1B}(T), 0\big) - F\Big), \tag{10.14}$$

and

$$\max_{I_{1B}} \int_0^T e^{-rt}\big(p_1 K_{1B} - C_1(I_{1B})\big)dt + e^{-rT} V_B^{m_2}\big(K_{1A}(T), K_{1B}(T), 0\big). \tag{10.15}$$

If an infinite time horizon is optimal, then the salvage value disappears and the value of the game is simply given by $V_f^{m_1}(\cdot)$ for $f = A, B$ and there is a unique stable steady state (see [25]).

As discussed above, we assume that the innovator announces the date of product introduction and has commitment power such that he cannot deviate from the announced date even though ex post it would be better to do so. Thus, the non-innovator takes T as given by the preannouncement and chooses his investment strategy in order to maximize the value of the game. Technically speaking, we employ Markov (feedback) strategies for the investment in capacities and open-loop strategies for the introduction time T.

Note that for any fixed T, the game in m_1 is still of linear quadratic structure. Since the problem in m_1 has a finite time horizon the coefficients in the value function depend on time and from the HJB-equations a set of Riccati equations for those coefficients is obtained. We solve this system using standard numerical solvers. The corresponding HJB-equations to be fulfilled are given in Appendix 2. Denote the value function of the game starting in m_1 and switching to m_2 at a fixed T by $V_f(K, t; T)$, $f = A, B$, and the corresponding profile of Markovian strategies in equilibrium by $\phi_{1f}(K, t; T)$, $f = A, B$.

Since the game is time-autonomous, i.e. t appears explicitly only in the discounting term e^{-rt}, we can consider equilibrium strategies which depend only on the remaining time till T. This then also hold for the value function and we have $V_f(K, t; T) = V_f(K, 0; T - t)$, $f = A, B \; \forall \; K$ and $t \le T$ (cf. [4]).

In particular, we have $\phi_{1f}(K, T; T) = \phi_{1f}(K, 0; 0)$ and for finite T, we denote the right hand side of the HJB-equation of firm A (Eq. (10.40) in Appendix 2) at the switching time by[4]

$$H(K) = p_1^{m_1}(K)K_{1A} - C(\phi_{1A}(K, 0; 0)) + V_{A,K_{1A}}^{m_2}(K, 0)(\phi_{1A}(K, 0; 0) - \delta K_{1A}) \\ + V_{A,K_{1B}}^{m_2}(K, 0)(\phi_{1B}(K, 0; 0) - \delta K_{1B}). \tag{10.16}$$

Note that the optimal strategies ϕ_{1A} and ϕ_{1B} stem from m_1 whereas derivatives of the value function of m_2 are considered. We assume that $V(K, t; T)$ is sufficiently smooth, i.e. let $V(K, t; T)$ be continuously differentiable in K and t for all T.

[4] Actually, $H(K)$ is the Hamiltonian where the co-state variable is replaced by the state derivatives of the scrap value (cf. Pontryagin's maximum principle with finite time horizon e.g. in [13]).

Then, the following lemma gives a sufficient condition for delaying the product introduction.

Lemma 10.1 *If for the initial capacities* $(K_{1A}(0), K_{1B}(0)) = K^{ini}$ *the inequality*

$$H(K^{ini}) > r(V_A^{m_2}(K^{ini}, 0) - F) \tag{10.17}$$

holds, then the optimal time of product introduction T^* *is positive, possibly infinite.*

Proof Consider the value for the innovator to stay for the duration of ϵ in m_1 and afterwards to switch to m_2 under the equilibrium strategy $\phi = (\phi_{1A}, \phi_{1B})$:

$$V_A(K(0), 0; \epsilon) = \int_0^{\epsilon} e^{-rs} F_A^{m_1}(K(s), \phi(K(s), s; \epsilon) ds + e^{-r\epsilon}(V_A^{m_2}(K(\epsilon), 0) - F). \tag{10.18}$$

where $F_A^{m_1}(\cdot)$ is the instantaneous profit function of the innovator in m_1. For a finite time horizon, since we consider non-stationary strategies, altering the terminal time would yield different investments in m_1 and hence different values for the terminal capacities. Thus, for the sake of clarity, here we denote the capacity at t for terminal time T by $K_{1f}(t, T)$, $f = A, B$. $K_{1A}(\epsilon, \epsilon)$ can then be derived via the initial value $K_{1A}(0, \epsilon)$ and the investments from 0 until ϵ:

$$K_{1A}(\epsilon, \epsilon) = K_{1A}(0, \epsilon) + \int_0^{\epsilon} (\phi_{1A}(K(\tau, \epsilon), \tau; \epsilon) - \delta K_{1A}(\tau, \epsilon)) d\tau. \tag{10.19}$$

Its derivative with respect to ϵ is then given by

$$\frac{\partial K_{1A}(\epsilon, \epsilon)}{\partial t} + \frac{\partial K_{1A}(\epsilon, \epsilon)}{\partial T} \tag{10.20}$$

$$= \phi_{1A}(K(\cdot), \tau; \epsilon) - \delta K_{1A}(\cdot) + \int_0^{\epsilon} \frac{\partial \phi_{1A}(K(\tau, \epsilon), \tau, \epsilon) - \delta K_{1A}(\tau, \epsilon)}{\partial T} d\tau. \tag{10.21}$$

In Eq. (10.18), subtracting $V_A(K(0), 0; 0)$ on both sides, dividing by ϵ and considering the limit $\epsilon \to 0$ yields

$$\begin{aligned} \frac{\partial V_A(K, 0, 0)}{\partial T} &= p_1^{m_1}(K) K_{1A}(0, 0) - C(\phi_{1A}(K, 0; 0)) \\ &+ V_{A,K_{1A}}^{m_2}(K, 0) \left(\dot{K}_{1A}(0, 0) + \frac{\partial K_{1A}(0, 0)}{\partial T} \right) \\ &+ V_{A,K_{1B}}^{m_2}(K, 0) \left(\dot{K}_{1B}(0, 0) + \frac{\partial K_{1B}(0, 0)}{\partial T} \right) \\ &- r \left(V_A^{m_2}(K_{1A}(0, 0), K_{1B}(0, 0), 0) - F \right) \end{aligned} \tag{10.22}$$

where no time derivatives of $V_A^{m_2}$ appear since we consider stationary strategies in m_2. Furthermore,

$$\frac{\partial K_{1f}(0,0)}{\partial T} = 0, \ f = A, B \tag{10.23}$$

and using inequality (10.17) we obtain

$$\frac{\partial V_A(K, 0, 0)}{\partial T} > 0, \tag{10.24}$$

which proves that delaying the product introduction marginally is better than introducing immediately.

It follows from Lemma 10.1 that (10.17) being violated is a necessary condition for immediate product introduction. It should be noted that it is, however, not possible to derive a (local) sufficient condition for immediate introduction since marginally being worse-off does not necessarily imply that immediate introduction is optimal. For some $T > 0$, the corresponding value might still outweigh immediate introduction's value.

From optimal control theory, it is known that for $H(K^{ini}) > r(V_A^{m_2}(K^{ini}, 0) - F)$, the innovator prefers not introducing the product immediately but introducing whenever $H = r(V_A^{m_2} - F)$ holds. Here, $H = r(V_A^{m_2} - F)$ is satisfied on a *switching line* (see Appendix 1). In an optimal control setting, the firm exerts control such that the state arrives at the switching line and the switch occurs. However, in a game, the other player can influence the time the switching line is reached because it controls the dynamics of its own capacity. In an equilibrium where the strategy determining when to introduce the new product is of feedback type, e.g. Markovian, this gives rise to intricate strategic effect to be considered. Here we assume however that firm A commits at $t = 0$ to the *time of product introduction* (which might be infinity if the firm decides not to introduce the product at all) rather than on a switching line in the state space and therefore these issues do not arise. Also, by choosing T, firm A influences the investment strategy of firm B. As it will turn out, this effect induces that in our setting in equilibrium the product in general is not introduced at the point in time when the state is on the switching line.

In order to characterize the optimal time of product introduction T, i.e. the choice of the time horizon of the game, which maximizes $V_A(K, 0, T)$, we proceed as follows. We consider a sufficiently large fixed time horizon and compute the optimal distance to the terminal time where the firm wants the game to start. For this, we use a *large* T, which is defined as follows.

Standard turnpike arguments (see [30] or [19]) yield that for $T \to \infty$, the change in the value function becomes small since it is converging to the (time-independent) value function of the infinite horizon game in mode m_1, $V_f^{m_1}$. For an ϵ with $0 < \epsilon \ll |V_A^{m_2}(K^{ini}, 0) - V_A^{m_1}(K^{ini})|$[5] and an initial capacity K^{ini}, a large T satisfies

$$\left| V_f(K^{ini}, 0; T) - V_f^{m_1}(K^{ini}) \right| \le \epsilon. \tag{10.25}$$

[5]Note that for higher choices of ϵ, inequality (10.25) might be satisfied for all T and hence would not yield a *large* T.

We denote by $T^l(\epsilon, K^{ini})$ the minimal T for which inequality (10.25) holds for all $T \geq T^l$. Among all capacities which yield positive prices, we select the maximal T^l which we denote by $T^L(\epsilon)$, i.e. $T^L(\epsilon) := T^l(\epsilon, K_{max})$ where $K_{max} = \arg\max_{K: p_1^{m_1}(K)\geq 0}(T^l(\epsilon, K))$.

Using this notation, in the following proposition we characterize firm A's choice of the optimal time of product introduction.

Proposition 10.1 *Let $V_f(K, t; T^L)$ be the value function of the game for a fixed large end time $T^L(\epsilon)$ for $f = A, B$. Let t^* be the time argument maximizing V_A for an initial pair $K^{ini} = (K_{1A}^{ini}, K_{1B}^{ini})$, i.e.*

$$t^*(K^{ini}) = \arg\max_{t\in[0,T^L]} V_A(K^{ini}, t; T^L). \tag{10.26}$$

If $t^*(K^{ini}) > 0$, then

$$T^*(K^{ini}) = T^L - t^*(K^{ini}), \tag{10.27}$$

is the optimal time of product introduction for $K(0) = K^{ini}$ and the value function in m_1 for $f = A, B$ and for initial capacities K^{ini} is given by

$$V_f^{opt}(K, 0, t, m_1) = V_f(K, t; T^*(K^{ini})). \tag{10.28}$$

Furthermore, if $t^*(K^{ini}) = 0$ for all $T \geq T^L(\epsilon)$ (i.e. for all $T^L(\tilde{\epsilon})$ with $\tilde{\epsilon} \leq \epsilon$), then

$$T^*(K^{ini}) = \infty, \tag{10.29}$$

is the optimal time of product introduction for $K(0) = K^{ini}$ and the value function is given by

$$V_f^{opt}(K, 0, t, m_1) = V_f^{m_1}(K), \quad f = A, B. \tag{10.30}$$

Proof Due to time invariance, the current value of the initial game defined on the time interval $[0, T^L]$ at t^* is equal to the current value at 0 of the game defined over $[0, T^*]$ where $T^* = T^L - t^*$. Hence, it is sufficient to derive the optimal distance to a fixed terminal time where the innovator wants the game to start.

If $t^*(K^{ini}) > 0$, i.e. $t^*(K^{ini})$ is interior in $[0, T^L]$, then for all $T \geq T^L$, according to inequality (10.25), $t^*(K^{ini})$ (shifted by $T - T^L$) is still an interior maximum. Hence, $T^L - t^*(K^{ini})$ is the optimal distance to the terminal time T^L.

If $t^*(K^{ini}) = 0$ for all $T \geq T^L(\epsilon)$, then the maximizing argument is at the left boundary. More precisely, for reducing ϵ and thereby increasing T^L, $t^* = 0$ remains optimal. Thus, $V_A(K^{ini}, t, T)$ is monotonously increasing in T. Hence, $T^* = \infty$ is optimal.

Essentially, from a family of value functions of the game for different values of T, i.e. for varying terminal times, the innovator has to select that one which maximizes his profits for the initial capacity. So, the optimal time of product introduction can be found via considering the value function for a fixed initial pair K^{ini} and a fixed sufficiently large terminal time and determining the optimal distance to the terminal time.[6] In the next corollary, we provide necessary conditions for the slope of the time derivative of the value function at the outset of the game.

Corollary 10.1 *(i) If immediate product introduction, i.e. a corner solution $T^* = 0$ is optimal, then*

$$\lim_{T\to 0}\Big(\lim_{t\to T^-} V_{A,t}(K^{ini}, t; T)\Big) \geq 0, \tag{10.31}$$

and

$$H(K^{ini}) \leq r(V_A^{m_2}(K^{ini}, 0) - F). \tag{10.32}$$

(ii) If no product introduction, i.e. $T^* = \infty$ is optimal, then

$$\lim_{T\to\infty} V_{A,t}(K^{ini}, 0; T) \leq 0, \tag{10.33}$$

(iii) For an interior solution, i.e. $0 < T^* < \infty$ to be optimal we must have

$$V_{A,t}(K^{ini}, 0; T^*) = 0. \tag{10.34}$$

Proof (i) For a corner solution $T^* = 0$, the maximizing argument of (10.26) is on the right boundary, i.e. $t^* = T^L$. Thus,

$$\lim_{t\to T^-} V_{A,t}(K^{ini}, t; T) \geq 0,$$

holds for all $T > 0$, which implies (10.31) . Furthermore, the HJB-equation under $T^* = 0$ yields

$$rS(K^{ini}) - V_{A,t}(K^{ini}, 0; 0) = H(K^{ini}). \tag{10.35}$$

As the limit of $V_{A,t}$ stays positive, we obtain (10.32).

(ii) For a corner solution $T^* = \infty$, the maximizing argument is on the left boundary, i.e. $t^* = 0$. This means $V_t(K^{ini}, 0; T^L) \leq 0$ for all T^L. Thus, $\lim_{T\to\infty} V_t(K^{ini}, 0; T) \leq 0$.

(iii) follows directly from the first order condition if firm A.

Note that Corollary 10.1 yields necessary conditions only. In particular, condition (10.34) might be satisfied for local maxima which are not globally optimal. To get an intuition for this necessary optimality condition, consider the difference of the value of the game for a fixed state variable vector when time moves from t to $t + \Delta$, $\Delta > 0$:

$$V(K^{ini}, t + \Delta; T^L) - V(K^{ini}, t; T^L). \tag{10.36}$$

---

[6]The idea of considering large values for the terminal time has been employed by several works, e.g. in [17].

Assuming firm A is free to choose between $t + \Delta$ and t, (10.36) measures the change in the value function in current- value terms. If (10.36) is positive, it is (locally) optimal for the firm to choose a later starting point than t, and an earlier starting point, else. As K^{ini} is not affected by the choice of T^*, maximizing with respect to the second argument of the value function yields for fixed T^L the (globally) optimal time-span $T^* = T^L - t$ for firm A between the initial time and the time of product introduction, which corresponds to the optimal time of product introduction of the free end time game. The first order condition of the optimization of $V(K^{ini}, t; T^L)$ with respect to t yields (10.34). Since it is not feasible to provide an analytical characterization of the globally optimal choice of T it s also not possible to to derive results about the dependence of the introduction time and investment patterns in equilibrium. In order to get a more complete picture of the dependence of the optimal introduction time, as well as of the resulting equilibrium capacity dynamics, from initial capacities and key model parameters, in the following section we compare the actual equilibrium solutions under different parameter constellations using numerical methods.

10.4 Dynamics

In this section, we first examine the behavior of the firms for an exogenously given product introduction time T. We then explore optimal timing and its dependence on adoption costs and initial capacities. In case of delay, we analyze how capacities evolve before introduction.

10.4.1 Exogenous Time Horizon

In order to depict optimal introduction time and the equilibrium investment paths, we use the following default parameter setting (similar to the parameter setting of [10])

$$r = 0.04, \ \delta = 0.2, \ \eta = 0.5, \ \theta = 0.1, \ \gamma_A = \gamma_B = 0.1, \tag{10.37}$$

We start by analyzing the equilibrium investment strategies $\phi_f(K, t; T)$, $f = A, B$, for a large fixed time horizon $T^L = 3$, fixed initial capacity $K^{ini} = (0.35, 0.35)$, and adoption costs $F = 1$. In Fig. 10.2 the investment strategies $\phi_{1f}(K^{ini}, 0, t, m_1)$ in mode m_1 are depicted as functions of $t \in [0, T^L]$ The dashed line corresponds to the infinite horizon case in m_1. Obviously, T^L is large enough to resemble the infinite horizon investment strategy at $t = 0$. In panel (a), we see that the innovator reduces his investments as time approaches T^L which is due to the decreased marginal value of the established capacity when the innovator introduces the new product. For the non-innovator, we have an interesting investment strategy which is non-

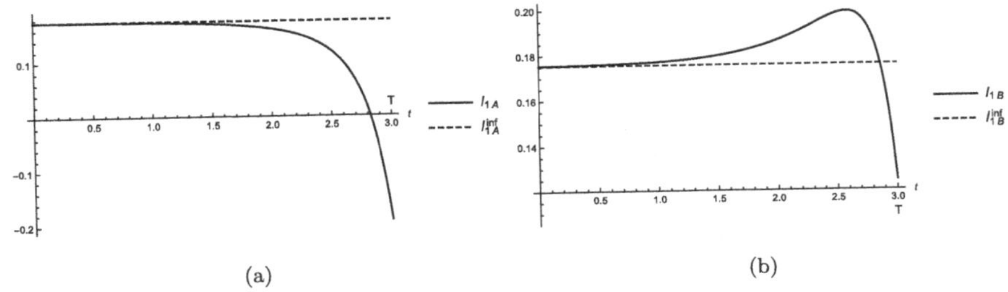

Fig. 10.2 Optimal investments of both firms in mode m_1 at a fixed capacity $K^{ini} = (0.35, 0.35)$ for $F = 1$ and $T^L = 3$

monotone in t. Note that the marginal value of its capacity is decreased in m_2 as well. Hence, eventually investments decline. The initial increase is due to the innovator's decreasing willingness to invest. Moreover, there is an intertemporal strategic effect, i.e. by increasing investment, via a higher capacity and lower price in the future, a firm can even further reduce the future investment of its competitor. As the innovator is affected on both markets by the established capacity while the non-innovator is affected only at the established market (since it is not producing product 2), the non-innovator has more influence on its competitor than the other way around.

Figure 10.2 is also suitable to assess the changes in investment incentives if the innovator (unexpectedly) preannounces the introduction of a new product at the capacity levels (0.35, 0.35). Comparing the solid lines with the dotted ones, which correspond to investment level if no introduction of a new product is expected, we see that for the innovator, the expectation of future product introduction yields a reduction of its investment in capacities for the established product. For the non-innovator, it depends on the length T till the announced time of product introduction. For $T \lessapprox 0.15$, there is a negative effect of the preannouncement on investment, whereas for higher T, investment of firm B increases.

10.4.2 Endogenous Time Horizon

Employing the approach described in Sect. 10.3 and using Proposition 10.1, we are able to derive the optimal T to be preannounced by the innovator. In particular, we calculate the value function of firm A for a sufficiently large T^L and then determine the optimal distance to the terminal time. The approach is illustrated in Fig. 10.3. It can be clearly seen that for the considered parameter the product is optimally introduced after a very short but positive delay of about $T^L - t^* \approx 0.12$. Furthermore, it can be seen that delaying the product introduction by more than 0.4 actually is dominated by immediate product introduction.

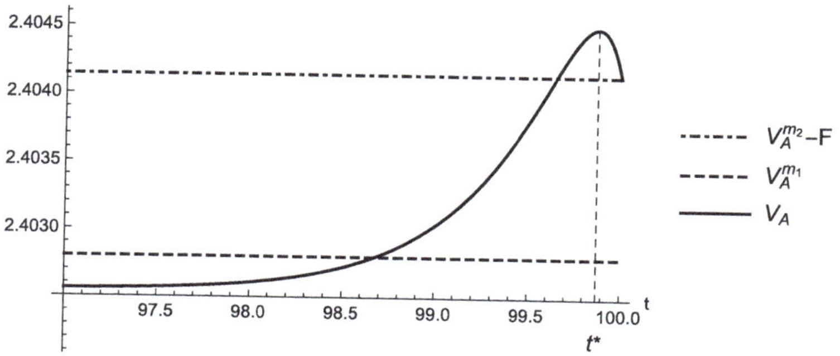

Fig. 10.3 Value function for the innovator for $F = 2.94$, $K^{ini} = (0.35, 0.35)$ and for $T^L = 100$

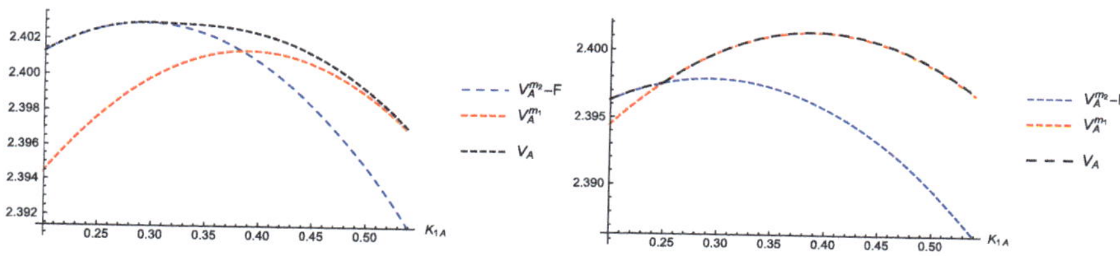

(a) $F = 2.94$.

(b) $F = 2.945$.

Fig. 10.4 Value function for $K_{1B} = K_{1B}^{m_1,ss} \approx 0.3697$

Using this approach we can obtain the equilibrium value of T for each pair of initial capacities and also the resulting value functions for both players. In Fig. 10.4 we show the equilibrium value function at $t = 0$ for firm A (black line) as well as the value obtained under immediate introduction (blue line) and no introduction (red line) as a function of K_{1A}^{ini} for a given value of K_{1B}^{ini}. More precisely, we set K_{1B}^{ini} to the steady state value of the infinite horizon game in m_1, which we denote by $K_{1B}^{m_1,ss}$. Assuming relatively low adoption costs (panel (a)) for low initial K_{1A}^{ini}, the innovator introduces immediately whereas for higher initial capacity, there is a gain by delaying the product introduction.[7] For even higher values of F not introducing becomes optimal for high capacities and hence infinite solutions for T occur. There arises an indifference point for the capacity K_{1A} (which depends on K_{1B}), where introducing (maybe after some delay) and not introducing at all yield the same value

[7]The value functions of immediate and no switching intersect at a point where the slopes of the value functions are very different and hence there is a kink. As usual in endogenous timing problems the option of delaying 'smoothes' the value function of firm A.

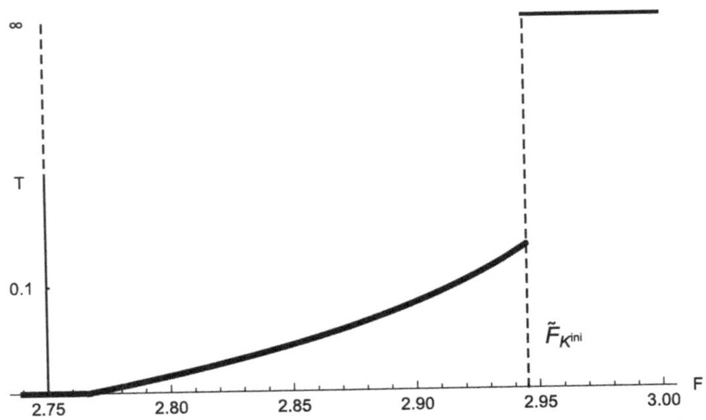

Fig. 10.5 Optimal time to switch to m_2 ($K^{ini} = (0.35, 0.35)$)

for the innovator.[8] If adoption costs F become too high firm A either introduces immediately or never (see Fig. 10.4b).

The pattern sketched above can be clearly seen in Fig. 10.5, which shows the optimal product introduction time T as a function of adoption costs for given initial capacities. For low adoption costs the firm wants to introduce the new product immediately, whereas above some threshold $\bar{F}_{K^{ini}}$, the firm chooses to introduce after some delay. This delay is higher the higher F is. There is another threshold $\tilde{F}_{K^{ini}}$ where the innovator abstains totally from product introduction and stays with its established product. Thus, there is a jump from some finite T to infinity at this threshold. Note that the thresholds depend on initial capacities.

A qualitative description of optimal timing for different levels of initial capacities of both firms is given in Fig. 10.6. Each arrow in the figure depicts the equilibrium trajectory of the capacities for the corresponding initial condition taking into account the time of product introduction chosen by firm A in equilibrium. Here, the steady state of m_1 lies in the interior of the area where firm A decides not to introduce the new product (i.e. $T = \infty$). Still, as can be seen in the figure, a trajectory starting in the area where in equilibrium we have $T = \infty$ might for some time run through the area in the state space where $T < \infty$ would be chosen if the state at $t = 0$ were in this area. Along such a trajectory, once the state $K_{1A}(t), K_{1B}(t))$ has entered the region $T < \infty$, it would be optimal for firm A to announce to introduce the new product after a finite delay, although at $t = 0$ the optimal choice was to commit not

[8]In the literature, such indifference points or curves, separating different basins of attraction of the dynamics under optimal investments, are called Skiba points (curves), see e.g. [20, 32]. For a discussion of issues related to the existence of Skiba points under Markov Perfect Equilibria of differential games see [7]. In general, the value functions have a kink at such Skiba points since, depending on whether firm A plans to introduce the new product or not, equilibrium investments are different. This means that equilibrium investment strategies exhibit jumps at this point in the state space and accordingly the value functions exhibit a kink. In Fig. 10.6 below, the Skiba curve can be seen as the line separating the area with $T = \infty$ from those with $T = 0$ respectively $T < \infty$.

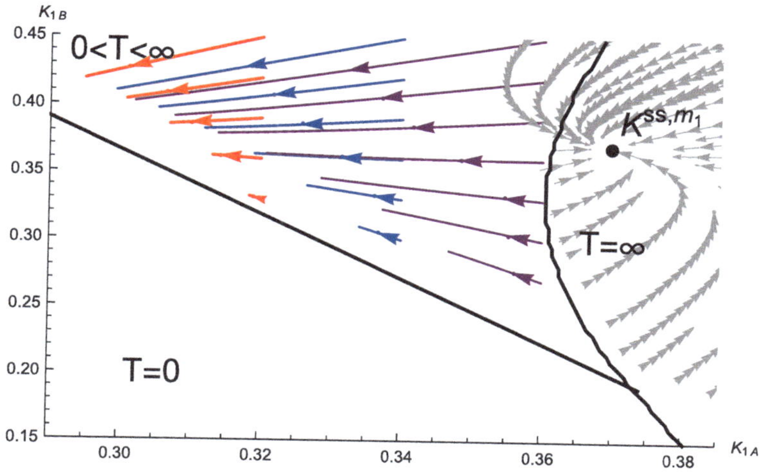

Fig. 10.6 Optimal trajectories for different initial capacities

to introduce the product. Hence, clearly this is a feature of the open-loop strategy for the timing choice, which requires commitment of the firm about the product introduction date at $t = 0$. Moreover, there are parameter settings where the steady state of m_1 does not lie in the corresponding area such that every trajectory starting in the $T = \infty$ area would end up in another $[0 < T < \infty]$ area where ex-post, the firm would like to introduce the product (possibly after some delay) if there were no commitment.

Furthermore, we are interested in how the optimal time of product introduction is influenced by the capacities of both firms. Regarding the capacity of the non-innovator, one might expect that if the non-innovator is stronger on the established market, the innovator has higher incentives to introduce the new product earlier in order to escape competition. But there is another effect as well, namely higher capacity of the non-innovator leads not only to a lower price of the established product but also to a lower price of the new product in m_2. In order to compensate for that, the innovator has incentives to decrease its own capacity on the established market in m_1 in order to be 'more prepared' when switching to m_2. Figure 10.6 suggests that the latter effect is stronger such that the stronger the competitor, the later the product introduction, i.e. T is increasing in K_{1B}. Moreover, the duration in m_1 is increasing in the innovator's capacity as well. Note that for the parameters considered here, the switching line, which separates the $0 < T < \infty$ from the $T = 0$ region, is never reached.

Another interesting observation is that for the innovator, for every initial capacity in the delaying region, it is optimal to reduce capacity whereas for the non-innovator, the dynamics of its capacity depends on initial capacities, in particular on K_{1B}. If K_{1B} is relatively low, then its capacity increases, otherwise it decreases as well. Note that the steady state value of the non-innovators capacity in m_2 is higher than in m_1.

Thus, it is very natural, that the non-innovator might increase its capacity already in m_1.

Considering initial conditions with $0 < T < \infty$ it could be expected that the change in mode at T leads to a discontinuous adjustment of the investment of firm B, since at $t = T$ the there is a discontinuous change in its instantaneous profit function and it has no influence on the choice of the product introduction time T. However, the investment trajectory of the non-innovator, is continuous at all $t \geq 0$, including at $t = T$, when the new product is introduced. Intuitively, in our setting, where the switching time T is fixed and known already at $t = 0$, firm B anticipates the marginal effect of investment on profits in m_2 even before T and therefore investment incentives do not jump at $t = T$.

Finally, we like to mention that in comparison to the monopoly case where the non-innovator does not exist, which has been analyzed in [16], we find the following interesting pattern: The innovator introduces earlier, i.e. the delay in product innovation is shorter but at the same time innovation occurs for a smaller range of costs of product introduction, i.e. for some F the innovator would innovate in monopoly but not in presence of a competitor even though the competitor is only active on the established market. Thus, we see a connection between the Schumpeterian and Arrowian perspective where market concentration facilitates innovation but decreases its speed.

10.5 Conclusion

In this paper, assuming commitment of the innovator with respect to the product introduction time, we have characterized how adoption costs and initial capacities for the established product influence the optimal timing of new product introduction in a dynamic duopoly market. In the interesting case of delay of product introduction, the innovator reduces capacities of the established product before the new product is introduced, whereas the dynamics of the non-innovator's capacity depends on initial capacities. Furthermore, in our setting the innovator would always like to further delay product introduction at the point in time where according to its initial commitment, the new product is brought to the market. More generally, our analysis indicates conditions for determining optimal mode transitions in multi-mode games under the assumption of open-loop determination of the transition times combined with Markov perfect equilibrium profiles within each mode. A challenging and interesting topic for future research clearly is the investigation of fully closed loop equilibria, where also the mode transitions are determined by feedback strategies of one (or several) players. The crucial difference between such a setting and the commitment scenario considered in the present paper is that any player whose controls influence the dynamics of the state(s) that enter the feedback strategy governing the mode transition can influence the timing of that transition, and maybe even prevent such a transition all-together. This feature, which is absent in our setting, makes the strategic interaction much more intricate and also might jeopardize the existence of a Markov-perfect equilibrium (see [6] for a discussion of similar technical issues arising in the context of Markov-perfect equilibria which exhibit Skiba points).

Acknowledgements The paper has profited from helpful comments from two anonymous referees. The authors gratefully acknowledge financial support from the German Research Foundation (DFG) through the Collaborative Research Centre 1283 "Taming uncertainty and profiting from randomness and low regularity in analysis, stochastics and their application".

Appendix

Appendix 1

As derived in Lemma 10.1, the innovator is indifferent between waiting marginally and introducing the new product if and only if $H = rS$, which reduces to

$$\frac{1}{2\gamma_2}\left(\frac{\partial V_A^{m_2}}{K_{2A}}\right)^2 = rF. \tag{10.38}$$

Rearranging Eq. (10.38) yields the switching line

$$K_{1B} = \frac{\sqrt{2r\gamma_2 F} - H_A^{m_2} - M_A^{m_2} K_{1A}}{N_A^{m_2}}. \tag{10.39}$$

Appendix 2

Given the terminal time T, the HJB-equations for non-stationary Markovian investment strategies are given by

$$\begin{aligned} rV_A(K_{1A}, K_{1B}, t) - \frac{\partial V_A(K_{1A}, K_{1B}, t)}{\partial t} = \max_{I_{1A}} \Big[p_1 K_{1A} - C_1(I_{1A}) &+ \frac{\partial V_A}{\partial K_{1A}}(I_{1A} - \delta K_{1A}) \\ &+ \frac{\partial V_A}{\partial K_{1B}}(I_{1B}^* - \delta K_{1B}) \Big] \end{aligned} \tag{10.40}$$

and

$$\begin{aligned} rV_B(K_{1A}, K_{1B}, t) - \frac{\partial V_B(K_{1A}, K_{1B}, t)}{\partial t} = \max_{I_{1B}} \Big[p_1 K_{1B} - C_1(I_{1B}) &+ \frac{\partial V_B}{\partial K_{1A}}(I_{1A}^* - \delta K_{1A}) \\ &+ \frac{\partial V_B}{\partial K_{1B}}(I_{1B} - \delta K_{1B}) \Big] \end{aligned} \tag{10.41}$$

with the transversality conditions

$$V_f(K_{1A}(T), K_{1B}(T), T) = V_f^{m_2}(K_{1A}(T), K_{1B}(T), T), f = A, B. \tag{10.42}$$

Maximizing the right hand side of the HJB-equations yields

$$I_{1f} = \frac{1}{\gamma}\frac{\partial V_f}{\partial K_{1f}}, \quad f = A, B. \tag{10.43}$$

Additionally, firm A has to select the optimal value for T maximizing its discounted stream of profits. Due to the linear-quadratic structure of the game, we impose the following form for the value function:

$$V_f = C_f(t) + D_f(t)K_{1A} + E_f(t)K_{1A}^2 + F_f(t)K_{1B} + G_f(t)K_{1B}^2 + L_f(t)K_{1A}K_{1B}, \; f = A, B. \tag{10.44}$$

Due to the finite time horizon, we consider non-stationary Markovian strategies and hence coefficients depend on time. Comparison of coefficients yields the following system of 12 riccati differential equations which are solved by standard numerical methods:

$$\begin{aligned}
rC_A(t) &= \frac{D_A(t)^2 + 2F_A(t)F_B(t) + 2\gamma_1 C'_A(t)}{2\gamma_1}\\
rD_A(t) &= \frac{\gamma_1 + D_A(t)(-\gamma_1\delta_1 + 2E_A(t)) + F_B(t)L_A(t) + F_A(t)L_B(t) + \gamma_1 D'_A(t)}{\gamma_1}\\
rE_A(t) &= \frac{2E_A(t)(-\gamma_1\delta_1 + E_A(t)) + L_A(t)L_B(t))}{\gamma_1} - 1 + E'_A(t)\\
rF_A(t) &= \frac{2F_B(t)G_A(t) + F_A(t)(-\gamma_1\delta_1 + 2G_B(t)) + D_A(t)L_A(t) + \gamma_1 F'_A(t))}{\gamma_1}\\
rG_A(t) &= \frac{G_A(t)(-4\gamma_1\delta_1 + 8G_B(t)) + L_A(t)^2 + 2\gamma_1 G'_A(t))}{2\gamma_1}\\
rL_A(t) &= \frac{2(-\gamma_1\delta_1 + E_A(t) + G_B(t))L_A(t) + 2G_A(t)L_B(t) + \gamma_1(-1 + L'_A(t))}{\gamma_1}\\
rC_B(t) &= \frac{2D_A(t)D_B(t) + F_B(t)^2 + 2\gamma_1 C'_B(t)}{2\gamma_1}\\
rD_B(t) &= \frac{D_B(t)(-\gamma_1\delta_1 + 2E_A(t)) + 2D_A(t)E_B(t) + F_B(t)L_B(t) + \gamma_1 D'_B(t)}{\gamma_1}\\
rE_B(t) &= \frac{(-4\gamma_1\delta_1 + 8E_A(t))E_B(t) + L_B(t)^2 + 2\gamma_1 E'_B(t))}{2\gamma_1}\\
rF_B(t) &= \frac{\gamma_1 + F_B(t)(-\gamma_1\delta_1 + 2G_B(t)) + D_B(t)L_A(t) + D_A(t)L_B(t) + \gamma_1 F'_B(t)}{\gamma_1}\\
rG_B(t) &= \frac{2G_B(t)(-\gamma_1\delta_1 + G_B(t)) + L_A(t)L_B(t))}{\gamma_1} - 1 + G'_B(t)\\
rL_B(t) &= \frac{2E_B(t)L_A(t) + 2(-\gamma_1\delta_1 + E_A(t) + G_B(t))L_B(t) + \gamma_1(-1 + L'_B(t))}{\gamma_1}
\end{aligned} \tag{10.45}$$

with transversality conditions $C_f(T) = C_f^{m_2}, D_f(T) = D_f^{m_2}, E_f(T) = E_f^{m_2}, F_f(T) = F_f^{m_2}, G_f(T) = G_f^{m_2}, L_f(T) = L_f^{m_2}, f = A, B$.

References

1. Åstebro, T. (2003). The return to independent invention: Evidence of unrealistic optimism, risk seeking or skewness loving? *Economic Journal, 113*(484), 226–239.
2. Åstebro, T., & Simons, K. (2003). Innovation exit: Why entrepreneurs pull the plug on their innovations. In *Proceedings of the DRUID Conference*, Copenhagen, June (pp. 12–14).
3. Bao, Y. K., Tellis, G. J., & Dawid, H. (2005). *Strategic Preannouncements of Innovations*. Center for Empirical Macroeconomics, Working Paper No. 90.
4. Caulkins, J. P., Feichtinger, G., Grass, D., Hartl, R. F., Kort, P. M., & Seidl, A. (2015). Skiba points in free end-time problems. *Journal of Economic Dynamics and Control, 51*, 404–419.
5. Chandy, R. K., & Tellis, G. J. (2000). The incumbent's curse? Incumbency, size, and radical product innovation. *Journal of Marketing, 64*(3), 1–17.
6. Dawid, H., Keoula, M., & Kort, P. (2016). Skiba phenomena in Markov perfect equilibria of asymmetric differential games. In H. Dawid, K. Dörner, G. Feichtinger, P. Kort, & A. Seidl (Eds.), *Dynamic perspectives on managerial decision making* (pp. 63–76).
7. Dawid, H., Keoula, M., & Kort, P. (2017). Numerical analysis of Markov-perfect equilibria with multiple stable steady states: A duopoly application with innovative firms. *Dynamic Games and Applications, 7*, 555–577.
8. Dawid, H., Keoula, M. Y., Kopel, M., & Kort, P. M. (2017). *Dynamic investment strategies and leadership in product innovation*. Bielefeld Working Papers in Economics and Management No. 03-2017.
9. Dawid, H., Kopel, M., & Dangl, T. (2009). Trash it or sell it? A strategic analysis of the market introduction of product innovations. *International Game Theory Review, 11*(03), 321–345.
10. Dawid, H., Kopel, M., & Kort, P. M. (2010). Dynamic strategic interaction between an innovating and a non-innovating incumbent. *Central European Journal of Operations Research, 18*(4), 453–463.
11. Dawid, H., Kopel, M., & Kort, P. M. (2013). R&D competition versus R&D cooperation in oligopolistic markets with evolving structure. *International Journal of Industrial Organization, 31*(5), 527–537.
12. Dixit, A. K., & Pindyck, R. S. (1994). *Investment under uncertainty*. Princeton: Princeton University Press.
13. Dockner, E. J., Jorgensen, S., Long, N. V., & Sorger, G. (2000). *Differential games in economics and management science*. Cambridge University Press.
14. Dutta, P. K., Lach, S., & Rustichini, A. (1995). Better late than early: Vertical differentiation in the adoption of a new technology. *Journal of Economics & Management Strategy*.
15. Farrell, J., & Saloner, G. (1986). Installed base and compatibility: Innovation, product preannouncements, and predation. *The American Economic Review, 76*(5), 940–955.
16. Gezer, S. (2019). Delaying product introduction: A dynamic analysis with endogenous time horizon. *Journal of Economic Dynamics and Control, 102*, 96–114.
17. Grass, D. (2012). Numerical computation of the optimal vector field: Exemplified by a fishery model. *Journal of Economic Dynamics and Control, 36*(10), 1626–1658.
18. Gromov, D., & Gromova, E. (2017). On a class of hybrid differential games. *Dynamic Games and Applications, 7*(2), 266–288.
19. Grüne, L., Semmler, W., & Stieler, M. (2015). Using nonlinear model predictive control for dynamic decision problems in economics. *Journal of Economic Dynamics and Control, 60*, 112–133.

20. Haunschmied, J., Kort, P., Hartl, R., & Feichtinger, G. (2003). A DNS-curve in a two-state capital accumulation model: A numerical analysis. *Journal of Economic Dynamics and Control, 27*, 701–716.
21. Heil, O., & Robertson, T. S. (1991). Toward a theory of competitive market signaling: A research agenda. *Strategic Management Journal, 12*(6), 403–418.
22. Hoppe, H. C., & Lehmann-Grube, U. (2005). Innovation timing games: A general framework with applications. *Journal of Economic Theory, 121*(1), 30–50.
23. Huberts, N., Dawid, H., Huisman, K. J. M., & Kort, P. M. (2019). Entry deterrence by timing rather than overinvestment in a strategic real options framework. *European Journal of Operational Research, 274*, 165–185.
24. Huisman, K. J. M., & Kort, P. M. (2015). Strategic capacity investment under uncertainty. *RAND Journal of Economics, 46*(2), 376–408.
25. Jun, B., & Vives, X. (2004). Strategic incentives in dynamic duopoly. *Journal of Economic Theory, 116*(2), 249–281.
26. Lilly, B., & Walters, R. (1997). Toward a model of new product preannouncement timing. *Journal of Product Innovation Management, 14*(1), 4–20.
27. Long, N. V., Prieur, F., Tidball, M., & Puzon, K. (2017). Piecewise closed-loop equilibria in differential games with regime switching strategies. *Journal of Economic Dynamics and Control, 76*, 264–284.
28. Mansfield, E. (1977). *The production and application of new industrial technology*. New York: Norton.
29. Masoudi, N., & Zaccour, G. (2013). A differential game of international pollution control with evolving environmental costs. *Environment and Development Economics, 18*(6), 680–700.
30. McKenzie, L. W. (1986). Optimal economic growth, turnpike theorems and comparative dynamics. In *Handbook of mathematical economics* (Vol. III, pp. 1281–1355). Amsterdam: North Holland.
31. Reddy, P. V., Schumacher, J. M., & Engwerda, J. C. (2015). Optimal management with hybrid dynamics-the shallow lake problem. *Lecture notes in control and information sciences* (Vol. 461, pp. 111–136). Cham: Springer.
32. Skiba, A. (1978). Optimal growth with a convex-concave production function. *Econometrica, 46*, 527–539.
33. Su, M., & Rao, V. R. (2010). New product preannouncement as a signaling strategy: An audience-specific review and analysis. *Journal of Product Innovation Management, 27*(5), 658–672.
34. Wang, Q. H., & Hui, K. L. (2012). Delayed product introduction. *Decision Support Systems, 53*, 870–880.
35. Yeung, D. W. (2000). Feedback solution of a class of differential games with endogenous horizons. *Journal of Optimization Theory and Applications, 106*(3), 657–675.

Chapter 11
On the Cournot-Ramsey Model with Non-linear Demand Functions

Luca Lambertini and George Leitmann

JEL Codes: C73 · L13

11.1 Introduction

The usual approach to oligopoly games relies on the assumption that firms face a linear demand function, for the sake of (i) ensuring concavity and (ii) making the game analytically solvable without resorting to numerical simulations. This, however, has the cost of leaving aside uncountably many scenarios potentially relevant from an empirical point of view, in which demand is non-linear. Indeed, very few attempts to treat the case of hyperbolic demand have been carried out [3, 14, 16, 18]. The same is true for the demand function with parametric curvature appearing in the Cournot game proposed by Anderson and Engers [1, 2], which hinges upon the ancillary but crucial assumption of costless production to keep the model analytically solvable, and has been used to study the impact of the curvature of market demand on cartel stability [15] and capacity accumulation in the Ramsey model [5], as well as to prove the existence of the best-response Hamiltonian potential function [11].

Here, we focus on three special cases of the demand structure introduced by Anderson and Engers [1, 2] which lend themselves to be treated in a fully analytical way in combination with a linear cost function, i.e. we look at three specifications

L. Lambertini (✉)
Department of Economics, University of Bologna, Strada Maggiore 45, 40125 Bologna, Italy
e-mail: luca.lambertini@unibo.it

G. Leitmann
College of Engineering, University of California at Berkeley, 320 McLaughlin Hall, Mailstop 1740, Berkeley 94720-1700, CA, USA
e-mail: gleit@berkeley.edu

J. L. Haunschmied et al. (eds.), *Dynamic Economic Problems with Regime Switches*, Dynamic Modeling and Econometrics in Economics and Finance 25,
https://doi.org/10.1007/978-3-030-54576-5_11

of the differential game appearing in Cellini and Lambertini [5]. These demand functions are used under open-loop information to solve the weakly and strongly time-consistent versions of the Cournot–Ramsey differential game, to characterise the impact of the curvature of market demand on the nature of the regime prevailing at the saddle-point equilibrium of the game, which can be either driven by market conditions or by technological efficiency and intertemporal parameters measuring discounting and capital depreciation.

Before dealing with the dynamic version of the model, we characterise the Nash equilibria of the static game, in correspondence of each of the three specifications of the demand function. Doing so, we single out a property of equilibrium output levels which is robust to the introduction of the Ramsey capacity accumulation dynamics, namely that the ranking of optimal market-drive quantities is sensitive to the level of the choke price (the vertical intercept of demand) as well as marginal cost. This, as we know from Anderson and Engers [1, 2], does not happen if production is costless, as in such a situation switching from convex to linear to concave demand functions implies a monotonic increase in individual and industry output.

This finding, in addition to telling that setting the marginal cost to zero, is not a neutral choice, has relevant bearings on the properties of the differential game and, in particular, on the attainment of the Ramsey rule (or, the lack thereof). Indeed, for instance, there are admissible ranges of the choke price where firms reach the steady state associated with the golden rule if demand is linear while being short of it if demand is convex or concave, which contrasts with the picture emerging from the costless case in Cellini and Lambertini [5]. Then, we also discuss the state-redundant solution, which, on the one hand, delivers strong time consistency of the market-driven output strategy but, on the other, causes the disappearance of the Ramsey solution and may even prevent firms to reach a steady state at all.

The remainder of the paper is organised as follows. The model is laid out in Sect. 11.2. Section 11.3 illustrates the static game, while Sect. 11.4 contains the analysis of the differential game, including both the regime switch between market-driven and Ramsey solutions and the state-redundant case. Concluding remarks are in Sect. 11.5.

11.2 Setup

A Cournot industry existing over continuous time $t \in [0, \infty)$ is populated by $\mathcal{N} = 1, 2, 3...n$ firms supplying a homogeneous good whose instantaneous demand function, as in Anderson and Engers [1, 2], is $Q(t) = a - p^{\alpha}(t)$, where $Q(t) = \sum_{i=1}^{n} q_i(t)$ is industry demand. Moreover, $a > 0$ is the choke price and parameter $\alpha > 0$ determines the shape of market demand. The demand function is convex (resp., concave) for all $\alpha \in (0, 1)$ (resp., $\alpha > 1$), and linear for $\alpha = 1$. Accordingly, the relevant inverse demand to be used in the Cournot game is $p(t) = [a - Q(t)]^{\frac{1}{\alpha}}$.

On the supply side, the instantaneous cost function of firm i is $C_i(t) = cq_i(t)$, with $c > 0$ measuring the time invariant marginal cost. Hence, the instantaneous individual profit function is $\pi_i(t) = [p(t) - c]\, q_i(t)$.

In order to produce and sell, firm i must build up physical capital (or, installed capacity) $k_i(t)$ through unsold output, according to the following Ramsey [21] dynamics:

$$\frac{dk_i(t)}{dt} \equiv \dot{k}_i = f(k_i(t)) - q_i(t) - \delta k_i(t) \tag{11.1}$$

in which $f(k_i(t))$ is a concave technology with $f'(k_i(t)) \equiv \partial f(k_i(t))/\partial k_i(t) > 0$ and $f''(k_i(t)) \equiv \partial^2 f(k_i(t))/\partial k_i(t)^2 < 0$, and $\delta > 0$ is the time-invariant decay rate of capacity, common to all firms. Capacity accumulates as a result of intertemporal relocation of unsold output, whenever $f(k_i(t)) > q_i(t)$. Each firm discounts future profits at the time-invariant rate $\rho > 0$ and has to choose its single control variable $q_i(t)$ so as to maximise the flow of discounted profits

$$\Pi_i = \int_0^\infty e^{-\rho t}\, [p(t) - c]\, q_i(t)dt, \tag{11.2}$$

under the set of n dynamic constraints (11.1), the initial conditions $\mathbf{k}(0) = [k_1(0), k_2(0), \ldots k_i(0), \ldots k_n(0)]$ and the transversality conditions

$$\lim_{t\to\infty} \mu_{ij}(t)\, k_j(t) = 0\; \forall\, i, j \tag{11.3}$$

where $\mu_{ij}(t)$ is the costate variable attached by firm i to the state variable k_j.

As stated in the introduction, thus far the impact of the above demand function has been exclusively investigated under the admittedly ad hoc assumption of costless production, to allow for the analytical solution of the first-order condition delivering the Cournot–Nash quantity, both in the static game [1, 2, 15] and in the dynamic one [5, 11].[1]

Indeed, neither the static nor the dynamic game lends themselves to a full analytical treatment for a generic value of α. However, there are three obvious values of the key parameter delivering solvable first-order conditions (FOCs), i.e. $\alpha \in \{1/2, 1, 2\}$. These are intuitive candidates for the solution of both the static and the dynamic version of the game since the resulting necessary conditions are either linear or quadratic in controls. Before delving into the details of the differential game, it is worth dwelling upon the features of the static Cournot–Nash equilibrium, whose implications are also relevant for the characterisation of the dynamic problem.

[1] In particular, in the costless case Dragone et al. [11] prove the existence of a best-response potential function for the Cournot–Ramsey differential game for all values of α (see also Dragone et al. [12]).

11.3 The Static Game

Leaving time aside for the moment, we may take a look at the static setting in which capacity accumulation is not an issue and firms solve a one-shot Cournot–Nash game in output levels. For any given α, the FOC for noncooperative profit maximisation is

$$\frac{\partial \pi_i}{\partial q_i} = \frac{(a-Q)^{\frac{1-\alpha}{\alpha}}\left[\alpha\left(a-q_i-\sum_{j\neq i} q_j\right)-q_i\right]}{\alpha} - c = 0 \tag{11.4}$$

and of course its solution, in general, would require the use of transcendental functions if $c > 0$, except in special cases. Clearly, if $\alpha = 1$, demand is linear and (11.4) yields the familiar Cournot–Nash output $q^{CN}_{\alpha=1} = (a-c)/(n+1)$. If $\alpha = 1/2$, whereby the demand function is convex, there exist two solutions:

$$q^{\pm}_{\alpha=1/2} = \frac{a(n+1) \pm \sqrt{a^2 + cn(n+2)}}{n(n+2)} \tag{11.5}$$

and the second-order condition for concavity selects $q^{CN}_{\alpha=1/2} = q^{-}_{\alpha=1/2}$ as the Cournot equilibrium quantity. If instead $\alpha = 2$, the demand function is concave and (11.4) has the following solutions:

$$q^{\pm}_{\alpha=2} = \frac{2\left[a(2n+1) - nc^2 \pm c\sqrt{a(2n+1) + c^2n^2}\right]}{4n(n+1)+1} \tag{11.6}$$

with $q^{CN}_{\alpha=2} = q^{-}_{\alpha=2}$ meeting the concavity requirement. For all $a > c$ and $n \geq 1$, the three equilibrium quantities are monotonically increasing in a, with[2]

$$\frac{\partial q^{CN}_{\alpha=2}}{\partial a} > \frac{\partial q^{CN}_{\alpha=1}}{\partial a} > \frac{\partial q^{CN}_{\alpha=1/2}}{\partial a} > 0 \tag{11.7}$$

Moreover, in correspondence of $a = c$,

$$\begin{aligned} sgn\left\{ q^{CN}_{\alpha=1/2}\Big|_{a=c} \right\} &= sgn\{c-1\} \\ q^{CN}_{\alpha=1}\Big|_{a=c} &= 0 \\ q^{CN}_{\alpha=2}\Big|_{a=c} &\gtreqless 0 \, \forall c \lesseqgtr 1 \end{aligned} \tag{11.8}$$

[2] Alternatively, one could look at the (negative) partial derivatives of equilibrium quantities q^{CN}_{α} and replicate the ensuing discussion in terms of the level of the common marginal cost.

Taken together, (11.7–11.8) imply that, for any $a > c > 1$, $q^{CN}_{\alpha=1/2}$, $q^{CN}_{\alpha=1}$ and $q^{CN}_{\alpha=2}$ will intersect as in Fig. 11.1.[3] In particular, the linear solution intersects the other two at

$$\begin{aligned}\widetilde{a} &= c + (n+1)\sqrt{c\,(c-1)} \\ \overline{a} &= 2c(n+1)\left(c + \sqrt{c\,(c-1)}\right) - c\,(2n+1)\end{aligned} \tag{11.9}$$

and $\widetilde{a}, \overline{a} \in \mathbb{R}^{+}$ iff $c \geq 1$. The expression of $\widehat{a}$ can only be approximated numerically.

While profits $\pi^{CN}_{\alpha=2}$ and $\pi^{CN}_{\alpha=1}$ generated by a concave or linear demand are always positive, the profits $\pi^{CN}_{\alpha=1/2}$ associated with a convex demand are non-negative for all[4]

$$a \geq \frac{c^3\,(n+2) - n}{2c} \equiv \underline{a} \in \left(\widetilde{a}, \widehat{a}\right) \tag{11.10}$$

Conversely, firms facing a convex demand with $a \in \left(c, \underline{a}\right)$ will shut down production because the choke price is too low. Hence, in order to ensure the viability of firms across the three market outcomes, we must confine our attention to the range $a \geq \underline{a}$. This condition identifies the relevant region of the space (a, q) in Fig. 11.1 in which firms produce positive outputs irrespective of the shape of market demand, to the right of the dashed line at $\underline{a}$. Accordingly, the intersection between $q^{CN}_{\alpha=1/2}$ and $q^{CN}_{\alpha=1}$ at $\widehat{a}$ is ruled out because $q^{CN}_{\alpha=1/2} = 0 < q^{CN}_{\alpha=2} < q^{CN}_{\alpha=1}$ for all $a \in \left(c, \underline{a}\right)$.

For any $a \geq 1 > c \geq 0$, $q^{CN}_{\alpha=2} > q^{CN}_{\alpha=1} > q^{CN}_{\alpha=1/2}$, i.e. if the marginal cost is sufficiently low, the sequence of equilibrium quantities reflects what we are accustomed with from the initial version of the model in which marginal production cost is nil. Indeed, this fact had to be expected at the very outset, on the basis of a continuity argument. That is, since $q^{CN}_{\alpha=2} > q^{CN}_{\alpha=1} > q^{CN}_{\alpha=1/2}$ if $c = 0$, then the same ranking can be expected to hold in the right neighbourhood of zero.

As we shall see in the dynamic version of the game, the impact of the demand curvature on the Cournot–Nash outcome may entail unusual consequences on the nature of the steady-state equilibrium.

---

[3] Also note that

$$\frac{\partial^2 q^{CN}_{\alpha=2}}{\partial a^2} > \frac{\partial^2 q^{CN}_{\alpha=1}}{\partial a^2} = 0 > \frac{\partial^2 q^{CN}_{\alpha=1/2}}{\partial a^2}$$

so that, while the Cournot–Nash quantity generated by a linear demand is indeed linear in the choke price, that generated by a concave (resp., convex) demand is convex (resp., concave) in the choke price.

[4] The reason is that if $\alpha = 1/2$ then the profit margin is negative for all $a \in \left(c, \underline{a}\right)$. So, the requirement is indeed about the positivity of the markup.

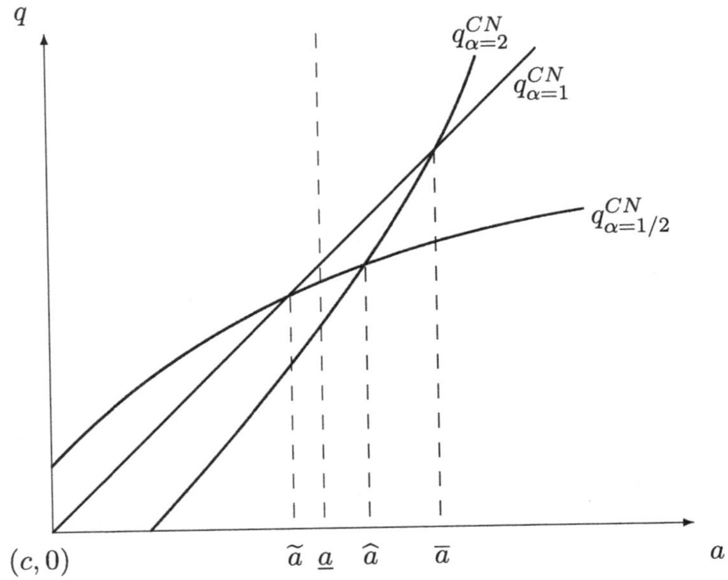

Fig. 11.1 Cournot–Nash quantities as functions of the choke price

11.4 The Differential Game

We are now in a position to investigate the differential game, in which the current value Hamiltonian function of the generic firm is

$$\mathcal{H}_i(t) = e^{-\rho t}\left\{[p(t) - c]\, q_i(t) + \lambda_{ii}(t)\,\dot{k}_i + \sum_{j \neq i} \lambda_{ij}(t)\,\dot{k}_j\right\} \tag{11.11}$$

in which the direct demand function takes one of the three forms considered above, alternatively. In (11.11), $\lambda_{ij}(t) = \mu_{ij}(t)\, e^{\rho t}$ is the *capitalised* costate variable that firm i attaches to state j; accordingly, the generic transversality condition can be written as $\lim_{t\to\infty} e^{-\rho t}\lambda_{ij}(t)\, k_j(t) = 0\,\forall\, i, j$. The game engendered by the linear demand coincides with Cellini and Lambertini [4, 6] and will not be illustrated in detail. In what follows, we dwell upon the scenarios in which demand is either convex ($\alpha = 1/2$) or concave ($\alpha = 2$).

11.4.1 Convex Demand

Suppose $\alpha = 1/2$. If so, the FOC taken on (11.11) is

$$\frac{\partial \mathcal{H}_i(t)}{\partial q_i(t)} = e^{-\rho t}\left\{\left[a - q_i(t) - Q_{-i}(t)\right]\left[a - 3q_i(t) - Q_{-i}(t)\right] - c - \lambda_{ii}(t)\right\} = 0 \tag{11.12}$$

where $Q_{-i}(t) \equiv \sum_{j \neq i} q_j(t)$, and the n costate equations are

$$\begin{aligned} \dot{\lambda}_{ii} &= \lambda_{ii}(t)\left[\delta + \rho - f'(k_i(t))\right] \\ \dot{\lambda}_{ij} &= \lambda_{ij}(t)\left[\delta + \rho - f'\left(k_j(t)\right)\right] \end{aligned} \tag{11.13}$$

Now, imposing the symmetry conditions $q_i(t) = q_j(t)$ and $\lambda_{ii}(t) = \lambda_{jj}(t)$ for all i and j, (11.12) yields

$$[a - nq(t)][a - (n+2)q(t)] - c - \lambda(t) = 0 \tag{11.14}$$

which can be solved to obtain the optimal value of the costate:

$$\lambda^*(t) = [a - nq(t)][a - (n+2)q(t)] - c \tag{11.15}$$

as well as the equilibrium individual quantity at any time t:

$$q^{CN}_{\alpha=1/2}(t) = \frac{a(n+1) - \sqrt{a^2 + [c + \lambda(t)]n(n+2)}}{n(n+2)} \tag{11.16}$$

and it is immediate that $q^{CN}_{\alpha=1/2}(t) = q^{CN}_{\alpha=1/2}$ iff $\lambda(t) = 0$, i.e. firms play the static Cournot–Nash quantity if and only if the value of the costate is zero. We will come back to this particular case below.[5]

Differentiating both sides of (11.16), w.r.t. time, we obtain the following control equation:

$$\frac{\partial q^{CN}_{\alpha=1/2}(t)}{\partial t} \equiv \dot{q}^{CN}_{\alpha=1/2} = -\frac{\dot{\lambda}}{2\sqrt{a^2 + [c + \lambda(t)]n(n+2)}} \tag{11.17}$$

The r.h.s. of (11.17) can be appropriately rewritten to obtain

$$\dot{q}^{CN}_{\alpha=1/2} = -\frac{\lambda(t)\left[\delta + \rho - f'(k(t))\right]}{2\sqrt{a^2 + [c + \lambda(t)]n(n+2)}} \tag{11.18}$$

[5] Also note that the $n-1$ costate variables $\lambda_{ij}(t)$ and their equations do not appear in (11.12) and therefore can be disregarded as they exert no impact on firm i's strategy.

in which the index of $k(t)$ has also disappeared in view of the symmetry condition. Substituting (11.15) into (11.18), the control equation takes its definitive shape:

$$\dot{q}^{CN}_{\alpha=1/2} = -\frac{[(a - nq(t))(a - (n+2)q(t)) - c]\left[\delta + \rho - f'(k(t))\right]}{2\sqrt{a^2 + [c + \lambda(t)]n(n+2)}} \tag{11.19}$$

and $\dot{q}^{CN}_{\alpha=1/2} = 0$ in correspondence of

$$q^{ss}_{\alpha=1/2} = \frac{a(n+1) \pm \sqrt{a^2 + cn(n+2)}}{n(n+2)} \qquad f'(k^{ss}) = \delta + \rho \tag{11.20}$$

where superscript ss mnemonics for *steady state*. For the same reasons as in the static case, the relevant Cournot–Nash solution is $q^{CN}_{\alpha=1/2} = q^{-}_{\alpha=1/2}$ as in (11.5), that is, the optimal market-driven output engendered by open-loop rules replicates the static Cournot–Nash quantity. The phase diagram appearing in Fig. 11.2 portrays the case in which the quantity q^{ss}_R associated with the Ramsey equilibrium condition, also known as Ramsey (modified) golden rule, $f'(k(t)) = \delta + \rho$ is higher than $q^{ss}_{\alpha=1/2}$. This situation yields three steady-state equilibria, E_1, E_2 and E_3, at the intersections between the linear loci (11.20) and the concave locus $dk(t)/dt = 0$. The relative positions of these three points depend upon the parameters of the model and the marginal productivity of physical capital. In particular, $q^{CN}_{\alpha=1/2}$ is strictly increasing in the choke price a and decreasing in the number of firms n, as is the case for any (quasi-)static Cournot–Nash quantity, and therefore the exact relative positions of E_1, E_2 and E_3 depend on the choke price and industry fragmentation, all else equal.

The stability properties of E_1, E_2 and E_3 can be established by looking at the Jacobian matrix of the state-control system composed by (11.1) and (11.19):

$$J = \begin{bmatrix} \dfrac{\partial \dot{k}}{\partial k} = f'(k(t)) - \delta & \dfrac{\partial \dot{k}}{\partial q} = -1 \\[2ex] \dfrac{\partial \dot{q}}{\partial k} = \dfrac{\Gamma f''(k(t))}{2\Lambda} & \dfrac{\partial \dot{q}}{\partial q} = \dfrac{(\Upsilon + \Xi)\left[\delta + \rho - f'(k(t))\right]}{2\Lambda} \end{bmatrix} \tag{11.21}$$

in which

$$\begin{gathered} \Gamma \equiv [(a - nq(t))(a - (n+2)q(t)) - c] \\ \Lambda \equiv a(n+1) - n(n+2)q(t) \\ \Upsilon \equiv a^2[n(n+2) + 2] - 2an(n+1)(n+2)q(t) \\ \Xi \equiv n(n+2)\left[n(n+2)q^2(t) + c\right] \end{gathered} \tag{11.22}$$

The trace and determinant of the above Jacobian matrix are

$$\mathcal{T}(J) = \frac{\partial \dot{k}}{\partial k} + \frac{\partial \dot{q}}{\partial q}$$
$$\Delta(J) = \frac{\partial \dot{k}}{\partial k} \cdot \frac{\partial \dot{q}}{\partial q} - \frac{\partial \dot{k}}{\partial q} \cdot \frac{\partial \dot{q}}{\partial k} \tag{11.23}$$

Substituting $q^{CN}_{\alpha=1/2} = q^{-}_{\alpha=1/2}$ in (11.23), we obtain

$$\mathcal{T}(J) = \rho;\ \Delta(J) = -\left[f'(k(t)) - \delta\right]\left[f'(k(t)) - \delta - \rho\right] \tag{11.24}$$

If instead one poses $f'(k(t)) = \delta + \rho$ so as to look at the case in which $q^{ss} = q^{ss}_R$, one obtains

$$\mathcal{T}(J) = \rho;\ \Delta(J) = \frac{f''(k(t)) \cdot \Gamma}{2\Lambda} \tag{11.25}$$

with $\Gamma < 0$ for all $q \in \left(q^{-}_{\alpha=1/2}, q^{+}_{\alpha=1/2}\right)$ and conversely outside this range; and $\Lambda > 0$ for all $0 < q < a(n+1)/[n(n+2)]$, which, in turn, belongs to the interval $\left(q^{-}_{\alpha=1/2}, q^{+}_{\alpha=1/2}\right)$.

All of this, considering the situation appearing in Fig. 11.2, implies that, if a steady-state point is unstable, it is necessarily a focus because $\mathcal{T}(J) = \rho$. Moreover,

Proposition 11.1 *Suppose $\alpha = 1/2$. If $q^{ss}_R > q^{CN}_{\alpha=1/2}$, then E_1 and E_3 are saddle points, while E_2 is an unstable focus. If instead $q^{ss}_R < q^{CN}_{\alpha=1/2}$, then E_1 is an unstable focus, while E_2 and E_3 are saddle points.*

Three special cases are worth mentioning. If $E_1 = E_2$, then this as well as E_3 are saddle points. If $E_1 = E_3$, then this as well as E_2 are saddle points. In this case, the horizontal locus is tangent to the concave locus at the peak of the latter. Finally, if the horizontal locus is strictly above the concave one, E_2 is the only steady state: it is engendered by the Ramsey rule and is a saddle point.

Under open-loop information, the saddle path actually taken by firms will depend on the initial condition on capacity, which, under symmetry, will be the same k_0 for all of them. Suppose, for instance, that the relevant situation is that represented in Fig. 11.2. In such a case, a sufficiently small k_0 will cause firms to follow the saddle path to E_1, which is drawn in the phase diagram, while if k_0 is large, they will converge to the saddle point equilibrium E_3. The latter, however, is inefficient, as the same equilibrium quantity (or, from the consumers' standpoint, the same level of consumption) is associated with a larger volume of installed capacity (or physical capital), as already stressed in Cellini and Lambertini [4, 6].

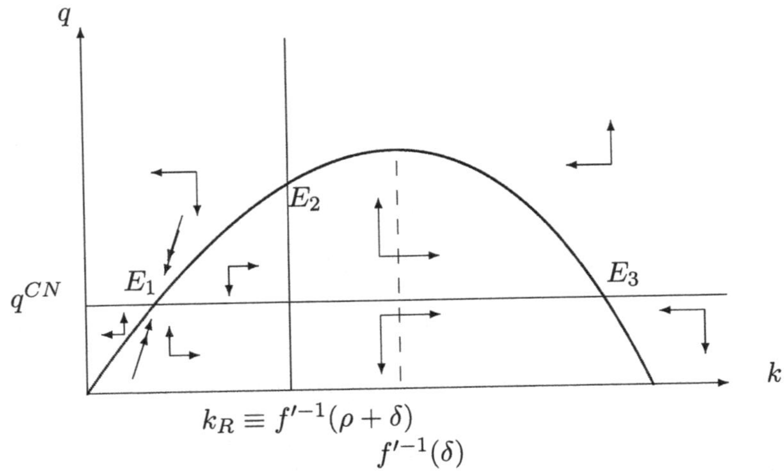

Fig. 11.2 The phase diagram with $\lambda_{ij}(t) \neq 0$

11.4.2 Concave Demand

Here, $\alpha = 2$ and the FOC taken on (11.11) is

$$\frac{\partial \mathcal{H}_i(t)}{\partial q_i(t)} = e^{-\rho t}\left\{\frac{2\left[a - Q_{-i}(t)\right] - 3q_i(t)}{2\sqrt{a - q_i(t) - Q_{-i}(t)}} - c - \lambda_{ii}(t)\right\} = 0 \tag{11.26}$$

while the n adjoint equations are the same as in (11.13). After imposing symmetry, (11.26) yields[6]

$$\lambda^* = \frac{2a - (2n+1)q - 2c\sqrt{a-q}}{2\sqrt{a-q}} \tag{11.27}$$

Then, differentiating both sides of (11.27) w.r.t. time and using (11.13), we have the following control equation:

$$\dot{q}^{CN}_{\alpha=2} = \frac{2\lambda(c+\lambda)\left[f'(k) - \delta - \rho\right]}{(2n+1)^2}\left[2n + \frac{a(2n+1) + 2n^2(c+\lambda)^2}{(c+\lambda)\sqrt{a(2n+1) + 2n^2(c+\lambda)^2}}\right] \tag{11.28}$$

The definitive formulation of the control equation obtains by substituting (11.27) into (11.28):

[6]Henceforth, the explicit indication of the time argument is omitted for the sake of brevity.

$$\dot{q}^{CN}_{\alpha=2} = \frac{2\left(a-nq\right)\left[2a-\left(2n+1\right)q\right]\left[2\left(a-c\sqrt{a-nq}-\left(2n-1\right)q\right)\right]\left[f'\left(k\right)-\delta-\rho\right]}{\sqrt{\Psi}} \tag{11.29}$$

where

$$\Psi \equiv \left[2a\left(n+1\right)-n\left(2n+1\right)q\right]\left[2a-\left(2n+1\right)q\right] \tag{11.30}$$

The control becomes stationary in correspondence of $q^{-}_{\alpha=2}$, $q^{+}_{\alpha=2}$ and

$$\widehat{q}_{\alpha=2} = \frac{a}{n};\ \widetilde{q}_{\alpha=2} = \frac{2a}{2n+1};\ f'\left(k^{ss}\right) = \delta+\rho \tag{11.31}$$

Note also that $\Psi \leq 0$ for all

$$q \in \left[\widetilde{q}_{\alpha=2}, \frac{2a\left(n+1\right)}{n\left(2n+1\right)}\right] \tag{11.32}$$

Therefore, since

$$0 < q^{-}_{\alpha=2} < \widetilde{q}_{\alpha=2} < q^{+}_{\alpha=2} < \widehat{q}_{\alpha=2} < \frac{2a\left(n+1\right)}{n\left(2n+1\right)} \tag{11.33}$$

we must concentrate on $q^{-}_{\alpha=2} = q^{ss}_{\alpha=2} = q^{CN}_{\alpha=2}$, which is the optimal quantity of the corresponding static game (see (11.6)), and the Ramsey quantity q^{ss}_{R} corresponding to the golden rule $f'\left(k^{ss}\right) = \delta+\rho$. Relying on the analysis of the stability properties of the state-control system (11.1) and (11.29), we may formulate the following Proposition, which shares its qualitative properties with Proposition 11.1:

Proposition 11.2 *Suppose $\alpha = 2$. If $q^{ss}_{R} > q^{CN}_{\alpha=2}$, then E_1 and E_3 are saddle points, while E_2 is an unstable focus. If instead $q^{ss}_{R} < q^{CN}_{\alpha=2}$, then E_1 is an unstable focus while E_2 and E_3 are saddle points.*

The details of the proof are in the Appendix. For obvious reasons, the resulting picture portraying the phase diagram in the situation in which the saddle-point equilibrium is market-driven is observationally equivalent to Fig. 11.2. From Cellini and Lambertini [4, 6], we also know that an analogous phase diagram emerges if $\alpha = 1$, i.e. in presence of a linear demand. We are now ready to comparatively assess the equilibrium configurations delivered by the three different demand functions considered in the foregoing analysis, in order to carry out a simple parametric analysis of the switch from the Cournot equilibrium to the Ramsey equilibrium (or the opposite).

11.4.3 Switching Regimes: Cournot Versus Ramsey

Were production costless, the static and dynamic versions of the game could be solved analytically to obtain the Cournot–Nash quantity $q^{CN}\left(\alpha\right) = a\alpha/\left(\alpha n+1\right)$ explicitly parametric in α, and we know from Cellini and Lambertini [5] that $\partial q^{CN}(\alpha)/\partial\alpha > 0$. This, in terms of the three specific examples considered in the present paper, would imply $q^{CN}_{\alpha=2} > q^{CN}_{\alpha=1} > q^{CN}_{\alpha=1/2}$.

If $c \in (0, a)$,, we cannot write any partial derivative w.r.t. the curvature of market demand because α necessarily takes numerical values, but from (11.7) and Fig. 11.1, we are aware that Cournot–Nash quantities (i) increase monotonically in the choke price, while (ii) their ranking changes as the choke price increases. The first property implies that, for sufficiently high levels of a, the Cournot–Nash strategy will cross the concave locus $\dot{k} = 0$ beyond the Ramsey equilibrium, and may even end up not intersecting it at all if a is very large. Yet, the second property entails that the identity of the Cournot–Nash quantity collapsing first into the Ramsey one is, *a priori*, undefined. With one exception, though, as we may exclude the case $q^{CN}_{\alpha=1/2} > q^{CN}_{\alpha=1} > q^{CN}_{\alpha=2}$, because $q^{CN}_{\alpha=1/2} > 0$ if and only if $a \geq \underline{a}$.

To see this, it suffices to observe that q^{ss}_R, being determined by $f'(k) = \delta + \rho$, is independent of demand conditions. Therefore, for instance, any $a \in (\underline{a}, \widehat{a})$, such that $q^{CN}_{\alpha=1} > q^{CN}_{\alpha=1/2} > q^{CN}_{\alpha=2}$, might indeed yield $q^{CN}_{\alpha=1} = q^{ss}_R$, which is the situation depicted in Fig. 11.3. This amounts to taking the conditions determining the Ramsey solution as given and asking oneself what happens if demand conditions change provoking a shift in the horizontal locus corresponding to the market-driven solution. But, in the opposite perspective, it might as well be the case that the marginal productivity of physical capital or decay and discount rates modify, triggering a change in the concave locus and a shift of the vertical locus (for instance, leftwards) so as to generate the very same consequence. This brief example suffices to suggest that changes in the curvature of demand and the choke price may push (pull) the industry into (out of) the Ramsey rule, and the same may happen due to changes in technological and intertemporal conditions.

Without delving in the details of all admissible cases, we may confine ourselves to formulate the following:

Proposition 11.3 *Suppose $f(k)$, δ and ρ are fixed. Then,*

- *for all $a \in (\underline{a}, \overline{a})$, $q^{CN}_{\alpha=1} > \max\left\{q^{CN}_{\alpha=1/2}, q^{CN}_{\alpha=2}\right\}$; therefore, the market-driven solution candidate to collapsing first into the Ramsey rule is the one generated by a linear demand;*
- *for all $a > \overline{a}$, $q^{CN}_{\alpha=2} > q^{CN}_{\alpha=1} > q^{CN}_{\alpha=1/2}$; therefore, the market-driven solution candidate to collapsing first into the Ramsey rule is the one generated by a concave demand.*

The above Proposition has a relevant Corollary:

Corollary 11.1 *For a sufficiently high value of the choke price a,*

$$q^{ss}_R \leq \min\left\{q^{CN}_{\alpha=1/2}, q^{CN}_{\alpha=1}, q^{CN}_{\alpha=2}\right\}$$

If so, then demand, cost and competitive conditions have no bearings at all on the equilibrium.

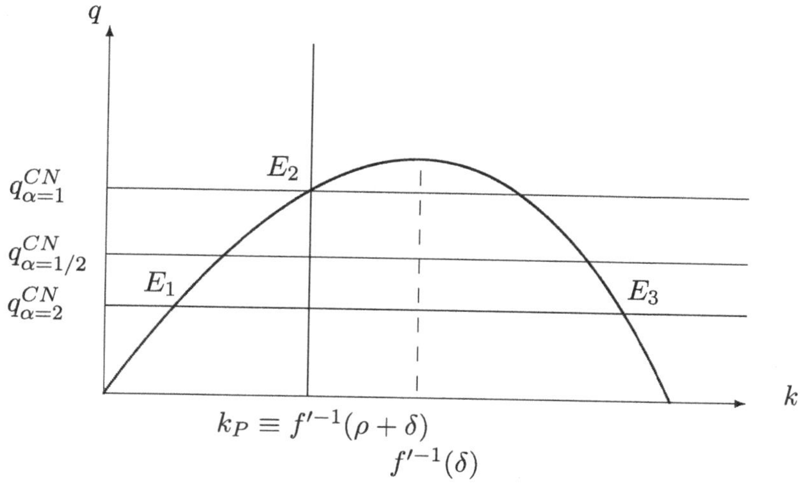

Fig. 11.3 The phase diagram with $\lambda_{ij}(t) \neq 0$

More explicitly, if the entire set of market-driven solutions collapses into the Ramsey one, then the curvature of market demand becomes immaterial as far as firms' production and sale decisions are concerned, since players only obey the Ramsey rule for any a, c and n, whereby also the level of marginal cost and the structure of the industry become irrelevant. This means that the equilibrium configurations generated by three very different markets are observationally equivalent, and therefore a casual observer looking at q_R^{ss} but uninformed about demand conditions would be altogether unable to formulate an educated guess about the shape of market demand.

All of this can be summarised in the following terms. As far as the issue at stake, namely regime switch, is concerned, the foregoing discussion amounts to saying that, if demand is low (or marginal cost is high) enough, the horizontal arm determines the capacity accumulation path as well as the steady state, which are indeed market-driven. It is also worth recalling that demand depends on both α and a. So, for instance, if the vertical intercept of demand increases sufficiently to determine $q_R^{ss} \leq \min\left\{q^{CN}_{\alpha=1/2}, q^{CN}_{\alpha=1}, q^{CN}_{\alpha=2}\right\}$, capacity accumulation and the resulting equilibrium just obey the Ramsey rule, which is independent of market behaviour. Consequently, in the latter case firms are not even aware of the specific pattern of strategic competition in output levels associated with the definition of their profit functions, the only residual trace of it being the fact that the equilibrium price level will be determined along a specific demand function after plugging the industry output Q_R^{ss} into it. This level of collective production, however, is entirely determined by depreciation and discounting, and therefore could as well be supplied by the same number of firms initially operating under Bertrand as well as monopolistic or perfect competition (cf. Cellini and Lambertini [4, 5]). In fact, the golden rule $f'(k^{ss}) = \delta + \rho$ is invariant across the three cases examined in this paper, while demand features are not. In some sense, the regime switch implies an analogous switch in the minds of entrepreneurs, who cannot tell what it would have been like to maximise profits in a proper Cournot game.

11.4.4 The State-Redundant Solution

There remains to examine the case in which $\lambda_{ii} = \lambda_{ij} = 0$ at all times, due to the fact that adjoint equations (11.13) are differential equations in separable variables admitting the nil solution. Taking this route, FOCs on controls become independent of costates and therefore also states, and this implies that the game is state-redundant and admits a strongly time-consistent solution represented by the static Cournot–Nash quantity.[7] Clearly, this holds true irrespective of the curvature of market demand.

More importantly, keeping in mind the above discussion, the additional implication of state redundancy is that the Ramsey equilibrium at $f'(k) = \delta + \rho$ disappears, and consequently the switch between the market-driven solution and the one determined by technology, decay and discounting never takes place, as firms systematically play q_α^{CN}, with α alternatively equal to 1/2, 1 or 2, forever. This delivers, for instance, the phase diagram appearing in Fig. 11.4, which is equivalent to Fig. 11.3 without the vertical locus; hence, the steady-state point E_2 disappears as well.

In the presence of the market-driven solution only, the dynamic properties of the game are illustrated by the horizontal arrows along q_α^{CN}, and there emerges that in absence of the Ramsey rule the saddle-point equilibrium is necessarily the inefficient solution corresponding to the rightwards intersection between the specific q_α^{CN} and $\dot{k} = 0$ in E_3, as already highlighted in Cellini and Lambertini [6].

Of course, if a is large enough, one of the horizontal loci (or even all of them) may not cross the concave locus. In such a case, the steady-state equilibrium will not exist for at least one of the three possible demand functions considered in this game. It is worth noting that this cannot happen if $\lambda_{ii} \neq 0$, as then the Ramsey solution indeed generates a saddle-point equilibrium whenever the relevant Cournot quantity lies above $\dot{k} = 0$.

This brief discussion shows that, in the Cournot–Ramsey game, strong time consistency under open-loop information comes at a twofold price, namely the certain lack of the Ramsey equilibrium and the possible disappearance of a steady-state equilibrium under a portion or all of the set of market demand schedules. Hence, we may formulate the following:

[7] The strand of research about the arising of degenerate feedback strategies under open-loop information dates back to the identification of strongly time-consistent solutions in trilinear games [8]. Later, the same result was shown to hold in exponential games [22] as well as in linear-state or, more generally, state-separable games [9], while Mehlmann and Willing [20] proved that differential game characterised by strongly time-consistent open-loop strategies is or can be reformulated as a state-redundant game. For more, see also Fershtman [13], Dockner et al. [10, Chap. 7], Cellini et al. [7] and Lambertini [17, Chap. 1].

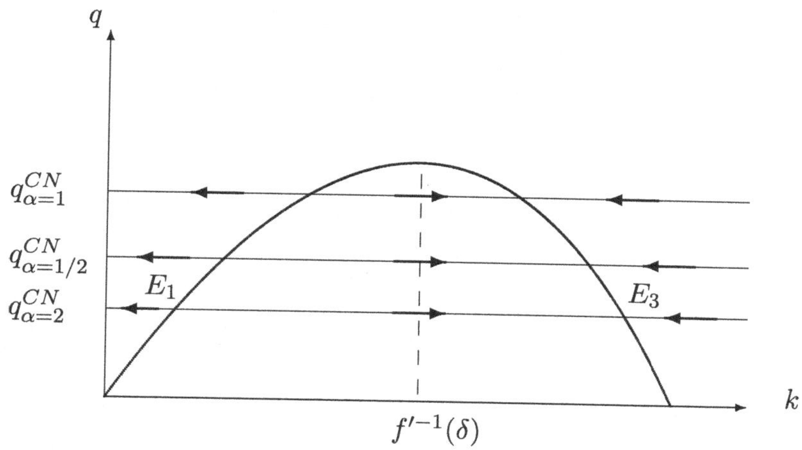

Fig. 11.4 The phase diagram with $\lambda_{ij}(t) = 0$

Proposition 11.4 *The state-redundant solution of the Cournot–Ramsey differential game admits the market-driven solution only, q_α^{CN}, for any curvature of market demand. This (i) excludes the attainment of the golden rule and (ii) may imply that there exists no steady state at all.*

Propositions 11.1–11.2 and 11.4 imply:

Corollary 11.2 *While the weakly time-consistent solution of the game yields at least one saddle-point equilibrium, irrespective of demand conditions, the strongly time-consistent solution does not ensure, in general, the existence of a steady state.*

Whenever the second case materialises, given the dynamics of k above the concave locus, the individual and aggregate capacity endowment is bound to shrink to zero. One might say that taking the route of state redundancy rules out regime switches by privileging the market-driven equilibrium, *provided the latter does exist*. That is, ruling out switches may take a heavy toll.

11.5 Concluding Remarks

We have investigated the impact of demand curvature on the equilibrium outcome of a Cournot–Ramsey industry in three special cases which are the representative of a class of games with analogous qualitative properties, to find that the choke price plays a key role in shaping the nature of the steady-state equilibrium. In particular, it determines whether and under what conditions the industry will switch from the market-driven Cournot-Nash solution to the golden rule one. We have also stressed that if the switch takes place, firms become unaware of the specific nature of strategic behaviour associated with the market-driven solution, as the Ramsey rule has no correlation with demand conditions. Additionally, we have also shown that the strongly time-consistent case excludes the attainment of the Ramsey solution and may also prevent firms from reaching a steady state, as the latter does not exist if the choke price is too high.

Acknowledgements We would like to thank two anonymous referees for helpful comments and suggestions. The usual disclaimer applies.

Appendix

To prove the claim in Proposition 11.2, it suffices to evaluate the sign of the trace and determinant of the Jacobian matrix associated with the state-control system made up by (11.1) and (11.29) at $f'(k) = \delta + \rho$:

$$\mathcal{T}(J) = \rho$$
$$\Delta(J) = \frac{2\sqrt{a-nq}\,[2a-(2n+1)\,q]\left[2a-(2n+1)\,q-2c\sqrt{a-nq}\right] f''(k)}{\sqrt{\Psi}} \tag{11.34}$$

Since $\mathcal{T}(J) > 0$, the Ramsey solution is either a saddle point or an unstable focus depending on the sign of the determinant. Knowing that $f''(k) < 0$, and verifying that

$$\frac{2\sqrt{a-nq}\,[2a-(2n+1)\,q]\left[2a-(2n+1)\,q-2c\sqrt{a-nq}\right]}{\sqrt{\Psi}} < 0 \tag{11.35}$$

for all $q \in \left(q^-_{\alpha=2}, q^+_{\alpha=2}\right)$ while the opposite holds outside this interval, we see that

$$q_R^{ss} \in \left(0, q^-_{\alpha=2}\right) \Leftrightarrow \Delta(J) < 0 \tag{11.36}$$

and

$$q_R^{ss} \in \left(q^-_{\alpha=2}, q^+_{\alpha=2}\right) \Leftrightarrow \Delta(J) > 0 \tag{11.37}$$

Hence, if $q_R^{ss} < q^{CN}_{\alpha=2} = q^-_{\alpha=2}$, the Ramsey rule produces a saddle-point equilibrium, while the market-driven solution is associated with an unstable focus. If instead $q_R^{ss} \in \left(q^-_{\alpha=2}, q^+_{\alpha=2}\right)$, the opposite applies.

References

1. Anderson, S. P., & Engers, M. (1992). Stackelberg vs Cournot oligopoly equilibrium. *International Journal of Industrial Organization, 10*, 127–135.
2. Anderson, S. P., & Engers, M. (1994). Strategic investment and timing of entry. *International Economic Review, 35*, 833–853.
3. Bischi, G. I., Chiarella, C., Kopel, M., & Szidarovszky, F. (2010). *Nonlinear oligopolies: Stability and bifurcations*. Heidelberg: Springer.
4. Cellini, R., & Lambertini, L. (1998). A dynamic model of differentiated oligopoly with capital accumulation. *Journal of Economic Theory, 83*, 145–155.
5. Cellini, R., & Lambertini, L. (2007). Capital accumulation, mergers, and the Ramsey golden rule. In M. Quincampoix, T. Vincent, & S. Jørgensen (Eds.), *Advances in dynamic game theory and applications*. Annals of the International Society of Dynamic Games (Vol. 8, pp. 487–505). Boston: Birkhäuser.
6. Cellini, R., & Lambertini, L. (2008). Weak and strong time consistency in a differential oligopoly game with capital accumulation. *Journal of Optimization Theory and Applications, 138*, 17–26.
7. Cellini, R., Lambertini, L., & Leitmann, G. (2005). Degenerate feedback and time consistency in differential games. In E. P. Hofer & E. Reithmeier (Eds.), *Modeling and control of autonomous decision support based systems. Proceedings of the 13th International Workshop on Dynamics and Control* (pp. 185–192). Shaker Verlag: Aachen.
8. Clemhout, S., & Wan, H. Y, Jr. (1974). A class of trilinear differential games. *Journal of Optimization Theory and Applications, 14*, 419–424.
9. Dockner, E. J., Feichtinger, G., & Jørgensen, S. (1985). Tractable classes of nonzero-sum open-loop Nash differential games: Theory and examples. *Journal of Optimization Theory and Applications, 45*, 179–197.
10. Dockner, E. J., Jørgensen, S., Long, N. V., & Sorger, G. (2000). *Differential games in economics and management science*. Cambridge: Cambridge University Press.
11. Dragone, D., Lambertini, L., & Palestini, A. (2012). Static and dynamic best-response potential functions for the non-linear Cournot game. *Optimization, 61*, 1283–1293.
12. Dragone, D., Lambertini, L., Leitmann, G., & Palestini, A. (2015). Hamiltonian potential functions for differential games. *Automatica, 62*, 134–138.
13. Fershtman, C. (1987). Identification of classes of differential games for which the open-loop is a degenerate feedback Nash equilibrium. *Journal of Optimization Theory and Applications, 55*, 217–231.
14. Lamantia, F. (2011). A nonlinear Duopoly with efficient production-capacity levels. *Computational Economics, 38*, 295–309.
15. Lambertini, L. (1996). Cartel stability and the curvature of market demand. *Bulletin of Economic Research, 48*, 329–334.
16. Lambertini, L. (2010). Oligopoly with hyperbolic demand: A differential game approach. *Journal of Optimization Theory and Applications, 145*, 108–119.
17. Lambertini, L. (2018). *Differential games in industrial economics*. Cambridge: Cambridge University Press.
18. Lambertini, L., & Palestini, A. (2014). On the feedback solution of differential oligopoly games with hyperbolic demand curve and capacity accumulation. *European Journal of Operational Research, 236*, 272–281.
19. Mehlmann, A. (1988). *Applied differential games*. New York: Plenum Press.
20. Mehlmann, A., & Willing, R. (1983). On nonunique closed-loop Nash equilibria for a class of differential games with a unique and degenerate feedback solution. *Journal of Optimization Theory and Applications, 41*, 463–472.
21. Ramsey, F. P. (1928). A mathematical theory of saving. *Economic Journal, 38*, 543–559.
22. Reinganum, J. (1982). A class of differential games for which the closed loop and open loop nash equilibria coincide. *Journal of Optimization Theory and Applications, 36*, 253–262.

Chapter 12
Optimal Taxation with Endogenous Population Growth and the Risk of Environmental Disaster

Tapio Palokangas

JEL Classification: J13 · O44 · Q53 · Q56 · Q58

12.1 Introduction

This document considers a market economy where firms produce goods from labor and capital and households save in capital, dictate the number of their children, and spend on health care to improve their survival. The economy contains the externality that population growth and capital accumulation boost pollution that threatens to trigger a lethal environmental disaster. Could this externality be eliminated by (linear) taxation? This research question is examined by a dynamic game where the benevolent government is the leader and the representative household the follower.

For the sake of clarity, the disaster is taken as a random regime shift that occurs only once, with the post-event regime holding indefinitely. As pointed out by de Zeeuw and Zemel [2], this restriction is not essential, and models of recurrent events, where several shifts occur at random times with independent intervals, can be analyzed using the same methodology. Because the construction of different mortality rates for different cohorts would excessively complicate the analysis, then, following Becker [1], it is assumed that the whole population has a uniform mortality rate, for simplicity.

Polasky et al. [12] analyze how the threat of future regime shift affects the optimal management of natural resources. They focus on harvesting a renewable resource (e.g., fishery), whose growth rate is dependent on the regime and whose stock can trigger a regime shift. They show that the possibility of the regime shift makes the

T. Palokangas (✉)
University of Helsinki, Helsinki GSE, IIASA, IZA,
P.O. Box 17, FIN-00014 Helsinki, Finland
e-mail: tapio.palokangas@helsinki.fi
URL: http://blogs.helsinki.fi/palokang/

J. L. Haunschmied et al. (eds.), *Dynamic Economic Problems with Regime Switches*, Dynamic Modeling and Econometrics in Economics and Finance 25,
https://doi.org/10.1007/978-3-030-54576-5_12

central planner *precautionary*, i.e., willing to maintain a larger stock of the resource. In this document, the government faces the risk of disaster due to pollution and behaves in a precautionary manner by keeping the damaging stock (pollution) at a lower level.

Many dynamic models of pollution control assume smooth convex damage functions (e.g., van der Ploeg and de Zeeuw [16], and Dockner and Long [4]), which ignores the effect of a potential regime shift on the optimal policy. Then, there is no need for precautionary measures against pollution: the policy maker should respond at the moment pollution occurs, but not beforehand. To examine the need for precautionary environmental policy, de Zeeuw and Zemel [2] consider the management of a system that is subject to the risk of an abrupt and random jump in pollution damage. This document applies the same idea for the management of a market economy when pollution-related mortality is the damage.

Haurie and Moresino [7], Polasky et al. [12], and de Zeeuw and Zemel [2] consider only the central planner that can fully control all resources of the economy. In contrast, this document examines the government in a market economy where microeconomic agents (households and firms) determine production, fertility, and capital accumulation, unintentionally generating lethal emissions, but the government can use only linear taxes. Public policy is then constructed as a dynamic Stackelberg game where the government is the leader. This approach has the benefit that the suggested policy rules can be presented directly in terms of observable variables (e.g., prices and the interest rate).

Tsur and Zemel [13, 14] ignore population growth, but examine the possibility of climate change in a market economy where firms employ labor, capital, and two energy inputs that are perfect substitutes: clean input that does not emit, and dirty input the emissions of which accumulate "hazardous" stock that threatens to trigger the climate change. As a result, they obtain a Pigouvian tax on the "hazardous" input. In contrast, this document considers endogenous population growth that may trigger the catastrophe. In order to avoid excessive complications in the model, the choice of energy inputs is ignored, and it is assumed that population growth and capital accumulation generate "hazardous" pollution as a by-product.

Harford [5, 6] addresses the issue of environmental and population externalities in a dynamic model where the individuals are altruistic toward their descendants and environmental pollution is a joint product of output. In his model, the social planner optimizes the utility of the representative individual. By comparing this optimum with the individuals' decisions, he shows that Pareto optimality requires both a pollution tax and a parental tax per child, because the former does not limit fertility enough to keep population stationary. To contribute to the discussion on this matter, this study adds Harford's parental tax into the potential tools of the government.

Palokangas [11] and Lehmijoki and Palokangas [9] examine optimal taxation in an economy where households dictate fertility and save for capital, while firms produce output from labor and capital, and population growth and capital accumulation generate pollution. In those studies, however, there is no precautionary motive, for the current damage (mortality) is a smooth function of current pollution. In this document, in contrast, there is a precautionary motive: pollution triggers the damage (i.e., lethal environmental disaster) randomly at any moment of time.

The remainder of this document is organized as follows. Section 12.2 presents the basic structure of the economy, including the behavior of competitive firms. Then, a stochastic Stackelberg game is defined, with the government as the leader and the representative micro-household as the follower. Section 12.3 considers the household's and Sect. 12.4 the government's behaviors. Section 12.5 presents optimal public policy and Sect. 12.6 summarizes the results.

12.2 The Economy as a Whole

12.2.1 Population and Labor Supply

In the model, time t is continuous. Population L grows at the rate that is equal to the fertility rate f minus the mortality rate m:

$$\frac{1}{L}\frac{dL}{dt} = f - m, \quad L(0) = L_0. \tag{12.1}$$

The units are normalized so that one unit of labor is needed to rear one newborn. Then, labor devoted to child rearing is equal to total fertility fL, and the remainder of the population, N, works in production:

$$N \doteq L - fL = (1 - f)L \;\; \Leftrightarrow \;\; n \doteq N/L = 1 - f. \tag{12.2}$$

12.2.2 The Goods Market

In the economy, there is only one good. The depreciation of capital is included in the production function of that good, so that the accumulation of capital K is given by $\frac{dK}{dt}$. Because (private) capital is the only asset in the model, private saving is equal to the accumulation of capital, $\frac{dK}{dt}$. The output of the good, Y, is used in consumption C, health care H, and investment $\frac{dK}{dt}$:

$$Y = C + H + \frac{dK}{dt}. \tag{12.3}$$

It is convenient to define output Y, consumption C, health care H, and capital K in proportion to population L:

$$y \doteq \frac{Y}{L}, \quad c \doteq \frac{C}{L}, \quad h \doteq \frac{H}{L}, \quad k \doteq \frac{K}{L}. \tag{12.4}$$

Noting (12.1), (12.3), and (12.4), *investment per head* in the economy is defined by

$$s \doteq \frac{dk}{dt} = \frac{d}{dt}\left(\frac{K}{L}\right) = \frac{1}{L}\frac{dK}{dt} - \frac{dL}{dt}\frac{K}{L^2} = \frac{1}{L}\frac{dK}{dt} + (m - f)k. \tag{12.5}$$

Because it is convenient to define investment per head $s \doteq \frac{dk}{dt}$ as a control in dynamic programming, then, by (12.5), private saving $\frac{dK}{dt}$ is given by

$$\frac{dK}{dt} = [s + (f - m)k]L. \tag{12.6}$$

12.2.3 Firms

The firms produce output Y from capital K and labor input N [cf. (12.2)] according to neoclassical technology:

$$Y = F(K, N), \quad F_K \doteq \frac{\partial F}{\partial K} > 0, \quad F_N \doteq \frac{\partial F}{\partial N} > 0, \quad F_{KK} \doteq \frac{\partial^2 F}{\partial K^2} < 0,$$
$$F_{NN} \doteq \frac{\partial^2 F}{\partial N^2} < 0, \quad F_{KN} \doteq \frac{\partial^2 F}{\partial K \partial N} > 0, \quad F \text{ linearly homogeneous.} \tag{12.7}$$

Noting (12.2), (12.4), and (12.7), output per head, y, can be defined as a function of capital per head, k, and the fertility rate, f, as follows:

$$1 - f = n \doteq N/L, \quad Y/L = F(k, n) = F(k, 1 - f) \doteq y(k, f),$$
$$y_k \doteq \frac{\partial y}{\partial k} = F_K(k, n) > 0, \quad y_f \doteq \frac{\partial y}{\partial f} = -F_N(k, n) < 0. \tag{12.8}$$

The representative firm maximizes its profit Π by capital input K and labor input N according to technology (12.7), given the wage w and the interest rate r. With (12.4) and (12.8), this implies

$$\Pi \doteq \max_{K,N}\left[F(K, N) - wN - rK\right] = L \max_{k,n}\left[F(k, n) - wn - rk\right]. \tag{12.9}$$

Because the production function F is subject to constant returns to scale (i.e., linearly homogeneous), then, in equilibrium, the marginal products of capital and labor, F_K and F_N, are equal to the interest rate r and the wage w, respectively, and total profit Π is equal to zero [cf. (12.8) and (12.9)]:

$$y(k, f) = F(k, n) = F_B n + F_K k = wn + rk = (1 - f)w + rk,$$
$$r = F_K(k, n) = y_k, \quad w = F_N(k, n), \quad y_f = -F_N(k, n) = -w. \tag{12.10}$$

12.2.4 Externality

It is assumed that aggregate capital K and aggregate population L pollute according to the geometric average $P = K^{\gamma}L^{1-\gamma} = k^{\gamma}L$, where $0 < \gamma < 1$ is a constant. Then, the change of pollution, $v \doteq \frac{dP}{dt}$, is obtained from (12.1) and (12.6) as follows:

$$\frac{1}{P}\frac{dP}{dt} = \frac{d\ln P}{dt} = \gamma\frac{d\ln k}{dt} + \frac{d\ln L}{dt} = \gamma\frac{1}{k}\frac{dk}{dt} + \frac{1}{L}\frac{dL}{dt} = \gamma\frac{s}{k} + f - m$$
$$\Leftrightarrow \quad v \doteq \frac{dP}{dt} = \left(\gamma\frac{s}{k} + f - m\right)P. \tag{12.11}$$

With result (12.11), population L can be replaced by pollution P as a predetermined state variable in the model.

The probability of the environmental disaster, π, is assumed to be an increasing function of pollution P. Then, the disaster can be considered as a random shock q with mean $\pi(P)$ as follows:

$$q = \begin{cases} 1 \text{ with probability } \pi(P), \\ 0 \text{ with probalility } 1 - \pi(P), \end{cases} \quad \text{where } \pi' > 0. \tag{12.12}$$

The externality in the economy is the following: the environmental shock q increases every individual's mortality rate m simultaneously, but each individual can decrease her personal mortality rate m by spending on her personal health care h with increasing marginal costs. This function is specified by

$$m = \chi(\delta q - h), \quad \chi' > 0, \quad \frac{d^2(-m)}{dh^2} = \chi'' > 0, \tag{12.13}$$

where the constant $\delta > 0$ is the effect of the shock q in terms of output per head (= in terms of health care per head, h) and

$$\left|\frac{dm}{dh}\right| = -\frac{dm}{dh} = \chi' \tag{12.14}$$

the marginal efficiency of personal health care h in decreasing the personal mortality rate m.

The household chooses its saving per head, $s = \frac{dk}{dt}$, fertility rate, f, and heath care per head, h. Because of the one-to-one correspondence between h and m through the function (12.13), health care h can be replaced by the mortality rate m as the household's control in the model, for convenience. Denoting the inverse function of χ by $z(m) \doteq \chi^{-1}(m)$ in (12.13) yields

$$h = \delta q - z(m), \quad z' \doteq \frac{1}{\chi'(m)} > 0, \quad z'' \doteq -\frac{\chi''}{(\chi')^2} < 0. \tag{12.15}$$

The factors affecting the mortality rate m [cf. (12.13)] affect also the level of health, ℓ, but in the opposite direction: the environmental shock q worsens every individual's health simultaneously, but each individual can improve her personal health ℓ by her personal health care h. Because the definition of health ℓ as a separate function of q and h would excessively complicate the analysis, and because it is technically convenient to handle the mortality rate as the household's control in the model, health ℓ and the mortality rate m are defined as negatively associated joint products of the same process[1]:

$$\ell(m), \quad \ell' < 0, \quad \ell'' \text{ exists.} \tag{12.16}$$

12.2.5 Public Policy

The government sets a poll tax $a \in \Re$ per head, the tax $\tau \in (-\infty, 1)$ on capital income rK, the parental tax $x \in \Re$ on the number of children, fL, and the tax $b \in (-1, \infty)$ on health care H. If a tax is negative, then it is a subsidy. Any set of linear taxes that support Pareto optimum in the model is equivalent to those taxes. The government's budget is [cf. (12.4)]:

$$aL + xfL + \tau rK + bH = 0 \;\Leftrightarrow\; a + xf + \tau rk + bh = 0. \tag{12.17}$$

In the model, the setup of public policy is a *Stackelberg game* as follows. The representative household is the *follower* that determines its consumption per head, c, its spending on health care per head, h, and its fertility rate, f, taking the taxes (a, x, τ, b) and the environmental shock q as given. The benevolent government is the *leader* that maximizes the representative household's utility by the taxes (a, x, τ, b), observing the follower's behavior, the behavior of the firms, (12.10), the budget constraint of its own, (12.17), and the risk of the regime shift (12.12, 12.15) due to pollution (12.11). The follower's and leader's behaviors are examined in Sects. 12.3 and 12.4.

12.3 The Household

12.3.1 Utility

According to Becker [1], an individual derives her utility $c(t)f(t)^{\alpha}$, where $\alpha > 0$ is a constant, from her consumption $c(t)$ and the fertility rate in her household, $f(t)$, at each time t. This study extends that framework by introducing personal health ℓ as

[1]The mortality rate m is introduced as a factor of utility through health ℓ [cf. (12.18)] only to ensure that the functions (12.26) and (12.42) can be strictly concave with respect to the mortality rate m for realistic values of consumption per head, c, capital per head, k, and the mortality rate m.

the third factor of individual utility. Consequently, noting (12.16), periodic utility u is a function of consumption per head, c, the fertility rate, f, and the mortality rate, m, as follows:

$$u(t) = c(t)f(t)^{\alpha}\ell\big(m(t)\big), \quad \alpha > 0, \quad \ell' < 0, \quad \ell'' \text{ exists.} \tag{12.18}$$

Let ρ be the constant rate of time preference for a hypothetical individual who could live forever. When an individual faces the mortality rate m, her probability of dying in a short time dt is equal to $m\,dt$. Then, the probability of her survival beyond the period $[\zeta, t]$ is given by $e^{m(\zeta-t)}$, and her expected periodic utility at time $t \geq \zeta$ is $e^{m(\zeta-t)}u(t)$. Consequently, noting (12.16), the representative member's utility for the whole period $t \in [\zeta, \infty)$ in the household is given by

$$\int_{\zeta}^{\infty} u(t)^{\sigma} e^{(\rho+m)(\zeta-t)}dt \ \text{ with } (12.18), \quad 0 < \sigma < 1, \tag{12.19}$$

where σ is a parameter and $\rho + m$ the effective rate of time preference with mortality. The closer σ is to one, the more eagerly the household transfers resources from present to future by saving.

12.3.2 Saving

Investment $\frac{dK}{dt}$ is equal to private saving:

$$\frac{dK}{dt} = wN + rK - C - hL - [a + xf + \tau rk + bh]L, \tag{12.20}$$

where w is the wage, r the interest rate, N labor supply, wN labor income, rK capital income, C consumption, h spending on health care per head, hL total spending on health care, and $[a + xf + \tau rk + bh]L$ tax expenditures [cf. (12.17)]. By (12.4), (12.6), (12.8), (12.15), and (12.20), consumption per head, c, can be defined as a function of the household's controls (s, f, m), capital per head, k, taxes (a, x, τ, b), the wage w, the interest rate r and the shock q as follows:

$$\begin{aligned}
&s + (f - m)k = \frac{1}{L}\frac{dK}{dt} = \frac{wN + rK - C}{L} - (1+b)h - a - xf - \tau rk \\
&= (1 - f)w + rk - c + (1+b)[z(m) - \delta q] - a - xf - \tau rk \ \Leftrightarrow \\
&c = \widetilde{c}(s, f, m, k, a, x, \tau, b, w, r, q) \doteq \\
&w + (m - f + r - \tau r)k - s + (1+b)[z(m) - \delta q] - (w + x)f - a.
\end{aligned} \tag{12.21}$$

12.3.3 Transformation from Real into Virtual Time

The mortality rate m can be eliminated from the discount factor of the utility function (12.19) by Uzawa's [15] transformation:

$$\theta(t) = (\rho + m)t \quad \text{with} \quad dt = \frac{d\theta}{\rho + m}. \tag{12.22}$$

Because $\theta(\zeta) = (\rho + m)\zeta$, $\theta(\infty) = \infty$, and $\frac{dt}{d\theta} = \frac{1}{\rho+m} > 0$ hold true, one can define $\theta(t)$ as an alternative time variable and set the variables in terms of it. Noting (12.10) and (12.22), the utility function (12.19) with (12.18) and the constraint $s = \frac{dk}{dt}$ can be transformed into virtual time θ as follows:

$$\int_{\zeta}^{\infty} \frac{c(\theta)^{\sigma} f(\theta)^{\alpha\sigma} \ell\big(m(\theta)\big)^{\sigma}}{\rho + m(\theta)} e^{\zeta-\theta} d\theta, \tag{12.23}$$

$$\frac{dk}{d\theta} = \frac{s(\theta)}{\rho + m(\theta)}, \quad k(0) = k_0. \tag{12.24}$$

12.3.4 Optimal Behavior

The household maximizes its utility (12.23) by investment per head, s, the fertility rate, f, and the mortality rate, m, subject to its consumption per head, (12.21), and its accumulation of wealth per head, (12.24), given the wage w, the interest rate r, the environmental shock q, and the taxes (a, x, τ, b). This defines the value function at initial time ζ as

$$\Phi(k, a, x, \tau, b, w, r, q, \zeta) \doteq \max_{(s, f, m) \text{ s.t. } (12.21), (12.24)} \int_{\zeta}^{\infty} \frac{c(\theta)^{\sigma} f(\theta)^{\alpha\sigma} \ell\big(m(\theta)\big)^{\sigma}}{\rho + m(\theta)} e^{\zeta-\theta} d\theta. \tag{12.25}$$

Following Dixit and Pindyck [3], and noting $s = \frac{dk}{dt}$ [cf. (12.5)], the Bellman equation for the household's program (12.25) is constructed as follows:

$$\Phi(k, a, x, \tau, b, w, r, q, \zeta) = \max_{(s,f,m)\, s.t. (12.21)} \Lambda(s, f, m, k, a, x, \tau, b, w, r, q, \zeta) \quad \text{with}$$

$$\Lambda(s, f, m, k, a, x, \tau, b, w, r, q, \zeta)$$
$$\doteq \frac{c^{\sigma} f^{\alpha\sigma} \ell}{\rho + m} + \frac{\partial \Phi}{\partial k} \frac{dk}{d\theta} = \frac{1}{\rho + m}\left[c^{\sigma} f^{\alpha\sigma} \ell(m)^{\sigma} + \frac{\partial \Phi}{\partial k} s\right]. \tag{12.26}$$

The first-order conditions for maximizing the function (12.26) by the controls (s, f, m) subject to (12.21) are given by

$$\frac{\partial\Lambda}{\partial s} = \frac{1}{\rho+m}\left(\sigma c^{\sigma-1} f^{\alpha\sigma}\ell^{\sigma}\frac{\partial\widetilde{c}}{\partial s} + \frac{\partial\Phi}{\partial k}\right) = \frac{1}{\rho+m}\left(\frac{\partial\Phi}{\partial k} - \sigma c^{\sigma-1} f^{\alpha\sigma}\ell^{\sigma}\right) = 0$$

$$\Leftrightarrow \quad \frac{\partial\Phi}{\partial k} = \sigma c^{\sigma-1} f^{\alpha\sigma}\ell^{\sigma}, \tag{12.27}$$

$$\frac{\partial\Lambda}{\partial f} = \frac{1}{\rho+m}\left(\alpha\sigma c^{\sigma} f^{\alpha\sigma-1}\ell^{\sigma} + \sigma c^{\sigma-1} f^{\alpha\sigma}\ell^{\sigma}\frac{\partial\widetilde{c}}{\partial f}\right) = \sigma\frac{c^{\sigma-1} f^{\alpha\sigma}\ell^{\sigma}}{\rho+m}\left(\alpha\frac{c}{f} - \frac{\partial\widetilde{c}}{\partial f}\right)$$

$$= \sigma\frac{c^{\sigma-1} f^{\alpha\sigma}\ell^{\sigma}}{\rho+m}\left(\alpha\frac{c}{f} - w - x - k\right) = 0 \quad \Leftrightarrow \quad f = \frac{\alpha c}{w+k+x}, \tag{12.28}$$

$$\frac{\partial\Lambda}{\partial m} = \frac{1}{\rho+m}\left[\sigma c^{\sigma-1} f^{\alpha\sigma}\ell^{\sigma}\frac{\partial\widetilde{c}}{\partial m} + \sigma c^{\sigma} f^{\alpha\sigma}\ell^{\sigma-1}\ell' - \frac{\Lambda}{\rho+m}\right]$$

$$= \frac{1}{\rho+m}\Bigg\{\sigma c^{\sigma-1} f^{\alpha\sigma}\ell(m)^{\sigma}[k+(1+b)z'(m)] + \sigma c^{\sigma} f^{\alpha\sigma}\ell(m)^{\sigma-1}\ell'(m)$$

$$- \frac{\Lambda}{\rho+m}\Bigg\} = 0 \quad \Leftrightarrow \quad k+(1+b)z' = \underbrace{\frac{\Lambda/(\rho+m)}{\sigma c^{\sigma-1} f^{\alpha\sigma}\ell^{\sigma}}}_{+} - c\underbrace{\frac{\ell'}{\ell}}_{-} > 0. \tag{12.29}$$

When the mortality rate m is held constant, the function Λ is strictly concave in controls s and f. To obtain a unique equilibrium for the household, the strict concavity of Λ must be extended for all controls (s, f, m). This is done by examining the second-order partial derivative of Λ with respect to m, which is obtained by (12.15), (12.19), (12.21), and (12.29) as follows:

$$\frac{\partial^2\Lambda}{\partial m^2} = \underbrace{\frac{1}{\rho+m}}_{+}\Bigg\{\underbrace{(\sigma-1)}_{-}\underbrace{\sigma c^{\sigma-2} f^{\alpha\sigma}\ell^{\sigma}}_{+}\underbrace{[k+(1+b)z']^2}_{+} + \underbrace{\sigma c^{\sigma-1} f^{\alpha\sigma}\ell^{\sigma}(1+b)}_{+}\underbrace{z''}_{-}$$

$$+ \underbrace{\sigma^2 c^{\sigma-1} f^{\alpha\sigma}\ell^{\sigma-1}}_{+}\underbrace{[k+(1+b)z']}_{+}\underbrace{\ell'}_{-} + \underbrace{(\sigma-1)}_{-}\underbrace{\sigma c^{\sigma} f^{\alpha\sigma}\ell^{\sigma-2}(\ell')^2}_{+}$$

$$+ \underbrace{\sigma c^{\sigma} f^{\alpha\sigma}\ell^{\sigma-1}}_{+}\ell'' + \underbrace{\frac{\Lambda}{(\rho+m)^2}}_{+}\Bigg\}.$$

If, in this equation, the negative effects of the mortality rate m *dominate over* the positive inter-temporal effect of the effective discount rate $\rho+m$ and the ambiguous effect of the mortality rate m through the second derivative ℓ'', then $\frac{\partial^2\Lambda}{\partial m^2} < 0$ holds true and the function Λ is strictly concave. Furthermore, by (12.29), one obtains

$$\frac{\partial^2\Lambda}{\partial m\partial b} = \frac{\sigma c^{\sigma-1} f^{\alpha\sigma}\ell^{\sigma}}{\rho+m} z' > 0.$$

Thus, differentiating equation (12.29) totally yields the mortality function

$$m = M(k, w, r, q, a, x, \tau, b, \zeta) \text{ with } \frac{\partial M}{\partial b} = -\frac{\partial^2 \Lambda}{\partial m \partial b} \Big/ \frac{\partial^2 \Lambda}{\partial m^2} > 0. \tag{12.30}$$

Results (12.28) and (12.30) can be explained as follows. An increase in the parental tax per child, x, decreases incentives to rear children (i.e., the fertility rate f falls relative to consumption c). When capital per head, k, increases, it is more difficult for the household to save that capital k for each newborn. This as well decreases incentives to rear children (i.e., f falls). An increase in the tax on health care, b, discourages health care, increasing the mortality rate m.

The solution of dynamic programming is based on finding a specification for the value function Φ. Then, one can use Merton's [10] Rule as follows. In the steady state $s = 0$, from the Bellman equation (12.26) it follows that

$$\left(\arg \max_{(s,f,m)\text{ s.t. (12.21)}} \Lambda\right)_{s=0} = \left(\arg \max_{(s,f,m)\text{ s.t. (12.21)}} \frac{c^\sigma f^{\alpha\sigma} \ell^\sigma}{\rho + m}\right)_{s=0}.$$

Thus, one can try the simplest case where the value function Φ is a positive constant ϑ times the maximized periodic utility in virtual time:

$$\Phi \doteq \vartheta \max_{(s,f,m)\text{ s.t. (12.21)}} \frac{c^\sigma f^{\alpha\sigma} \ell^\sigma}{\rho + m}. \tag{12.31}$$

Plugging (12.31) into the Bellman equation (12.26) in the steady state $s = 0$ yields

$$\vartheta = 1. \tag{12.32}$$

Thus, the value function (12.31) becomes

$$\Phi \doteq \max_{(s,f,m)\text{ s.t. (12.21)}} \frac{c^\sigma f^{\alpha\sigma} \ell^\sigma}{\rho + m}. \tag{12.33}$$

Inserting this into the first-order condition (12.27) and noting (12.21) yield

$$\sigma c^{\sigma-1} f^{\alpha\sigma} \ell^\sigma = \frac{\partial \Phi}{\partial k} = \sigma \frac{c^{\sigma-1} f^{\alpha\sigma} \ell^\sigma}{\rho + m} \frac{\partial \widetilde{c}}{\partial k} \Leftrightarrow \rho + m = \frac{\partial \widetilde{c}}{\partial k} = m - f + r - \tau r$$
$$\Leftrightarrow f = (1 - \tau) r - \rho. \tag{12.34}$$

The household's response functions are the fertility rate relative to consumption, (12.28), and the fertility rate (12.34). The government takes these together with the firm's responses (12.10) as constraints in its optimization.

12.4 The Government

12.4.1 Setup for Public Policy

The definition of pollution, (12.11), determines the fertility rate f as a function of the controls (s, v, m) and state variables (k, P):

$$f(s, v, m, k, P) \doteq m + \frac{v}{P} - \gamma\frac{s}{k}, \quad \frac{\partial f}{\partial s} = -\frac{\gamma}{k} < 0, \quad \frac{\partial f}{\partial v} = \frac{1}{P} > 0, \quad \frac{\partial f}{\partial m} = 1,$$
$$\frac{\partial f}{\partial k} = \gamma\frac{s}{k^2}, \quad \frac{\partial f}{\partial P} = -\frac{v}{P^2}. \tag{12.35}$$

The government balances its budget (12.17) by the poll tax a. Because there is one-to-one correspondence from the other taxes (τ, x, b) to (s, v, m) through the system (12.21) [with (12.34)], (12.35) [with (12.28)], and (12.30), investment per head, $s = \frac{dk}{dt}$, the change of pollution, $v = \frac{dP}{dt}$, and the mortality rate m can replace the taxes (τ, x, b) as the government's controls in the model.

Inserting the production function (12.10), the government's budget (12.17), and the fertility function (12.35) into the household's consumption per head, (12.21), it is possible to define consumption per head, c, as a function of the government's controls (s, v, m) and the state variables (k, P) as follows:

$$c = \widehat{c}(s, v, m, k, P, q) = \widetilde{c} = y(k, f) + (m - f)k - s - \delta q + z(m), \tag{12.36}$$

with the partial derivatives

$$\frac{\partial \widehat{c}}{\partial s} = (y_f - k)\frac{\partial f}{\partial s} - 1, \quad \frac{\partial \widehat{c}}{\partial v} = (y_f - k)\frac{\partial f}{\partial v} = \frac{y_f - k}{P},$$
$$\frac{\partial \widehat{c}}{\partial m} = k + z' + (y_f - k)\frac{\partial f}{\partial m}, \quad \frac{\partial \widehat{c}}{\partial q} = -\delta, \quad \frac{\partial \widehat{c}}{\partial P} = (y_f - k)\frac{\partial f}{\partial P},$$
$$\frac{\partial \widehat{c}}{\partial k} = y_k + m - f + (y_f - k)\frac{\partial f}{\partial k}. \tag{12.37}$$

By (12.22), the constraint $v = \frac{dP}{dt}$ can be written in virtual time θ as follows:

$$\frac{dP}{d\theta} = \frac{v(\theta)}{\rho + m(\theta)}, \quad P(0) = P_0. \tag{12.38}$$

12.4.2 Optimization

The government maximizes the representative household's welfare (12.23) by its controls (s, v, m) subject to the occurrence of the environmental shock, (12.12), the accumulation of capital per head and aggregate pollution, (12.24) and (12.38), and the determination of the fertility rate and consumption per head, (12.35) and (12.36). Thus, its value function at initial time ζ is defined by

$$\Psi(k, P, q, \zeta) \doteq \max_{\substack{(s(\zeta), v(\zeta), m(\zeta)) \\ \text{s.t.}(12.24),(12.35),(12.36),(12.38)}} \int_{\zeta}^{\infty} \frac{c(\theta)^{\sigma} f(\theta)^{\alpha\sigma} \ell\big(m(\theta)\big)^{\sigma}}{\rho + m(\theta)} e^{\zeta-\theta} d\theta, \quad (12.39)$$

where $q = 0$ holds true before and $q = 1$ after the shock. Noting (12.39), one can define the *relative damage* of the shock in terms of welfare as follows:

$$D(k, P, \zeta) \doteq \frac{\Psi(k, P, 0, \zeta) - \Psi(k, P, 1, \zeta)}{\Psi(k, P, 0, \zeta)}. \quad (12.40)$$

The following result is proven in Appendix:

Proposition 12.1 *If the loss of income due to the shock, δ, is small relative to consumption per head before the shock, $c|_{q=0}$, (e.g., if $\frac{\delta}{c|_{q=0}}$ is less than 10%), then the relative damage of the shock in terms of welfare,* (12.40)*, is approximately in fixed proportion σ to it, $D(k, P, \zeta) \approx \sigma \frac{\delta}{c|_{q=0}}$.*

The parameter $\sigma \in (0, 1)$ tells how willing the households are to save for future in capital, i.e., $\frac{1}{1-\sigma}$ is the elasticity of inter-temporal substitution. [cf. (12.19)]. If σ is close to zero, then relative damage of the shock is insignificant in terms of current consumption. The closer σ is to one, the greater the relative damage D is in terms of current consumption.

At the occurrence of the environmental shock, q jumps permanently from 0 to 1 [cf. (12.12)], changing welfare (12.39) from $\Psi(k, P, 0, \zeta)$ into $\Psi(k, P, 1, \zeta)$. Thus, by Kamien and Schwartz [8] and Dixit and Pindyck [3], the Bellman equation for the government's program is [cf. (12.12) and (12.39)]

$$\Psi = \max_{(s(\zeta), v(\zeta), m(\zeta)) \text{ s.t.}(12.24),(12.35),(12.36),(12.38)} \Upsilon(s, f, m, k, P, q, \zeta) \text{ with} \quad (12.41)$$

$$\begin{aligned}
\Upsilon(s, f, m, k, P, q, \zeta) &\doteq \frac{c^{\sigma} f^{\alpha\sigma} \ell^{\sigma}}{\rho + m} + \frac{\partial \Psi}{\partial k}(k, P, q, \zeta)\frac{dk}{d\theta} + \frac{\partial \Psi}{\partial P}(k, P, q, \zeta)\frac{dP}{d\theta} \\
&\quad + \pi(P)\big[\Psi(k, P, 1, \zeta) - \Psi(k, P, q, \zeta)\big] \\
&= \frac{1}{\rho + m}\left[c^{\sigma} f^{\sigma\alpha} \ell^{\sigma} + \frac{\partial \Psi}{\partial k}(k, P, 1, \zeta)s + \frac{\partial \Psi}{\partial P}(k, P, 1, \zeta)v\right] \\
&\quad + \pi(P)\big[\Psi(k, P, 1, \zeta) - \Psi(k, P, q, \zeta)\big], \quad (12.42)
\end{aligned}$$

where the fertility rate, f, and consumption per head, c, are determined by (12.35) and (12.36), $\pi(P)$ is the probability of the environmental shock [cf. (12.12)], and the difference $\Psi(k, P, 1, \zeta) - \Psi(k, P, q, \zeta)$ is the immediate change of welfare due to that shock. Note that the latter term $\pi\big[\Psi(k, P, 1, \zeta) - \Psi(k, P, q, \zeta)\big]$ vanishes entirely after the shock when $q = 1$ holds true.

From the equilibrium condition $y_f = -w$ [cf. (12.10)] and the household's first-order condition (12.28), it follows that

$$x = \alpha c/f - w - k = \alpha c/f + y_f - k. \tag{12.43}$$

Noting (12.35), (12.37), and (12.43), the first-order conditions for the maximization (12.41) subject to (12.42) are obtained as follows:

$$\begin{aligned}
0 = \frac{\partial \Upsilon}{\partial s} &= \frac{1}{\rho + m}\left[c^\sigma f^{\alpha\sigma} \ell^\sigma \left(\frac{\sigma}{c}\frac{\partial \widehat{c}}{\partial s} + \frac{\sigma\alpha}{f}\frac{\partial f}{\partial s}\right) + \frac{\partial \Psi}{\partial k}\right] \\
&= \frac{1}{\rho + m}\left\{\sigma c^{\sigma-1} f^{\alpha\sigma} \ell^\sigma \left[\left(y_f - k + \alpha\frac{c}{f}\right)\frac{\partial f}{\partial s} - 1\right] + \frac{\partial \Psi}{\partial k}\right\} \\
&= \frac{1}{\rho + m}\left[\sigma c^{\sigma-1} f^{\alpha\sigma} \ell^\sigma \left(x\frac{\partial f}{\partial s} - 1\right) + \frac{\partial \Psi}{\partial k}\right] \\
&= \frac{1}{\rho + m}\left[-\sigma c^{\sigma-1} f^{\alpha\sigma} \ell^\sigma \left(x\frac{\gamma}{k} + 1\right) + \frac{\partial \Psi}{\partial k}\right],
\end{aligned} \tag{12.44}$$

$$\begin{aligned}
0 = \frac{\partial \Upsilon}{\partial v} &= \frac{1}{\rho + m}\left[c^\sigma f^{\alpha\sigma} \ell^\sigma \left(\frac{\sigma}{c}\frac{\partial \widehat{c}}{\partial v} + \frac{\sigma\alpha}{f}\frac{\partial f}{\partial v}\right) + \frac{\partial \Psi}{\partial P}\right] \\
&= \frac{1}{\rho + m}\left[\sigma c^{\sigma-1} f^{\alpha\sigma} \ell^\sigma \left(y_f - k + \alpha\frac{c}{f}\right)\frac{\partial f}{\partial v} + \frac{\partial \Psi}{\partial P}\right] \\
&= \frac{1}{\rho + m}\left(\sigma c^{\sigma-1} f^{\alpha\sigma} \ell^\sigma x\frac{\partial f}{\partial v} + \frac{\partial \Psi}{\partial P}\right) = \frac{1}{\rho + m}\left(\sigma c^{\sigma-1} f^{\alpha\sigma} \ell^\sigma \frac{x}{P} + \frac{\partial \Psi}{\partial P}\right),
\end{aligned} \tag{12.45}$$

$$\begin{aligned}
0 = \frac{\partial \Upsilon}{\partial m} &= \frac{1}{\rho + m}\left[c^\sigma f^{\alpha\sigma} \ell^\sigma \left(\frac{\sigma}{c}\frac{\partial \widehat{c}}{\partial m} + \frac{\sigma\alpha}{f}\frac{\partial f}{\partial m} + \sigma\frac{\ell'}{\ell}\right) - \frac{\Upsilon}{\rho + m}\right] \\
&= \frac{1}{\rho + m}\left\{\sigma c^{\sigma-1} f^{\alpha\sigma} \ell^\sigma \left[\left(y_f - k + \alpha\frac{c}{f}\right)\frac{\partial f}{\partial m} + k + z' + c\frac{\ell'}{\ell}\right] - \frac{\Upsilon}{\rho + m}\right\} \\
&= \frac{1}{\rho + m}\left[\sigma c^{\sigma-1} f^{\alpha\sigma} \ell^\sigma \left(x + k + z' + c\frac{\ell'}{\ell}\right) - \frac{\Upsilon}{\rho + m}\right] \Leftrightarrow \\
&\frac{1}{\sigma c^{\sigma-1} f^{\alpha\sigma} \ell^\sigma}\frac{\Upsilon}{\rho + m} - c\frac{\ell'}{\ell} = x + k + z'.
\end{aligned} \tag{12.46}$$

The function Υ [cf. (12.42)] is strictly concave in (s, v). To ensure that the government's equilibrium is unique, this property is extended by assuming that the function Υ as well is strictly concave in its arguments (s, f, m).

12.4.3 Solution

In a steady state with $s = v = 0$, from the Bellman equation (12.41) with (12.42), it follows that

$$\left(\arg \max_{\substack{(s(\zeta),v(\zeta),m(\zeta)) \text{ s.t.} \\ (12.24),(12.35),(12.36),(12.38)}} \Upsilon\right)_{s=v=0} = \left(\arg \max_{\substack{(s(\zeta),v(\zeta),m(\zeta)) \text{ s.t.} \\ (12.24),(12.35),(12.36),(12.38)}} \frac{c^\sigma f^{\alpha\sigma} \ell^\sigma}{\rho + m}\right)_{s=v=0}.$$

Thus, one can try the simplest case where the value function Ψ is a positive constant ϖ times the maximized periodic utility in virtual time:

$$\Psi(k, P, q, \zeta) = \varpi \max_{\substack{(s(\zeta),v(\zeta),m(\zeta)) \text{ s.t.} \\ (12.24),(12.35),(12.36),(12.38)}} \frac{c^\sigma f^{\alpha\sigma} \ell^\sigma}{\rho + m} > 0,$$
$$\varpi > 0, \quad \overline{\Psi}(k, P, \zeta) \doteq \Psi(k, P, 1, \zeta), \tag{12.47}$$

where $\overline{\Psi}$ is the value after the disaster when $q = 1$. Noting (12.35), (12.37), (12.43), and (12.47), the partial derivatives of the value function (12.47) with respect to the state variables (k, P) are obtained as follows:

$$\begin{aligned}
\frac{\partial \Psi}{\partial k} &= \Psi \frac{\partial \ln \Psi}{\partial k} = \Psi \frac{\sigma}{c} \frac{\partial \widehat{c}}{\partial k} + \frac{\sigma\alpha}{f} \frac{\partial f}{\partial k} = \Psi \frac{\sigma}{c}\left(\frac{\partial \widehat{c}}{\partial k} + \alpha \frac{c}{f} \frac{\partial f}{\partial k}\right) \\
&= \Psi \frac{\sigma}{c}\left[y_k + m - f + \left(y_f - k + \alpha \frac{c}{f}\right)\frac{\partial f}{\partial k}\right] \\
&= \Psi \frac{\sigma}{c}\left(y_k + m - f + x\frac{\partial f}{\partial k}\right) = \Psi \frac{\sigma}{c}\left(y_k + m - f + x\gamma \frac{s}{k^2}\right),
\end{aligned} \tag{12.48}$$

$$\begin{aligned}
\frac{\partial \Psi}{\partial P} &= \Psi \frac{\partial \ln \Psi}{\partial P} = \Psi \frac{\sigma}{c}\left(\frac{\partial \widehat{c}}{\partial P} + \alpha \frac{c}{f} \frac{\partial f}{\partial P}\right) = \Psi \frac{\sigma}{c}\left(y_f - k + \alpha \frac{c}{f}\right)\frac{\partial f}{\partial P} = \Psi \frac{\sigma}{c} x \frac{\partial f}{\partial P} \\
&= -\Psi \frac{\sigma}{c} x \frac{v}{P^2}.
\end{aligned} \tag{12.49}$$

Dividing the Bellman equation (12.41) and (12.42) by the value function (12.47) and noting the definition of the relative damage, (12.40), yield

$$\begin{aligned}
1 = \frac{\Upsilon}{\Psi} &= \frac{c^\sigma f^{\alpha\sigma}}{\rho + m}\frac{1}{\Psi} + \frac{1}{\Psi}\frac{\partial \Psi}{\partial k}\frac{dk}{d\theta} + \frac{1}{\Psi}\frac{\partial \Psi}{\partial P}\frac{dP}{d\theta} + \pi(P)\frac{\Psi(k, P, 1, \zeta) - \Psi(k, P, q, \zeta)}{\Psi(k, P, q, \zeta)} \\
&= \begin{cases} \frac{1}{\varpi} + \frac{1}{\Psi}\frac{\partial \Psi}{\partial k}\frac{dk}{d\theta} + \frac{1}{\Psi}\frac{\partial \Psi}{\partial P}\frac{dP}{d\theta} - \pi(P)D(k, P, \zeta) & \text{for } q = 0, \\ \frac{1}{\varpi} + \frac{1}{\Psi}\frac{\partial \Psi}{\partial k}\frac{dk}{d\theta} + \frac{1}{\Psi}\frac{\partial \Psi}{\partial P}\frac{dP}{d\theta} & \text{for } q = 1. \end{cases}
\end{aligned} \tag{12.50}$$

In this study, the steady-state value of a variable is denoted by superscript (*). There are different steady states before ($q = 0$) and after ($q = 1$) the shock. Because (12.50) holds in both of these steady states where $\frac{dP}{d\theta} = \frac{dk}{d\theta} = 0$ hold true, the multiplier ϖ is piecewise constant as follows:

$$\varpi\big|_{q=0} = \frac{1}{1+\pi^* D^*} < 1, \quad \varpi|_{q=1} = 1, \tag{12.51}$$

where $\pi^* \doteq \pi(P^*|_{q=0})$ is the probability of the disaster [cf. (12.12)], $D^* \doteq D(k^*|_{q=0}, P^*|_{q=0}, \xi)$ the relative damage [cf. (12.40)], and $\pi^* D^*$ the expected relative damage in the steady state before the occurrence of the shock.

12.5 Optimal Policy

12.5.1 The Parental Tax per Child

It is assumed that the relative change of pollution, $\frac{v}{P}$, is either negative or positive, but small enough for $\frac{v}{P} < \frac{\rho+m}{\varpi}$. Inserting the value function (12.47) and its partial derivative (12.49) into the government's first-order condition (12.45) and noting the government's fertility function (12.35) yield

$$\begin{aligned} 0 &= (\rho+m)\frac{\partial \Upsilon}{\partial v} = \sigma c^{\sigma-1} f^{\alpha\sigma} \ell^{\sigma} \frac{x}{P} + \frac{\partial \Psi}{\partial P} = \sigma c^{\sigma-1} f^{\alpha\sigma} \ell^{\sigma} \frac{x}{P} - \Psi \frac{\sigma}{c} x \frac{v}{P^2} \\ &= \frac{\Psi}{P}\frac{\sigma}{c} x \left(\frac{c^{\sigma} f^{\alpha\sigma} \ell^{\sigma}}{\Psi} - \frac{v}{P} \right) = \frac{\Psi}{P}\frac{\sigma}{c} x \underbrace{\left(\frac{\rho+m}{\varpi} - \frac{v}{P} \right)}_{+} \Leftrightarrow x = 0. \end{aligned} \tag{12.52}$$

Thus, in contrast to Harford (1997, 1998), the parental tax per child is not positive in this case:

Proposition 12.2 *The parental tax per child can be eschewed, $x = 0$.*

Because the other taxes eliminate the externality through pollution and mortality, this tax is unnecessary.

12.5.2 Taxing Capital Income

Plugging $x = 0$ [cf. (12.52)], the profit maximization condition $y_k = r$ [cf. (12.10)], and the value function (12.47) into the partial derivative (12.48) and the first-order condition (12.44), one obtains

$$
\begin{aligned}
0 &= \frac{\partial \Psi}{\partial k} - \sigma c^{\sigma-1} f^{\alpha\sigma} \ell^{\sigma} \left(x \frac{\gamma}{k} + 1 \right) \\
&= \Psi \frac{\sigma}{c} \left(y_k + m - f + x\gamma \frac{s}{k^2} \right) - \sigma c^{\sigma-1} f^{\alpha\sigma} \left(x \frac{\gamma}{k} + 1 \right) \\
&= \Psi \frac{\sigma}{c} \left(r + m - f \right) - \frac{\sigma}{c} \frac{\Psi}{\varpi} (\rho + m) = \Psi \frac{\sigma}{c} \left(r + m - f - \frac{\rho + m}{\varpi} \right) \\
&\Leftrightarrow \; r = f - m + \frac{\rho + m}{\varpi}.
\end{aligned}
\tag{12.53}
$$

Because the ratio of the difference between the fertility and mortality rates to the sum of the rate of time preference and the mortality rate, $\frac{f-m}{\rho+m}$, is insignificant, one can approximate the first-order condition (12.53) as follows:

$$
\frac{r}{\rho + m} = \underbrace{\frac{f - m}{\rho + m}}_{\approx 0} + \frac{1}{\varpi} \approx \frac{1}{\varpi} \;\Leftrightarrow\; \frac{\rho + m}{r} \approx \varpi. \tag{12.54}
$$

Plugging (12.51), (12.53), and (12.54) into the household's response (12.34) yields the optimal tax

$$
\begin{aligned}
\tau &= \frac{r - f - \rho}{r} = \frac{1}{r} \left(\frac{\rho + m}{\varpi} - m - \rho \right) = \frac{\rho + m}{r} \left(\frac{1}{\varpi} - 1 \right) = \varpi \left(\frac{1}{\varpi} - 1 \right) \\
&= \begin{cases} \varpi|_{q=0} \left(\frac{1}{\varpi|_{q=0}} - 1 \right) = \frac{\pi^* D^*}{1+\pi^* D^*} & \text{for } q = 0, \\ 0 & \text{for } q = 1. \end{cases}
\end{aligned}
$$

This result can be rephrased by the following proposition:

Proposition 12.3 *Before the disaster, the optimal tax on capital income is an increasing function of the expected relative damage $\pi^* D^*$ as follows:*

$$
\tau\Big|_{q=0} \approx \frac{\pi^* D^*}{1 + \pi^* D^*}.
$$

After the disaster, that tax can be eschewed, $\tau|_{q=1} = 0$.

12.5.3 Taxing Health Care

Because optimal public policy leads to the Pareto optimum, where consumption per head, c, the fertility rate, f, and the mortality rate, m, are equal in the household's and the government's problems, then, by (12.26), (12.32) and (12.41), the ratio of the household's and the government's value functions, (12.33) and (12.47), is $\Upsilon/\Lambda = \Psi/\Phi = \varpi/\vartheta = \varpi$. From this, (12.13), (12.15), (12.51), and the comparison of the household's and the government's first-order conditions, (12.29) and (12.46), it follows that

$$\begin{aligned}
0 &= (\rho+m)\frac{\partial\Upsilon}{\partial m} = \sigma c^{\sigma-1} f^{\alpha\sigma}\ell^{\sigma}(k+z') - \frac{\Upsilon}{\rho+m} \\
&= \sigma c^{\sigma-1} f^{\alpha\sigma}\ell^{\sigma}(k+z') - \frac{\varpi\Lambda}{\rho+m} \\
&\Leftrightarrow\ k+z' = \frac{1}{\sigma c^{\sigma-1} f^{\alpha\sigma}\ell^{\sigma}}\frac{\varpi\Lambda}{\rho+m} = \varpi[k+(1+b)z'] \ \Leftrightarrow \\
b &= \left(\frac{1}{\varpi}-1\right)\left(\frac{k}{z'}+1\right) = \left(\frac{1}{\varpi}-1\right)(k\chi'+1) \\
&= \begin{cases} (k\chi'+1)\pi^{*}D^{*} > 0 & \text{for } q=0, \\ 0 & \text{for } q=1. \end{cases}
\end{aligned} \tag{12.55}$$

Noting (12.15), the result (12.55) can be rephrased as follows.

Proposition 12.4 *Before the disaster, the tax on health care should be in proportion* $(k\chi'+1)$ *to the expected relative loss for the disaster* $\pi^{*}D^{*}$,

$$b|_{q=0} = (k\chi'+1)\pi^{*}D^{*} > 0,$$

where k *is capital per head and* χ' *the marginal efficiency of personal health care* h *in decreasing the mortality rate* m [*cf.* (12.14)]. *After the disaster, that tax can be eschewed,* $b|_{q=1} = 0$.

Because a single household ignores the effect of its health care h on the other households' mortality rate m through the increase of population L and pollution P, its demand for health care exceeds the socially optimal level before the occurrence of the disaster. Thus, the demand for health care must be discouraged by the tax b. The more efficiently personal health care decreases mortality (i.e., the greater χ'), or the more capital k each surviving person needs, the higher the tax b must be. If health care is very inefficient in decreasing mortality (i.e., χ' is close enough to 0), then the tax is roughly equal to the expected relative loss for the disaster, $b \approx \pi^{*}D^{*}$.

12.6 Conclusions

This study examines the optimal management of a market economy where (i) households decide on saving, health care, and the number of their children; (ii) the government controls their activity only by linear taxes; and (iii) population growth and capital accumulation generate pollution, increasing the risk of a lethal environmental disaster. In this situation, it turns out that a rational government should perform the following precautionary policy.

To implement Pareto optimality—i.e., to internalize the external link from population growth and capital accumulation to welfare though pollution and mortality—it is necessary to set precautionary taxes (i.e., taxes prior to the disaster) on capital income and the demand for health care. These are increasing functions of the expected relative damage of the disaster. The specific tax rules are given by Propositions 12.2, 12.3, and 12.4. In particular, Harford's (1997, 1998) parental tax is wholly unnecessary in this setup. In addition, only the revenue raising-poll tax is needed.

There are two reasons for this sharp result. First, because there is no incremental contribution of pollution to the mortality rate, there is only the precautionary, but no maintenance motive for the government to intervene. Second, because the mortality rate can be decreased by spending on health care, the mortality shock turns into an increase in the cost of health care, which has the same effect as an exogenous fall of income.

Acknowledgements The author thanks the two anonymous reviewers and the editors for constructive comments and IIASA (Laxenburg, Austria) for hospitality during my visit in November 2018 when this paper was started.

Appendix: The Approximation of the Relative Damage *D*

Because, by (12.35) and (12.37), the fertility rate f doesn't, but consumption per head, c, does depend on the shock q, the partial derivative of the value function (12.47) with respect to the shock q is negative:

$$\frac{\partial \Psi}{\partial q}(k, P, q, \zeta) = \frac{\partial}{\partial q} \max_{s,v,m} \frac{\varpi c^{\sigma} f^{\alpha\sigma} \varrho^{\sigma}}{\rho + m} = \sigma \frac{\Psi}{c} \frac{\partial c}{\partial q} = -\frac{\sigma \delta}{c} \Psi(k, P, q, \zeta) < 0. \tag{12.56}$$

Consider now what happens for the value function (12.47) if q jumps discretely from 0 to 1. Applying the mean value theorem to (12.47), and noting (12.56), one obtains the following: there exists a value $\xi \in (0, 1)$ so that

$$\Psi(k, P, 1, \zeta) - \Psi(k, P, 0, \zeta) = \frac{\partial \Psi}{\partial q}(k, P, \xi, \zeta) = -\frac{\sigma \delta}{c|_{q=\xi}} \Psi(k, P, \xi, \zeta) < 0. \tag{12.57}$$

Furthermore, from (12.37) it follows that $c|_{q=\xi} = c|_{q=0} - \delta\xi$. Given this, (12.47) and (12.57), the relative damage (12.40) can be approximated by

$$D(k,P,\zeta) \doteq \frac{\Psi(k,P,0,\zeta)-\Psi(k,P,1,\zeta)}{\Psi(k,P,0,\zeta)} = \frac{\sigma\delta}{c|_{q=\xi}}\frac{\Psi(k,P,\xi,\zeta)}{\Psi(k,P,0,\zeta)}$$
$$= \frac{\sigma\delta}{c|_{q=\xi}}\left(\frac{c|_{q=\xi}}{c|_{q=0}}\right)^{\sigma} = \frac{\sigma\delta}{c|_{q=0}}\left(\frac{c|_{q=\xi}}{c|_{q=0}}\right)^{\sigma-1} = \frac{\sigma\delta}{c|_{q=0}}\left(1-\frac{\xi\delta}{c|_{q=0}}\right)^{\sigma-1} > 0$$
$$\text{with } \lim_{\frac{\delta}{c|_{q=0}}\to 0}\left[\frac{c|_{q=0}}{\sigma\delta}D(k,P,\zeta)\right] = \lim_{\frac{\delta}{c|_{q=0}}\to 0}\Big(1-\xi\underbrace{\frac{\delta}{c|_{q=0}}}_{\to 0}\Big)^{\sigma-1} = 1$$
$$\text{and } \lim_{\frac{\delta}{c|_{q=0}}\to 0} D(k,P,\zeta) = \frac{\sigma\delta}{c|_{q=0}}. \tag{12.58}$$

Noting (12.19), the result (12.58) leads to the approximation $D(k,P,\zeta) \approx \sigma\frac{\delta}{c|_{q=0}}$.

References

1. Becker, G. S. (1981). *A treatise on the family.* Cambridge (Mass.): Harvard University Press.
2. de Zeeuw, A., & Zemel, A. (2012). Regime shifts and uncertainty in pollution control. *Journal of Economic Dynamics and Control, 36*, 939–950.
3. Dixit, A., & Pindyck, K. (1994). *Investment under uncertainty*. Princeton: Princeton University Press.
4. Dockner, E. J., & Long, N. V. (1993). International pollution control: Cooperative versus noncooperative strategies. *Journal of Environmental Economics and Management, 24*, 13–29.
5. Harford, J. D. (1997). Stock pollution, child-bearing externalities, and the social discount rate. *Journal of Environmental Economics and Management, 33*, 94–105.
6. Harford, J. D. (1998). The ultimate externality. *The American Economic Review, 88*, 260–265.
7. Haurie, A., & Moresino, F. (2006). A stochastic control model of economic growth with environmental disaster prevention. *Automatica, 42*, 1417–1428.
8. Kamien, M. I., & Schwartz, N. (1981). *Dynamic optimization. The calculus of variations and optimal control in economics and management.* New York: North-Holland.
9. Lehmijoki, U., & Palokangas, T. (2018) Optimal population policy with deadly emissions. Paper presented in the 5*th* International Workshop on Economic Growth, Environment and Natural Resources (2019). *European University at St.* Russia: Petersburg.
10. Merton, R. C. (1971). Optimum consumption and portfolio rules in a continuous-time model. *Journal of Economic Theory, 3*, 373–413.
11. Palokangas, T. (2018). Optimal taxation with endogenous fertility and health-damaging emissions. *IFAC PapersOnLine, 51*(32), 79–83.
12. Polasky, A., de Zeeuw, A., & Wagener, F. (2011). Optimal management with potential regime shifts. *Journal of Environmental Economics and Management, 62*, 229–240.
13. Tsur, Y., & Zemel, A. (2008). Regulating environmental threats. *Environmental and Resource Economics, 39*, 297–310.
14. Tsur, Y., & Zemel, A. (2009). Endogenous discounting and climate policy. *Environmental and Resource Economics, 44*, 507–520.
15. Uzawa, H. (1968) Time preference, the consumption function, and optimum asset holdings. In: J Wolfe (Ed.) *Value, Capital, and Growth* Aldine, Chicago (Chapter 21).
16. Van der Ploeg, F., & de Zeeuw, A. (1992). International aspects of pollution control. *Environmental and Resource Economics, 2*, 117–139.

Chapter 13
A Regime-Switching Model with Applications to Finance: Markovian and Non-Markovian Cases

E. Savku and G.-W. Weber

13.1 Introduction

A Markov regime-switching model is a continuous-time process with discrete components. Hence, this type of stochastic processes is applied to finance, psychology, automotive, aircraft traffic, etc., where a hybrid nature is required to catch a real-life phenomenon effectively. While the continuous-time process evolves according to a stochastic differential equation, the discrete component belongs to a finite or a countable set, by the way, such systems can be considered as an interleaving among a finite or countable family of diffusion or jump-diffusion processes.

Regime-switching models were first studied by Quandt [48] to derive a method to estimate the parameters of a linear regression system with two different regimes. Hamilton [22] followed Goldfeld's and Quandt's Markov regime-switching regression work [19] and investigated whether the business cycle between a recessionary state and a growth state is better denoted by such discrete components. The results of this investigation led researchers to focus on regime-switching models more and more in financial applications.

Generally, regime-switching models are seen as proxies of the different states of the economy, such as a gross domestic product, a purchase management index, and a sovereign credit rating. A discrete shift among different regimes can be observed

E. Savku (✉)
Ecole Polytechnique, CMAP, 91128 Palaiseau, France
e-mail: esavku@gmail.com

E. Savku · G.-W. Weber
Middle East Technical University, IAM, Dumlupinar Bulvari, Universiteler Mahallesi, No:1, 06800 Cankaya, Ankara, Turkey
e-mail: gerhard-wilhelm.weber@put.poznan.pl

G.-W. Weber
Faculty of Management Engineering, Poznan University of Technology, ul. Strzelecka 11, 60-965 Poznan, Poland

J. L. Haunschmied et al. (eds.), *Dynamic Economic Problems with Regime Switches*, Dynamic Modeling and Econometrics in Economics and Finance 25,
https://doi.org/10.1007/978-3-030-54576-5_13

after a change in a monetary or an exchange rate policy or after a new regulation. If we take into account that the plot programs, even just in tick sizes, like MiFID (Markets in Financial Instruments Directive) I and II in 2004–2018, see [31], and similar ones in the USA by the Security and Exchange Commission since 2016 and the Tokyo Stock Exchange in 2014–2015 [24] deeply modified the microstructure of financial markets, a very realistic and great importance of regime-switching systems in economy arises brightly. Moreover, we may describe not only a day-by-day change in a financial market (from a bullish day to a bearish one), but also, in some instants, some major events, such as the bankruptcy of Lehman Brothers in September 2008, or the 1973 oil crisis can be represented. Hence, many authors worked on regime-switches from different perspectives: option pricing and risk minimization [11, 12, 60], consumption [34, 51], determining optimal selling rules [58], optimal asset allocation [59], and more can be found in [5, 32, 37, 44, 55].

There are several models, which describe the hybrid nature of regimes mathematically, see [1, 18, 23]. Here, for this study, we will present some constructions and financial applications based on the semimartingale representation of Markov chains established by Elliott et al. [10]. We approach to this very large area of research by the tools of stochastic optimal control, both by the Dynamic Programming Principle (DPP) and the stochastic Maximum Principle (MP). Furthermore, we give examples for both diffusion and jump-diffusion processes with regimes. It can be considered that a Brownian motion describes the random shock of stock prices, a Poisson random measure interprets larger price fluctuations of the stock as a consequence of sudden changes in the market, and finally, Markov regime-switches carry the uncertainty of the macroeconomic indicators.

It is well known that in the past thirty years, game theory has been universally highlighted to explain the strategic interactions in economics, behavioral and social sciences. In this sense, the existence of an optimal strategy against others', called *Nash equilibrium*, becomes a cornerstone in modern economics. Hence, first, we summarize some financial applications within the framework of stochastic differential games in a Markov regime-switching diffusion system by DPP [13, 35]. We present corresponding Hamilton-Jacobi-Bellman–Isaacs (HJBI) equations and the analytical solutions for portfolio optimization problems. Moreover, several authors focused on stochastic game theory without regimes; see [8, 21, 26, 27] and the references therein.

Second, we concentrate on another main tool, the stochastic Maximum Principle both with regimes and time delay. In real-life events, market participants pay a deep attention to the historical performance of the risky assets for pricing options, making optimal investment decisions, getting better and better portfolios, etc. This main concern is carried out as memory in the dynamics of the stochastic processes with paying the price of violating the Markov property. This kind of systems are called Stochastic Differential Delay Equations (SDDEs). The first existence-uniqueness result of time-delayed diffusion processes was driven by Itô and Nisio [25], which was followed by Kushner [29]. A fundamental reference is Mohammed [39], in which he introduces and presents a very detailed theory of SDDEs. The difference between the methods of obtaining a solution of a system with and without delay component arises at the first sight. Let us explain this for a diffusion process as in Mohammed [39].

Let $(\Omega, \mathbb{F}, (\mathscr{F}_t)_{t\geq 0}, \mathbb{P})$ be a probability space satisfying that $(\mathscr{F}_t)_{t\geq 0}$ is a right-continuous filtration and for each $t \geq 0$, $\mathscr{F}_t$ contains all $\mathbb{P}$-null sets in $\mathbb{F}$. Here, we define

$$dX(t) = \sigma X(t - \delta) dW(t), \qquad t \geq 0. \tag{13.1}$$

In Eq. (13.1), if $\delta = 0$, then the closed-form solution is

$$X(t) = X(0)e^{\sigma W(t) - \sigma^2/2}, \qquad t \geq 0.$$

If we assume $\delta > 0$, we need an initial path $\theta(\cdot)$ to solve Eq. (13.1) such that

$$X(t) = \theta(t), \qquad -\delta \leq t \leq 0.$$

Then, by recursive Itô integration over steps of length δ, we observe that there is no closed-form solution if terminal time T is not finite:

$$\begin{aligned}
X(t) &= \theta(0) + \sigma \int_0^t \theta(u - \delta) dW(u), \qquad 0 \leq t \leq \delta, \\
X(t) &= X(r) + \sigma \int_r^t \left[\theta(0) + \sigma \int_0^{v-\delta} \theta(u - \delta) dW(u) \right] dW(v), \qquad \delta < t \leq 2\delta, \\
\ldots &= \ldots, \qquad 2\delta < t \leq 3\delta, \\
&\vdots
\end{aligned}$$

The solution process $\{X(t) : t \geq -\delta\}$ is still an $\mathscr{F}_t$-martingale but it is not Markovian any more. Let us define the segment $X_t : [-\delta, 0] \to \mathbb{R}^n$ by

$$X_t(s) = X(t + s) \quad \text{a.s.} \quad t \geq 0,\ s \in [-\delta, 0].$$

Then, a general representation for an SDDE is as follows:

$$\begin{aligned}
dX(t) &= h(t, X_t)dt + g(t, X_t)dW(t), \qquad t \geq 0, \\
X_0 &= \theta(t), \qquad t \in [-\delta, 0],
\end{aligned}$$

where the initial path $\theta(\cdot) \in C([-\delta, 0], \mathbb{R}^n)$ is an $\mathscr{F}_0$-measurable process.

However, including memory into the dynamics of the system provides a strong tool to capture the real-life phenomenon of finance and economics in a much more realistic way, applying DPP becomes more and more complicated (see [14–17, 30, 45]). While the requirement of a Markovian structure in DPP can not be omitted, with its infinite-dimensional nature, the stochastic MP arises as a more practical way to formulate financial problems for a non-Markovian process.

Cadenillas and Karatzas [3] provided the first use of stochastic MP, which states that an optimal control process maximizes a functional, called *Hamiltonian*, and satisfies the *optimality system*, described by forward–backward stochastic differential equations. Later on, several authors gave valuable applications [34, 41, 42, 54, 60]. Here, in this setting, the counterpart of the Partial Differential Equations (PDEs) in DPP can be seen as Backward Stochastic Differential Equations (BSDEs) in MP. A systematic work of BSDEs was first established by Pardoux and Peng [46], and their connection to financial mathematics attracted many authors very rapidly; see [4, 6, 9, 20, 27, 33, 52].

When memory is taken into account in a problem formulation, the corresponding adjoint equations appear in their new forms, called Anticipated (time-advanced) BSDEs (ABSDEs). This type of equations was developed originally in a diffusion setting. The pioneering study of Peng and Yang [47] introduced this new type of BSDEs, proved the existence-uniqueness and comparison theorems, and, moreover, constructed the duality between SDDEs and ABSDEs.

Peng and Yang [47] presented the ABSDEs as follows:

$$
\begin{aligned}
-dY(t) &= f(t, Y(t), Z(t), Y(t+\delta_1(t)), Z(t+\delta_2(t)))ds - Z(t)dW(t), \;\; t \in [0, T], \\
Y(t) &= \xi(t) \text{ and } Z(t) = \psi(t), \qquad t \in [T, T+K],
\end{aligned}
$$

where $\delta_i(\cdot)$, $i = 1, 2$, be $\mathbb{R}^+$-valued continuous functions on $[0, T]$.

In the sequel, the existence-uniqueness and duality theorems, which support the underlying theory of an optimal control problem with delay in MP, were extended to a jump-diffusion process in [42, 56] and to a Markov regime-switching jump-diffusion model by Savku and Weber [49].

This work is organized as follows: In Sect. 13.2, we briefly introduce a Markov regime-switching model, which is established by Elliott, Aggoun, and Moore [10]. In Sect. 13.3, we illustrate zero-sum and nonzero-sum stochastic differential game applications by the methods of DPP. In Sect. 13.4, we present the necessary and sufficient Maximum Principles with time delay and regimes and give an example of optimal consumption with memory. The last section is devoted to a conclusion and an outlook.

13.2 Preliminaries

Let us explain the main result to include Markov regime-switches to the dynamics of stochastic differential equations with and without delay. Throughout this chapter, we work on a finite time horizon $T > 0$ and $t \in [0, T]$.

Let $(X(t) : t \in [0, T])$ be a continuous-time, finite-state, and observable Markov chain. The finite-state space of the homogenous and irreducible Markov chain $X(t)$, $S = \{e_1, e_2, \ldots, e_D\}$, is called a canonical state space, where $D \in \mathbb{N}$, $e_i \in \mathbb{R}^D$ and the jth component of e_i is the Kronecker delta δ_{ij} for each pair of $i, j = 1, 2, \ldots, D$.

The generator of the chain under $\mathbb{P}$ is defined by $\Lambda(t) := [\lambda_{ij}(t)]_{i,j=1,2,\ldots,D}$. For each $i, j = 1, 2, \ldots, D$, $\lambda_{ij}(t)$ is the transition intensity of the chain from each state e_i to state e_j at time t. For $i \neq j$, $\lambda_{ij}(t) \geq 0$ and $\sum_{j=1}^{D} \lambda_{ij}(t) = 0$; hence, $\lambda_{ii}(t) \leq 0$. We suppose that for each $i, j = 1, 2, \ldots, D$, with $i \neq j$, $\lambda_{ij}(t) > 0$ and $\lambda_{ii}(t) < 0$, $t \in [0, T]$.

Elliott et al. [10] proved the following semimartingale representation for a Markov chain α:

$$X(t) = X(0) + \int_0^t \Lambda^T(u)X(u)du + M(t), \tag{13.2}$$

where $(M(t) : t \in [0, T])$ is an $\mathbb{R}^D$-valued $(\mathbb{F}, \mathbb{P})$-martingale and Λ^T represents the transpose of the matrix.

For the following sections, we present a survey of applications in finance with Markov regime-switches based on this representation.

13.3 Stochastic Differential Games with Regimes

The idea of combining game theory, DPP, and regime-switches, which are very powerful tools of mathematics, attracted many authors, [2, 7, 12, 50, 55, 57]. In this section, we present two financial examples of zero-sum and nonzero-sum stochastic games by summarizing the results of Elliott and Siu [13] and Ma et al. [35].

13.3.1 A Zero-Sum Game Application

Elliott and Siu [13] seek a robust optimal portfolio strategy under model uncertainty in a continuous-time Markov-modulated financial market. The states of the Markov chain interpret the different states of an economy as macroeconomic indicators. The problem is formulated as a Markov regime-switching version of a two-player, zero-sum stochastic differential game between the agent and the market. The agent maximizes the minimal expected utility of terminal wealth over a family of probability measures in a finite time horizon by DPP.

Let $(\Omega, \mathbb{F}, \mathscr{P})$ be a complete probability space, where $\mathscr{P}$ is generated by a family of absolutely continuous real-world probability measures. Elliott and Siu [13] presents a market with a risk-free bond $B(t)$ and a risky asset $S(t)$, which evolve as follows:

$$B(t) = \exp\left(\int_0^t r(u)du\right), \qquad B(0) = 1,$$

and

$$dS(t) = \mu(t)S(t)dt + \sigma(t)S(t)dW(t),$$
$$S(0) = s > 0$$

correspondingly, where $\{W(t) : t \in [0, T]\}$ is a Brownian motion. Here, r, μ and σ are defined by

$$r(t) = \langle r, X(t) \rangle, \quad \mu(t) = \langle \mu, X(t) \rangle, \quad \sigma(t) = \langle \sigma, X(t) \rangle,$$

where $r = (r_1, r_2, \ldots, r_D)^T \in \mathbb{R}^D$, $\mu = (\mu_1, \mu_2, \ldots, \mu_D)^T \in \mathbb{R}^D$, and $\sigma = (\sigma_1, \sigma_2, \ldots, \sigma_D)^T \in \mathbb{R}^D$, i.e., r, μ and σ get constant real values at each state. The authors work on a complete probability space with an enlarged σ-field $\mathscr{G}(t)$, generated by $\mathscr{F}^X(t)$ and $\mathscr{F}^S(t)$, $t \in [0, T]$. Hence, the wealth process is governed by

$$\begin{aligned} dV(t) &= V^\pi(t)([r(t) + \pi(t)(\mu(t) - r(t))]dt + \pi(t)\sigma(t)dW(t)), \\ V^\pi(0) &= v > 0, \end{aligned} \tag{13.3}$$

where π is self-financing and

$$\int_0^T \pi^2(t)dt < \infty, \; \mathscr{P} - a.s.$$

Then, π is called admissible, and let $\mathscr{A}$ be the class of such processes.
Let $\theta := \{\theta(t) : t \in [0, T]\}$ represent a Markovian regime-switching process such that

$$\theta(t) = \langle \theta(t), X(t) \rangle,$$

where $\theta(t) := (\theta_1(t), \theta_2(t), \ldots, \theta_D(t))^T \in \mathbb{R}^D$, $\theta_i(t) \geq 0$ for all $i = 1, 2, \ldots, D$ and $\theta_{(N)}(t) := \max_{1 \leq i \leq N} \theta_i(t) < \infty$, $t \in [0, T]$. Let Θ be the space of all such θ processes.

For each $\theta(t) \in \Theta$, the density process for the Brownian motion associated with θ is defined as

$$\alpha_1^\theta(t) := \exp\left(- \int_0^t \theta(s)dW(s) - \frac{1}{2}\int_0^t \theta^2(s)ds \right),$$

where $\alpha_1^\theta := \left\{\alpha_1^\theta(t) : t \in [0, T]\right\}$ is a $(\mathscr{G}, \mathscr{P})$-martingale, see Elliott and Siu [13] for technical details.

Let $\Lambda^\theta(t) := [\lambda_{ij}^\theta(t)]_{i,j=1,2,\ldots,D}$ represent a second family of generators for the Markov chain X such that

$$\lambda_{ij}^\theta(t) = \theta\lambda_{ij}(t), \quad t \in [0, T]. \tag{13.4}$$

We also define $D^{\theta}(t) = \left[\lambda_{ij}^{\theta}(t)/\lambda_{ij}(t)\right]_{i,j=1,2,\ldots,D}$.

Lemma 13.1 (Elliott and Siu [13]) *Let* $\Lambda_0(t) := \Lambda(t) - diag(\lambda(t))$, *where* $\lambda(t) := (\lambda_{11}(t), \lambda_{22}(t), \ldots, \lambda_{DD}(t))^T$, *for each* $t \in [0, T]$. *Suppose*

$$\tilde{N} := N(t) - \int_0^t \Lambda_0(u)X(u)du, \qquad t \in [0, T], \tag{13.5}$$

where

$$N(t) = \int_0^t (I - diag(X(u-)))dX(u), \qquad t \in [0, T],$$

counts the number of times the chain X *jumps to the state* e_i *in the time interval* $[0, t]$, *for each* $i = 1, 2, \ldots, D$. *I denotes the* $(D \times D)$*-identity matrix.*

Let us introduce another density process as follows:

$$\alpha_2^{\theta}(t) = 1 + \int_0^t \alpha_2^{\theta}(u-)\left[D_0(u)X(u-) - 1\right]^T (dN(u) - \Lambda_0(u)X(u)du),$$

where α_2^{θ} is an $(\mathscr{F}^X, \mathscr{P})$-martingale. Hence, the authors define

$$\alpha^{\theta}(t) := \alpha_1^{\theta}(t)\alpha_2^{\theta}(t),$$

which is a $(\mathscr{G}, \mathscr{P})$-martingale.

Then, authors define a real-world probability measure $Q^{\theta} \sim \mathscr{P}$ as

$$\frac{dQ^{\theta}}{d\mathscr{P}} := \alpha^{\theta}(T), \quad \text{for each } \theta \in \Theta.$$

Finally, there is a family of real-world probability measures $\mathscr{P}_{\lambda} := \mathscr{P}_{\lambda}(\Theta) = \{Q^{\theta} : \theta \in \Theta\}$.

Therefore,

$$\tilde{N}^{\theta} := N(t) - \int_0^t \Lambda_0^{\theta}(u)X(u)du, \qquad t \in [0, T], \quad \theta \in \Theta,$$

is an $(\mathscr{F}^X, Q^{\theta})$-martingale.

Here, X, which is a Markov chain with a family of generators $\alpha^{\theta}(t)$, $t \in [0, T]$, under Q^{θ}, is represented by

$$X(t) = X(0) + \int_0^t \Lambda^{\theta}(u)X(u)du + M^{\theta}(t),$$

where $M^{\theta} := \{M^{\theta}(t) : t \in [0, T]\}$ is an $(\mathscr{F}^X, Q^{\theta})$-martingale. Moreover, by Girsanov's theorem,

$$W^{\theta}(t) = W(0) + \int_0^t \theta(s)ds$$

is a standard Brownian motion under Q^{θ}.

The problem is formulated as in Mataramvura and Øksendal [36] within the framework of Markov regime-switches as follows:

$$\begin{aligned} dZ(t) &= (dZ_0(t), dZ_1^{\pi,\theta}(t), dZ_2^{\theta}(t))^T \\ &= (dt, dV^{\pi,\theta}(t), dX(t))^T, \\ Z(0) &= z = (s, z_1, z_2)^T, \end{aligned}$$

where

$$\begin{aligned} &dZ_0(t) = dt, \qquad Z_0(0) = s \in [0, T], \\ &dZ_1(t) = Z_1(t)\Big([r(t) + \pi(t)(\mu(t) - r(t)) - \theta(t)\pi(t)\sigma(t)]dt + \pi(t)\sigma(t)dW^{\theta}(t) \Big), \\ &\ Z_1(0) = z_1 > 0, \\ &dZ_2(t) = \Lambda^{\theta}(t)Z_2(t)dt + dM^{\theta}(t), \quad Z_2(0) = z_2. \end{aligned}$$

Now, conditioning on $Z(0) = z$, the robust utility maximization problem can be formulated as a two-player, zero-sum Markovian regime-switching stochastic differential game as follows:

$$\begin{aligned} \Phi(z) &= \sup_{\pi \in \mathscr{A}} \left(\inf_{\theta \in \Theta} E_{\theta}^{z} \left[U(V^{\pi,\theta}(T)) \right] \right) \\ &= E_{\hat{\theta}}^{z} \left[U(V^{\hat{\pi},\hat{\theta}}(T)) \right]. \end{aligned}$$

Let $\mathscr{H}$ denote the space of functions $h(\cdot,\cdot,\cdot) : [0, T] \times \mathbb{R}^+ \times \mathscr{E} \to \mathbb{R}$ such that for each $x \in \mathscr{E}$, $h(\cdot,\cdot,x) \in C([0, T] \times \mathbb{R}^+)$.
Then, we introduce

$$H(s, z_1) := (h(s, z_1, e_1), h(s, z_1, e_2), \ldots, h(s, z_1, e_D))^T \in \mathbb{R}^D.$$

Hence, the Markovian regime-switching generator $\mathscr{L}^{\theta,\pi}$ is represented as follows:

$$\begin{aligned} \mathscr{L}^{\theta,\pi}[h(s, z_1, x)] =& \frac{\partial h}{\partial s} + z_1[r(s) + (\mu(s) - r(s))\pi(z) - \theta(z)\pi(z)\sigma(s)]\frac{\partial h}{\partial z_1} \\ &+ \frac{1}{2}z_1^2\pi^2(z)\sigma^2(s)\frac{\partial^2 h}{\partial z_1^2} + \theta(z)\left\langle H(s, z_1), \Lambda(s)x \right\rangle. \end{aligned}$$

Now, we can present the main theorem of this work.

Theorem 13.1 (Elliott and Siu [13]) *Let $\bar{\mathscr{O}}$ denote the closure of $\mathscr{O}$. Suppose there exists a function ϕ such that for each $x \in \mathscr{E}$, $\phi(\cdot, \cdot, x) \in Crm^2(O) \cap C(\hat{\mathscr{O}})$ and a Markovian control $(\hat{\theta}(t), \hat{\pi}(t)) \in \Theta \times \mathscr{A}$, such that*

1. $\mathscr{L}^{\theta,\hat{\pi}}[\phi(s, z_1, x)] \geq 0$, *for all* $\theta \in \Theta$ *and* $(s, z_1, x) \in \mathscr{O} \times \mathscr{E}$,
2. $\mathscr{L}^{\hat{\theta},\pi}[\phi(s, z_1, x)] \leq 0$, *for all* $\pi \in \mathscr{A}$ *and* $(s, z_1, x) \in \mathscr{O} \times \mathscr{E}$,
3. $\mathscr{L}^{\hat{\theta},\hat{\pi}}[\phi(s, z_1, x)] = 0$, *for all* $(s, z_1, x) \in \mathscr{O} \times \mathscr{E}$,
4. *for all* $(\theta, \pi) \in \Theta \times \mathscr{A}$,

$$\lim_{t \to T^-} \phi(t, Z_1^{\theta,\pi}(t), X(t)) = U(Z_1^{\theta,\pi}(T)).$$

5. *let $\mathscr{K}$ denote the set of stopping times $\tau \leq T$. The family $\left\{\phi(Z_1^{\theta,\pi}(\tau)) : \tau \in \mathscr{K}\right\}$ is uniformly integrable.*

For each $z \in \mathscr{O} \times \mathscr{E}$ and $(\theta, \pi) \in \Theta \times \mathscr{A}$, we write

$$J^{\theta,\pi}(z) := E_\theta^z[U(Z_1^{\theta,\pi}(T))].$$

Then,

$$\begin{aligned}\phi(z) =& \Phi(z) \\ =& \inf_{\theta \in \Theta}\left(\sup_{\pi \in \mathscr{A}} J^{\theta,\pi}(z)\right) = \sup_{\pi \in \mathscr{A}}\left(\inf_{\theta \in \Theta} J^{\theta,\pi}(z)\right) \\ =& \sup_{\pi \in \mathscr{A}} J^{\hat{\theta},\pi}(z) = \inf_{\theta \in \Theta} J^{\theta,\hat{\pi}}(z) = J^{\hat{\theta},\hat{\pi}}(z), \qquad z \in \mathscr{O} \times \mathscr{E},\end{aligned}$$

and $(\hat{\theta}, \hat{\pi})$ is an optimal Markovian control.

Let us summarize the results of this problem formulation for a power utility function as in [13]:

$$U(v) = \frac{v^{1-\gamma}}{1-\gamma},$$

where $v \in [0, \infty)$ and γ is the risk-aversion coefficient.

Now, an ansatz form for the function ϕ is as follows:

$$\phi(z) = \frac{z_1^{1-\gamma}(g(s, x))^{1-\gamma}}{1-\gamma}, \quad \forall z \in \mathscr{O} \times \mathscr{E},$$

where for each $(s, x) \in (0, T) \times \mathscr{E}$, $g(s, x)$ does not vanish, $g(T, x) = 1$, and the authors define $G(s, \gamma)$ as follows:

$$G(s, \gamma) = (g^{1-\gamma}(s, e_1), g^{1-\gamma}(s, e_2), \ldots, g^{1-\gamma}(s, e_D))^T.$$

Let us represent $g_i(s) := g(s, e_i)$, $\mu(s) = \mu_i$, $\sigma(s) = \sigma_i$ and $r(s) = r_i$, for each $i = 1, 2, \ldots, D$, when $x = e_i$. Then, g satisfies the following system of first-order nonlinear ODEs:

$$\frac{dg_i(s)}{ds} + \left(r_i + \frac{\gamma}{2}(\hat{\pi}^*(s, e_i)^2\sigma_i^2\right)g_i(s) + \hat{\theta}^*(s, e_i)\hat{\pi}^*(s, e_i)\sigma_i = 0,$$

where

$$\hat{\pi}^*(s, e_i) = \frac{\langle G(s, \gamma), \Lambda(s)e_i\rangle}{(1-\gamma)\sigma_i g_i^{1-\gamma}(s)},$$

and

$$\hat{\theta}^*(s, e_i) = \frac{\mu_i - r_i - \gamma\hat{\pi}^*(s, e_i)\sigma_i^2}{\sigma_i}, \quad i = 1, 2, \ldots, D.$$

13.3.2 A Nonzero-Sum Game Application

Moreover, Ma et al. [35] illustrated a nonzero-sum game application within the framework of a diffusion regime-switching system by using Eq. (13.2). There are two risky assets and a risk-free asset, and each investor may invest in just one of the risky assets. The dynamics of the risky assets is defined as follows:

$$dS_k(t) = S_k(t)\left(\mu_k(t)dt + \sigma_k(t)dW_k(t)\right), \qquad k = 1, 2,$$

where $\mu_k(t) = \mu_k(t, X(t)) = \langle \mu_k, X(t)\rangle$, and similar for $\sigma(t)$ and $r(t)$. The authors assume that the interest rate r, the appreciation rate μ, and the volatility σ have constant real values at each state as in [13], i.e., $\mu_k = (\mu_k^1, \mu_k^1, \ldots, \mu_k^D) \in \mathbb{R}^D$, $\sigma_k(t) = (\sigma_k^1, \sigma_k^2, \ldots, \sigma_k^D) \in \mathbb{R}^D$, for each $k = 1, 2$. Here, in this setup, Brownian motions $W_k(t)$, $k = 1, 2$, are correlated with correlation coefficient ρ.

The authors formalize a kind of collaboration between two investors by the maximization of the sum of their wealth processes under a stochastic optimal control problem construction. Let us represent the wealth processes of each investor:

$$\begin{aligned}
dY_1(t) &= [r(t, X(t))Y_1(t) + (\mu_1(t, X(t)) - r(t, X(t)))\pi_1(t)]\,dt + \sigma_1(t)\pi_1(t)dW_1(t),\\
Y_1(0) &= y_1(0),\\
dY_2(t) &= [r(t, X(t))Y_2(t) + (\mu_2(t, X(t)) - r(t, X(t)))\pi_2(t)]\,dt + \sigma_2(t)\pi_2(t)dW_2(t),\\
Y_2(0) &= y_2(0),
\end{aligned}$$

where π_1 and π_2 represent the amount invested to the risky assets S_1 and S_2, correspondingly. Hence, the wealth process for this nonzero-sum stochastic differential game is

$$
\begin{aligned}
dZ^{\pi_1,\pi_2} &= \Big(r(t, X(t))Z(t) + (\mu_1(t, X(t)) - r(t, X(t)))\pi_1(t) \\
&\quad + (\mu_2(t, X(t)) - r(t, X(t)))\pi_2(t)\Big)dt \\
&\quad + \sigma_1(t)\pi_1(t)dW_1(t) + \sigma_2(t)\pi_2(t)dW_2(t), \\
Z(0) &= y_1(0) + y_2(0).
\end{aligned}
$$

If a Nash equilibrium exists, this implies that each player's strategy is a best response against the other one. Furthermore, there can be no unilateral profitable deviation for each player's action. Since a Nash equilibrium is self-enforcing, at the equilibrium, each player knows that moving brings a worse payoff.

Subsequently, we give a mathematical representation for the Nash equilibrium of such a stochastic differential game.

Definition 13.1 Let us assume that for the optimal strategy of Investor 2, $\pi_2^* \in \Theta_2$, the best response of Investor 1 satisfies

$$
J_1^{\pi_1,\pi_2^*}(t, z, e_i) \leq J_1^{\pi_1^*,\pi_2^*}(t, z, e_i) \text{ for all } \pi_1 \in \Theta_1,\ e_i \in S,\ z \in K,
$$

and for the optimal strategy of Investor 1, $\pi_1^* \in \Theta_1$, the best response of Investor 2 satisfies

$$
J_2^{\pi_1^*,\pi_2}(t, z, e_i) \leq J_2^{\pi_1^*,\pi_2^*}(t, z, e_i) \text{ for all } \pi_2 \in \Theta_2,\ e_i \in S,\ z \in K.
$$

Here, for $k = 1, 2$ J_k are the payoff functionals of each investor, Θ_k represents the set of admissible controls, and $K \subset \mathbb{R}$, which is an open set, symbolizes the solvency region. Then, the pair of optimal control processes $(\pi_1^*, \pi_2^*) \in \Theta_1 \times \Theta_2$ is called a *Nash equilibrium* for such a stochastic differential game.

The authors solve the problem for exponential utility functions for each investor:

$$
U_k(y) = \frac{-1}{\gamma_k} e^{-\gamma_k y}, \qquad k = 1, 2,
$$

where γ_k, $k = 1, 2$, are the coefficients of absolute risk-aversion (CARA) of each investor.

The ansatz forms of the value functions are as follows:

$$
V_1(t, z, e_i) = \frac{-1}{\gamma_1} e^{-\gamma_1 z e^{-\int_t^T r(s)ds}} f(t, e_i),
$$

where $f(t, e_i)$ is a suitable positive function with the boundary condition $f(T, e_i) = 1$ for all $e_i \in S$, and

$$V_2(t, z, e_i) = \frac{-1}{\gamma_2} e^{-\gamma_2 z e^{\int_t^T r(s)ds}} g(t, e_i),$$

where $g(t, e_i)$ is a suitable positive function with the boundary condition $g(T, e_i) = 1$ for all $e_i \in S$. The problem can be formulated as

$$\begin{aligned}\mathscr{L}^{(\pi_1,\pi_2)} V_k(t, z, e_i) =& [r(t)z + (\mu_1(t) - r(t))\pi_1(t) + (\mu_2(t) - r(t))\pi_2(t)] \frac{\partial V_k^i}{\partial z} \\ &+ \frac{1}{2}\left[\sigma_1^2(t)\pi_1^2 + \sigma_2^2(t)\pi_2^2 + 2\rho\sigma_1(t)\pi_1\sigma_2(t)\pi_2\right] \frac{\partial^2 V_k^i}{\partial z^2} \\ &+ \sum_{j=1}^{D} \lambda_{ij}\left[V_k(t, z, e_j) - V_k(t, z, e_i))\right], \qquad i = 1, 2, \ldots, D.\end{aligned}$$

Hence, for each $k = 1, 2$, the HJBI equations can be described as follows:

$$\begin{aligned}\frac{\partial V_1^i}{\partial t} + \sup_{\pi_1 \in \Theta_1} \mathscr{L}^{(\pi_1,\pi_2^*)}(t, z, e_i) = 0, \\ \frac{\partial V_2^i}{\partial t} + \sup_{\pi_2 \in \Theta_2} \mathscr{L}^{(\pi_1^*,\pi_2)}(t, z, e_i) = 0.\end{aligned}$$

Finally, we have the optimal portfolios for each investor as

$$\begin{aligned}\pi_1^*(t) = \frac{1}{1-\rho^2}\Bigg(&\frac{\mu_1(t) - r(t)}{\gamma_1\sigma_1^2(t)} e^{-\int_t^T r(s)ds} \\ &- \frac{\mu_2(t) - r(t)}{\gamma_2\sigma_1(t)\sigma_2(t)} e^{-\int_t^T r(s)ds}\Bigg)\end{aligned}$$

and

$$\begin{aligned}\pi_2^*(t) = \frac{1}{1-\rho^2}\Bigg(&\frac{\mu_2(t) - r(t)}{\gamma_2\sigma_2^2(t)} e^{-\int_t^T r(s)ds} \\ &- \frac{\mu_2(t) - r(t)}{\gamma_1\sigma_1(t)\sigma_2(t)} e^{-\int_t^T r(s)ds}\Bigg).\end{aligned}$$

Moreover, the Feyman–Kac representations of $f(t, e_i)$ and $g(t, e_i)$ can be obtained as follows:

$$f(t, e_i) = E^{t,i}\left[\int_t^T a(u, X(u)))du\right],$$

and

$$g(t, e_i) = E^{t,i}\left[\int_t^T b(u, X(u)))du\right],$$

where

$$a(t, e_i) = \frac{\gamma_1^2}{2(1-\rho^2)} \Bigg[\frac{(\mu_2(t) - r(t))^2}{\gamma_2^2 \sigma_2^2(t)} - \frac{(\mu_1(t) - r(t))^2}{\gamma_1^2 \sigma_1^2(t)} \\ - 2\frac{(\mu_2(t) - r(t))^2}{\gamma_1 \gamma_2 \sigma_2^2(t)} + 2\rho \frac{(\mu_1(t) - r(t))(\mu_2(t) - r(t))}{\gamma_1^2 \sigma_1(t) \sigma_2(t)} \Bigg]$$

and

$$b(t, e_i) = \frac{\gamma_2^2}{2(1-\rho^2)} \Bigg[\frac{(\mu_1(t) - r(t))^2}{\gamma_1^2 \sigma_1^2(t)} - \frac{(\mu_2(t) - r(t))^2}{\gamma_2^2 \sigma_2^2(t)} \\ - 2\frac{(\mu_1(t) - r(t))^2}{\gamma_1 \gamma_2 \sigma_1^2(t)} + 2\rho \frac{(\mu_1(t) - r(t))(\mu_2(t) - r(t))}{\gamma_2^2 \sigma_1(t) \sigma_2(t)} \Bigg].$$

By the way, the value function of this nonzero-sum stochastic differential game is provided.

13.4 Stochastic Maximum Principle with Memory and Regimes

The Stochastic Maximum Principle is one of the main foundations of stochastic optimal control theory; hence, many authors focused on this tool, see some recent approaches and applications in [34, 37, 38, 40, 43, 53, 54, 60]. Øksendal et al. [42] proved some theorems of necessary and sufficient maximum principles within the framework of a delayed jump-diffusion model, and Savku and Weber [51] carried these results over to a Markov regime-switching setup and illustrated their results by an application in finance for an optimal consumption rate derived from a cash flow with a delay effect. In this section, we prefer to present the results of this very general setup [51].

Let $(\Omega, \mathbb{F}, (\mathscr{F}_t)_{t\geq 0}, \mathbb{P})$ be a complete probability space, where $\mathbb{F} = (\mathscr{F}_t : t \in [0, T])$ and $(\mathscr{F}_t)_{t\geq 0}$ is a right-continuous, $\mathbb{P}$-completed filtration generated by a Brownian motion $W(\cdot)$, a Poisson random measure $N(\cdot, \cdot)$ and a Markov chain $\alpha(\cdot)$ with constant transition intensities. We assume that these processes are independent of each other and adapted to $\mathbb{F}$. Let $\mathscr{B}_0$ be the Borel σ-field generated by an open subset O of $\mathbb{R}_0 := \mathbb{R} \setminus \{0\}$, whose closure does not contain the point 0. Let

$$(N(dt, dz) : t \in [0, T], z \in \mathbb{R}_0)$$

be the Poisson random measure on $([0, T] \times \mathbb{R}_0, \mathscr{B}([0, T]) \otimes \mathscr{B}_0)$. Let $\tilde{N}(dt, dz) := N(dt, dz) - \nu(dz)dt$ be the compensated Poisson random measure, where ν is the Lévy measure of the jump measure $N(\cdot, \cdot)$ such that

$$\nu(\{0\}) = 0 \quad \text{and} \quad \int_{\mathbb{R}} (1 \wedge |z|^2)\nu(dz) < \infty.$$

We introduce a set of Markov jump martingales associated with the chain α as in [60] by using Eq. (13.2).

Let $J^{ij}(t)$ represent the number of the jumps from the state e_i to the state e_j up to time t for each $i, j = 1, 2, \ldots, D$, with $i \neq j$ and $t \in [0, T]$. Then,

$$\begin{aligned}
J^{ij}(t) &:= \sum_{0<s\leq t} \langle \alpha(s-), e_i \rangle \left\langle \alpha(s), e_j \right\rangle \\
&= \sum_{0<s\leq t} \langle \alpha(s-), e_i \rangle \left\langle \alpha(s) - \alpha(s-), e_j \right\rangle \\
&= \int_0^t \langle \alpha(s-), e_i \rangle \left\langle d\alpha(s), e_j \right\rangle \\
&= \int_0^t \langle \alpha(s-), e_i \rangle \left\langle \Lambda^T \alpha(s), e_i \right\rangle ds + \int_0^t \langle \alpha(s-), e_i \rangle \left\langle dM(s), e_j \right\rangle \\
&= \lambda_{ij} \int_0^t \langle \alpha(s-), e_i \rangle \, ds + m_{ij}(t),
\end{aligned}$$

where the processes m_{ij}s are $(\mathbb{F}, \mathbb{P})$-martingales and called the basic martingales associated with the chain α. For each fixed $j = 1, 2, \ldots, D$, let Φ_j be the number of the jumps into state e_j up to time t. Then,

$$\begin{aligned}
\Phi_j(t) &:= \sum_{i=1, i\neq j}^{D} J^{ij}(t) \\
&= \sum_{i=1, i\neq j}^{D} \lambda_{ij} \int_0^t \langle \alpha(s-), e_i \rangle \, ds + \tilde{\Phi}_j(t).
\end{aligned}$$

Let us define $\tilde{\Phi}_j(t) := \sum_{i=1, i\neq j}^{D} m_{ij}(t)$ and $\lambda_j(t) := \sum_{i=1, i\neq j}^{D} \lambda_{ij} \int_0^t \langle \alpha(s-), e_i \rangle \, ds$; then for each $j = 1, 2, \ldots, D$,

$$\tilde{\Phi}_j(t) = \Phi_j(t) - \lambda_j(t)$$

is an $(\mathbb{F}, \mathbb{P})$-martingale. By $\tilde{\Phi}(t) = (\tilde{\Phi}_1(t), \tilde{\Phi}_2(t), \ldots, \tilde{\Phi}_D(t))^T$, we represent a compensated random measure on $([0, T] \times S, \mathscr{B}([0, T]) \otimes \mathscr{B}_S)$, where $\mathscr{B}_S$ is a σ-field of S.

Now, let us represent the model:

$$
\begin{aligned}
dX(t) &= b(t, X(t), Y(t), A(t), \alpha(t), u(t))dt \\
&\quad + \sigma(t, X(t), Y(t), A(t), \alpha(t), u(t))dW(t) \\
&\quad + \int_{\mathbb{R}_0} \eta(t, X(t), Y(t), A(t), \alpha(t), u(t), z)\tilde{N}(dt, dz) \\
&\quad + \gamma(t, X(t), Y(t), A(t), \alpha(t), u(t))d\tilde{\Phi}(t), \qquad t \in [0, T], \qquad (13.6) \\
X(t) &= x_0(t), \qquad t \in [-\delta, 0],\ x_0 \in C([-\delta, 0]; \mathbb{R}),
\end{aligned}
$$

where

$$
Y(t) = X(t-\delta) \text{ and } A(t) = \int_{t-\delta}^{t} e^{-\rho(t-r)} X(r)dr \text{ for } t \in [0, T].
$$

Here, $\delta > 0$, $\rho \geq 0$ and $T > 0$ are given constants and b, σ, η and γ are $\mathscr{F}_t$-measurable functions for all t. Let $\mathscr{U}$ be a nonempty, closed, convex subset of $\mathbb{R}$. An admissible control is a $\mathscr{U}$-valued, $\mathscr{F}_t$-measurable, càdlàg process $u(t)$, $t \in [0, T]$, such that SDDEJR (13.6) has a unique solution and

$$
E\left[\int_0^T |u(t)|^2\, dt\right] < \infty.
$$

We denote the set of all admissible controls by $\mathscr{A}$. Our problem is to find an optimal control $\hat{u} \in \mathscr{A}$ such that

$$
\begin{aligned}
J(\hat{u}) &= \sup_{u \in \mathscr{A}} J(u(\cdot)) \\
&= \sup_{u \in \mathscr{A}} E\left[\int_0^T f(t, X(t), Y(t), A(t), \alpha(t), u(t))dt + g(X(T), \alpha(T))\right],
\end{aligned}
$$

where $J(u(\cdot))$ represents the performance criterion (objective functional) of the problem. The Hamiltonian is defined as follows:

$$
\begin{aligned}
&H(t, x, y, a, e_i, u, p, q, r, w) = f(t, x, y, a, e_i, u) + b(t, x, y, a, e_i, u)p \\
&+ \sigma(t, x, y, a, e_i, u)q + \int_{\mathbb{R}_0} \eta(t, x, y, a, e_i, u, z)r(t, z)\nu(dz) \\
&+ \sum_{j=1}^{D} \gamma^j(t, x, y, a, e_i, u)w^j(t)\lambda_{ij}.
\end{aligned}
$$

The adjoint equations corresponding to u and $X(t) := X^{(u)}(t)$ for the unknown, adapted processes $p(t) \in \mathbb{R}$, $q(t) \in \mathbb{R}$, $r(t, z) \in \mathscr{R}$ and $w(t) \in \mathbb{R}^D$, $t \in [0, T]$, are given by the following Anticipated Backward Stochastic Differential Equation with Jumps and Regimes (ABSDEJR):

$$
\begin{aligned}
dp(t) &= E[\mu(t)|\mathscr{F}_t]dt + q(t)dW(t) + \int_{\mathbb{R}_0} r(t,z)\tilde{N}(dt,dz) + w(t)d\tilde{\Phi}(t), \\
p(T) &= g_x(X(T),\alpha(T)),
\end{aligned} \tag{13.7}
$$

where

$$
\begin{aligned}
\mu(t) = &-\frac{\partial H}{\partial x}(t, X(t), Y(t), A(t), \alpha(t), u(t), p(t), q(t), r(t,\cdot), w(t)) \\
&-\frac{\partial H}{\partial y}(t+\delta, X(t+\delta), Y(t+\delta), A(t+\delta), \alpha(t+\delta), \\
&u(t+\delta), p(t+\delta), q(t+\delta), r(t+\delta,\cdot), w(t+\delta))\mathbf{1}_{[0,T-\delta]}(t) \\
&-e^{\rho t}\left(\int_t^{t+\delta} \frac{\partial H}{\partial a}(s, X(s), Y(s), A(s), \alpha(s), u(s), p(s), q(s), \right. \\
&\left. r(s,\cdot), w(s))e^{-\rho s}\mathbf{1}_{[0,T]}(s)ds\right).
\end{aligned} \tag{13.8}
$$

As seen in Eq. (13.8), we observe not only the present values but also the future values of the processes. In this sense, such equations are called time advanced or anticipated.

Now let us establish the required assumptions for the Necessary Maximum Principle.

- Let $\hat{u}$ be an optimal control process and $\beta \in \mathscr{A}$ be such that $\hat{u} + \beta = v \in \mathscr{A}$. Since $\mathscr{U}$ is a convex set, for any $v \in \mathscr{A}$ the perturbed control process $u^s = \hat{u} + s(v - \hat{u}),\ 0 < s < 1$, is also in $\mathscr{A}$.
- The directional derivative of the performance criterion $J(\cdot)$ at $\hat{u}$ in the direction of $v - \hat{u}$ is given by

$$
\frac{d}{ds}J(\hat{u} + s(v-\hat{u}))|_{s=0} := \lim_{s\to 0^+} \frac{J(\hat{u} + s(v-\hat{u})) - J(\hat{u})}{s}.
$$

- Since $\hat{u}$ is an optimal control, then a necessary condition for optimality is

$$
\frac{d}{ds}J(\hat{u} + s(v-\hat{u}))|_{s=0} \le 0.
$$

Now, we can present the following theorem of the Necessary Maximum Principle.

Theorem 13.2 (Savku and Weber [51]) *Let $\hat{u} \in \mathscr{A}$ be an optimal control with the corresponding trajectories $\hat{X}(t)$, $\hat{Y}(t)$, $\hat{A}(t)$, and $(\hat{p}(t), \hat{q}(t), \hat{r}(t,z), \hat{w}(t))$ be the unique solution of the corresponding adjoint equation. Under some technical conditions (see [51]), for any $v \in \mathscr{U}$, we have*

$$
\frac{\partial H}{\partial u}(t, \hat{X}(t), \hat{Y}(t), \hat{A}(t), \alpha(t), \hat{u}(t), \hat{p}(t), \hat{q}(t), \hat{r}(t,\cdot), \hat{w}(t))(v - \hat{u}(t)) \le 0
$$

$dt - a.e,\ \mathbb{P} - a.s..$

Savku and Weber [51] proved the necessary and maximum principles under full and partial information, here, Theorem 13.2 and the following Theorem 13.3 are the versions under full information.
Let us present a sufficient maximum principle for the system (13.6) as follows.

Theorem 13.3 (Savku and Weber [51]) *Let $\hat{u} \in \mathscr{A}$ with corresponding state processes $\hat{X}(t), \hat{Y}(t), \hat{A}(t)$ and adjoint processes $\hat{p}(t), \hat{q}(t), \hat{r}(t,z), \hat{w}(t)$, assumed to satisfy the ABSDEJR (13.7)–(13.8). Suppose that the following assertions hold:*

$$E\Big[\int_0^T \hat{p}(t)^2\Big((\sigma(t) - \hat{\sigma}(t))^2 + \int_{\mathbb{R}_0} (\eta(t,z) - \hat{\eta}(t,z))^2 \nu(dz) + \sum_{j=1}^{D} (\gamma^j(t) - \hat{\gamma}^j(t))^2 \lambda_j(t)\Big) dt\Big] < \infty$$

and

$$E\Big[\int_0^T (X(t) - \hat{X}(t))^2\Big(\hat{q}^2(t) + \int_{\mathbb{R}_0} \hat{r}^2(t,z)\nu(dz) + \sum_{j=1}^{D} (\hat{w}^j)^2(t)\lambda_j(t)\Big) dt\Big] < \infty.$$

Furthermore, we assume that the following conditions are fulfilled:

1. *For almost all $t \in [0,T]$,*

$$\begin{aligned} &H(t, \hat{X}(t), \hat{Y}(t), \hat{A}(t), \alpha(t), \hat{u}(t), \hat{p}(t), \hat{q}(t), \hat{r}(t,\cdot), \hat{w}(t)) \\ &= \max_{u \in \mathscr{U}} H(t, \hat{X}(t), \hat{Y}(t), \hat{A}(t), \alpha(t), u(t), \hat{p}(t), \hat{q}(t), \hat{r}(t,\cdot), \hat{w}(t)). \end{aligned}$$

2. *H is a concave function of x, y, a, u for all $(t, e_i) \in [0,T] \times S$.*
3. *$g(x, e_i)$ is a concave function of x for each $e_i \in S$.*

Then $\hat{u}(t)$ is an optimal control process and $\hat{X}(t), \hat{Y}(t), \hat{A}(t)$ are the corresponding controlled state processes.

Here, the concavity condition of the Hamiltonian is a cornerstone of the proof of Theorem 13.3 (see Savku and Weber [51]). Pamen [44] proved a more general stochastic maximum principle, which can be applied to a nonconcave Hamiltonian without time delay in the dynamics of a Markov regime-switching jump-diffusion process.

13.4.1 An Optimal Consumption Problem

In this subsection, we apply the results of Theorems 13.2 and 13.3 to a delayed cash flow with jumps and regime-switches. Let us introduce the optimal consumption problem as follows:

Let $b(t, \alpha(t))$, $\sigma(t, \alpha(t))$, $\eta(t, \alpha(t))$, and $\gamma(t, \alpha(t))$ be given bounded adapted processes and the consumption rate $c(t) \geq 0, t \in [0, T]$, be a càdlàg adapted process. Then the corresponding net cash flow $X(t) = X^c(t)$ is defined by Eq. (13.9):

$$\begin{aligned} dX(t) &= (X(t-\delta)b(t, \alpha(t)) - c(t))dt + X(t-\delta)\sigma(t, \alpha(t))dW(t) \\ &\quad + X(t-\delta)\int_{\mathbb{R}_0} \eta(t, \alpha(t), z)\tilde{N}(dt, dz) \\ &\quad + X(t-\delta)\gamma(t, \alpha(t))d\tilde{\Phi}(t), \qquad t \in [0, T], \\ X(t) &= x_0(t), \ t \in [-\delta, 0], \ x_0 \in C([-\delta, 0]; \mathbb{R}). \end{aligned} \tag{13.9}$$

Let $U(t, c, e_i), i = 1, 2, \ldots, D$, be a given function such that

- $t \mapsto U(t, c, e_i)$ is $\mathscr{F}_t$-adapted for each $c \geq 0$ and $e_i \in S$,
- $c \mapsto U(t, c, e_i)$ is $\mathscr{C}^1$, $\frac{\partial U}{\partial c}(t, c, e_i) > 0$ for each $e_i \in S$,
- $c \mapsto \frac{\partial U}{\partial c}(t, c, e_i)$ is strictly decreasing for each $e_i \in S$,
- $\lim_{c \to \infty} \frac{\partial U}{\partial c}(t, c, e_i) = 0$ for all $t \in [0, T]$ and $e_i \in S$.
- Let $v_0(t, e_i) := \frac{\partial U}{\partial c}(t, 0, e_i)$ and define

$$I(t, v, e_i) := \begin{cases} 0, & \text{if } v \geq v_0(t, e_i), \\ (\frac{\partial U}{\partial c}(t, \cdot, e_i))^{-1}(v), & \text{if } 0 \leq v < v_0(t, e_i). \end{cases}$$

Our problem is to find the consumption rate $\hat{c}(t)$ such that

$$J(\hat{c}) = \sup_{c \in \mathscr{A}} E\left[\int_0^T U(t, c(t), \alpha(t))dt + g(X(T), \alpha(T))\right].$$

Consequently, in such a case, the Hamiltonian gets the form:

$$\begin{aligned} H(t, x, y, a, e_i, c, p, q, r(\cdot), w) &= U(t, c, e_i) + (b(t, e_i)y - c)p + y\sigma(t, e_i)q \\ &\quad + y\int_{\mathbb{R}_0} \eta(t, e_i, z)r(t, z)\nu(dz) + y\sum_{j=1}^{D} \gamma^j(t, e_i)w^j(t)\lambda_{ij}. \end{aligned}$$

Hence, the corresponding adjoint equations are defined by

$$
\begin{aligned}
dp(t) = & -E[(b(t+\delta,\alpha(t+\delta))p(t+\delta)+\sigma(t+\delta,\alpha(t+\delta))q(t+\delta) \\
& +\int_{\mathbb{R}_0}\eta(t+\delta,\alpha(t+\delta),z)r(t+\delta,z)\nu(dz) \\
& +\sum_{j=1}^{D}\gamma^j(t,\alpha(t+\delta))w^j(t+\delta)\lambda_j(t))\mathbf{1}_{[0,T-\delta]}(t)|\mathcal{F}_t]dt \\
& +q(t)dW(t)+\int_{\mathbb{R}_0}r(t,z)\tilde{N}(dt,dz)+w(t)\tilde{\Phi}(t),\ t\in[0,T]; \\
p(T) = & \ g_x(X(T),\alpha(T)).
\end{aligned}
$$

Here, we observe that maximizing H with respect to c gives

$$\frac{\partial U}{\partial c}(t,\hat{c}(t),\alpha(t)) = p(t).$$

Therefore, we may summarize our findings by the following proposition.

Proposition 13.1 (Savku and Weber [51]) *Let $p(t), q(t), r(t,z)$ and $w(t)$ be the solution of the corresponding ABSDEJR, and suppose that $0 \le p(t) \le v_0(t,\alpha(t))$ holds for all $t \in [0,T]$.*

Then, the optimal consumption rate, $\hat{c}(t)$, and the corresponding optimal terminal wealth, $\hat{X}(t)$, are given implicitly by the coupled equations

$$\hat{c}(t) = I(t,p(t),\alpha(t))$$

and Eq. (13.9), respectively.

Now, in order to finalize our optimal consumption problem, we assume that $b(t,\alpha(t))$ is deterministic and $g(x,e_i)=kx,\ k>0$.

Hence, by the Martingale Representation Theorem for regime-switching jump-diffusions, Crépéy and Matoussi [6], we can choose $q=r=w=0$. Then the ABSDEJRs (13.7)–(13.8) become

$$
\begin{aligned}
dp(t) &= -b(t+\delta)p(t+\delta)\mathbf{1}_{[0,T-\delta]}(t)dt, \qquad t<T, \\
p(t) &= k, \qquad t\in[T-\delta,T].
\end{aligned}
$$

To solve this, we introduce

$$h(t) = p(T-t), \qquad t\in[0,T].$$

Then, we obtain the DDEs:

$$
\begin{aligned}
dh(t) &= b(T-t+\delta)h(t-\delta)dt, \qquad t\in[\delta,T], \\
h(t) &= k, \qquad t\in[0,\delta].
\end{aligned}
$$

Finally, we can determine $h(t)$ inductively on each interval as follows:

$$h(t) = h(j\delta) + \int_{j\delta}^{t} h'(s)ds = h(j\delta) + \int_{j\delta}^{t} b(T - t + \delta)h(s - \delta)ds \qquad (13.10)$$

for $t \in [j\delta, (j+1)\delta]$.

Whereas we may apply any utility function, in order to obtain a more specific solution form we prefer to use $U(t, c, e_i) = \phi(t, e_i)\ln(1 + c)$ for all $i = 1, 2, \dots, D$, where $\phi(t, e_i)$ is an $\mathbb{R}^+$-valued, càdlàg and $\mathscr{F}_t$-adapted function such that

$$E\left[\int_0^t |\phi(t, \alpha(t))|^2 \, dt\right] < \infty.$$

Then, following the theorem presents the solution of our problem.

Theorem 13.4 (Savku and Weber [51]) *The optimal consumption rate* $\hat{c}(t)$ *under the above construction is explicitly given by*

$$\begin{aligned}\hat{c}(t) &= I(t, h_\delta(T - t), \alpha(t)|_{\alpha(t)=e_i}) \\ &= \begin{cases} 0, & \text{if} \quad h_\delta(T - t) \geq \phi(t, e_i), \\ \frac{\phi(t,e_i)}{h_\delta(T-t)} - 1, & \text{if} \quad 0 \leq h_\delta(T - t) < \phi(t, e_i), \end{cases}\end{aligned}$$

where $h(\cdot) = h_\delta(\cdot)$ *is determined by Eq. (13.10).*

13.5 Conclusion and Outlook

In this survey study, we introduced mathematical the structure of Markov regime-switches and combined it with stochastic differential games and a time delay under the headline of the stochastic optimal control theory. Moreover, we provided optimal portfolio and consumption formulas for the systems of stochastic differential equations modulated by Markov regime-switches.

Mathematically, regime-switches should not be seen just as an additional jump part with carrying very similar properties of a compensated Poisson random measure. Many fundamental results, such as existence-uniqueness theorems of BSDE, ABSEDs, and SDDE with regimes, comparison theorems, duality, etc, have to be extended for a mathematically concrete and satisfactory scientific work. Furthermore, these additional jump parts, generated by a compensating counting measure of a Markov chain bring several challenges, not only in theory, but also in getting analytical results. Consequently, still there are many mathematical problems waiting to be solved with these settings.

Furthermore, the shifts between the states of a Markov chain may characterize some of the psychological phenomena from the perspective of macroeconomic indicators, for example, the currency risk or the country risk may affect the market psychology and the investors' preferences easily.

Markov-switching models, particularly with a delay, have gained a growing interest recently in so many disciplines, both theoretical and applied, e.g., in finance and economics, in medicine and neuroscience. Herewith, they demonstrate their unifying and pioneering scientific potentials and their future promise for a high-quality modeling, a balance of interests, and a decision-making in a world characterized by the highest complexity, stochastic uncertainties, and "human factors" [28]. We, the authors of this chapter, express our hope to having generated an interest in this emerging research agenda and its positive impact in the world of tomorrow.

References

1. Bensoussan, A., & Menaldi, J. L. (2003). Stochastic hybrid control. *Journal of Mathematical Analysis and Applications, 249*, 261–288.
2. Bensoussan, A., Siu, C. C., Yamd, S. C. P., & Yange, H. (2014). A class of nonzero-sum stochastic differential investment and reinsurance games. *Automatica, 50*(6), 2025–2037.
3. Cadenillas, A., & Karatzas, I. (1995). The stochastic maximum principle for linear, convex optimal control with random coefficients. *SIAM Journal of Control and Optimization, 33*(2), 590–624.
4. Cohen, S. N., & Elliott, R. J. (2010). Comparisons for backward stochastic differential equations on Markov chains and related no-arbitrage conditions. *Annals of Applied Probability, 20*(1), 267–311.
5. Crepey, S. (2010). About the pricing equations in finance. *Paris-Princeton Lectures on Mathematical Finance*, 63–203.
6. Crepey, S., & Matoussi, A. (2008). Reflected and doubly reflected BSDEs with jumps: A priori estimate and comparison. *The Annals of Applied Probability, 18*(5), 2041–2069.
7. Deng, C., Zeng, X., & Zhu, H. (2018). Nonzero-sum stochastic differential reinsurance and investment games with default risk. *European Journal of Operational Research, 264*(3), 1144–1158.
8. El-Karoui, N., & Hamadene, S. (2003). BSDEs and risk-sensitive control, zero-sum and nonzero-sum game problems of stochastic functional differential equations. *Stochastic Processes and their Applications, 107*(1), 145–169.
9. El-Karoui, N., Peng, S., & Quenez, M. C. (1997). Backward stochastic differential equations. *Mathematical Finance, 7*(1), 1–71.
10. Elliott, R. J., Aggoun, L., & Moore, J. B. (1995). *Hidden Markov models: Estimation and control*. New York: Springer.
11. Elliott, R. J., Chan, L., & Siu, T. K. (2005). Option pricing and Esscher transform under regime switching. *Annals of Finance, 1*(4), 423–432.
12. Elliott, R. J., & Siu, T. K. (2008). On risk minimizing portfolios under a Markovian regime-switching Black-Sholes economy. *Annals of Operations Research, 176*(1), 271–291.
13. Elliott, R. J., & Siu, T. K. (2009). Robust optimal portfolio choice under Markovian regime-switching model. *Methodology and Computing in Applied Probability, 11*(2), 145–157.
14. Elsanosi, I., Øksendal, B., & Sulem, A. (2000). Some solvable stochastic control problems with delay. *Stochastics and Stochastic Reports, 71*, 69–89.
15. Federico, S. (2011). A stochastic control problem with delay arising in a pension fund model. *Finance and Stochastics, 15*(3), 421–459.
16. Federico, S., Goldys, B., & Gozzi, F. (2010). HJB equations for the optimal control of differential equations with delays and state constraints, i: Regularity of viscosity solutions. *SIAM Journal on Control and Optimization, 48*(8), 4910–4937.

17. Federico, S., Goldys, B., & Gozzi, F. (2011). HJB equations for the optimal control of differential equations with delays and state constraints, ii: Verification and optimal feedbacks. *SIAM Journal on Control and Optimization*, *49*(6), 2378–2414.
18. Ghosh, M. K., & Bagchi, A. (2003). Modeling stochastic hybrid systems. In *21st IFIP TC7 Conference on System Modelling and Optimization*.
19. Goldfeld, S. M., & Quandt, R. E. (1973). A Markov model for switching regressions. *Journal of Econometrics*, *1*(1), 3–16.
20. Hamadene, S. (1998). Backward-forward SDE's and stochastic differential games. *Stochastic Processes and their Applications*, *77*(1), 1–15.
21. Hamadene, S., & Hassani, M. (2006). BSDEs with two reflecting barriers driven by a Brownian and a Poisson noise and related Dynkin game. *http://www.math.washington.edu/ > ejpecp/ 11*(5), 121–145.
22. Hamilton, J. D. (1989). A new approach to the economic analysis of nonstationary time series and business cycle. *Econometrica*, *57*(2), 357–384.
23. Hu, J., Lygeros, J., & Sastry, S. (2000). Towards a theory of stochastic hybrid systems. In N. Lynch & B. H. Krogh (Eds.), *Hybrid systems: Computation and control*, 1790 in LNCS pp. 160–173.
24. Huang, W., Lehalle, C.-A., & Rosenbaum, M. (2016). How to predict the consequences of a tick value change? Evidence from the Tokyo stock exchange pilot program. *Market Microstructure and Liquidity* *2*(03n04), 1750001.
25. Itô, K., & Nisio, M. (1964). On stationary solutions of a stochastic differential equation. *Journal of Mathematics*, *4*(1), 1–75.
26. Karatzas, I., & Li, Q. (2011). BSDE approach to nonzero-sum stochastic differential games of control and stopping. In *Stochastic processes, finance and control: A festschrift in honor of Robert J. Elliott*. World Scientific Publishing Co. Pte. Ltd.
27. Korn, R., & Menkens, O. (2005). Worst-case scenario portfolio optimization: A new stochastic control approach. *Mathematical Methods of Operations Research*, *62*(1), 123–140.
28. Kropat, E., Turkay, M., & Weber, G.-W. (2020). Fuzzy analytics and stochastic methods in neurosciences. *IEEE Transactions on Fuzzy Systems (IEEE TFS), Special issue (to appear)*, *28*(1), 1–121.
29. Kushner, H. J. (1968). On the stability of processes defined by stochastic differential difference equations. *Journal of Differential Equations*, *4*(3), 424–443.
30. Laruelle, S., Rosenbaum, M., & Savku, E. (2020). Assessing MiFID 2 regulation on tick sizes: A transaction costs analysis viewpoint. *Market Microstructure and Liquidity*, *5*(1), 2050003. World Scientific.
31. Laruelle, S., Rosenbaum, M., & Savku, E. (2018). Assessing MiFID 2 regulation on tick sizes: A transaction costs analysis viewpoint. *Available at SSRN, 3256453*.
32. Li, C.-Y., Chen, S.-N., & Lin, S.-K. (2016). Pricing derivatives with modeling co2 emission allowance using a regime-switching jump diffusion model: With regime-switching risk premium. *The European Journal of Finance*, *22*(10), 887–908.
33. Lin, X., Zhang, C., & Siu, T. K. (2012). Stochastic differential portfolio games for an insurer in a jump-diffusion risk process. *Mathematical Methods of Operations Research*, *75*(1), 83–100.
34. Lv, S., Tao, R., & Wu, Z. (2016). Maximum principle for optimal control of anticipated forward-backward stochastic differential delayed systems with regime-switching. *Optimal Control Applications and Methods*, *37*(1), 154–175.
35. Ma, C., Wu, H., & Lin, X. (2015). Nonzero-sum stochastic differential portfolio games under a Markovian regime switching model. *Hindawi Publishing Corporation Mathematical Problems in Engineering*.
36. Mataramvura, S., & Øksendal, B. (2008). Risk minimizing portfolios and HJBI equations for stochastic differential games. *Stochastics: An International Journal of Probability and Stochastic Processes*, *80*(4), 317–337.
37. Menoukeu-Pamen, O., & Momeya, R. H. (2017). A maximum principle for Markov regime-switching forward-backward stochastic differential games and applications. *Mathematical Methods of Operations Research*, *85*(3), 349–388.

38. Meyer-Brandis, T., Øksendal, B., & Zhou, X. Y. (2012). A mean-field stochastic maximum principle via Malliavin calculus. *Stochastics: An International Journal of Probability and Stochastic Processes 84*, 5–6 (201), 643–666.
39. Mohammed, S. E. A. (1984). *Stochastic functional differential equations*. Pitman.
40. Øksendal, B., & Sulem, A. (2001). A maximum principle for optimal control of stochastic systems with delay, with applications to finance. *Optimal Control and Partial Differential Equations* (01 2000).
41. Øksendal, B., & Sulem, A. (2011). *A maximum principle for optimal control of stochastic systems with delay, with application to finance*. Amsterdam: IOS press.
42. Øksendal, B., Sulem, A., & Zhang, T. (2011). Optimal control of stochastic delay equations and time-advanced backward stochastic differential equations. *Advances in Applied Probability, 43*(2), 572–596.
43. Pamen, O. M. (2015). Optimal control for stochastic delay systems under model uncertainty: A stochastic differential game approach. *Journal of Optimization Theory and Applications, 167*(3), 998–1031.
44. Pamen, O. M. (2017). Maximum principles of Markov regime-switching forward-backward stochastic differential equations with jumps and partial information. *Journal of Optimization Theory and Applications, 175*(2), 373–410.
45. Pang, T., & Hussain, A. (2016). An infinite time horizon portfolio optimization model with delays. *Mathematical Control & Related Fields, 6*(4), 629–651.
46. Pardoux, E., & Peng, S. (1990). Adapted solution of a backward stochastic differential equation. *Systems and Control Letters, 14*(1), 55–61.
47. Peng, S., & Yang, Z. (2009). Anticipated backward stochastic differential equations. *The Annals of Probability, 37*(3), 877–902.
48. Quandt, R. E. (1958). The estimation of the parameters of a linear regression system obeying two seperate regimes. *Journal of the American Association, 53*(284), 873–880.
49. Savku, E. (2017). *Advances in optimal control of Markov regime-switching models with applications in finance and economics*. Ph.D. thesis, IAM, METU, Turkey.
50. Savku, E., & Weber, G.-W. (2020). Stochastic differential games for optimal investment problems in a Markov regime-switching jump-diffusion market. *Annals of Operations Research*, 1–26. Springer.
51. Savku, E., & Weber, G.-W. (2018). A stochastic maximum principle for a markov regime-switching jump-diffusion model with delay and an application to finance. *Journal of Optimization Theory and Applications, 179*(2), 696–721.
52. Seifried, F. T. (2010). Optimal investment for worst-case crash scenarios: A martingale approach. *Mathematics of Operations Research, 35*(3), 559–579.
53. Shen, Y., Meng, Q., & Shi, P. (2014). Maximum principle for mean-field jump-diffusion stochastic delay differential equations and its application to finance. *Automatica, 50*(5), 1565–1579.
54. Shen, Y., & Siu, T. K. (2013). The maximum principle for a jump-diffusion mean-field model and its application to the mean-variance problem. *Nonlinear Analysis, 86*(2), 58–73.
55. Shen, Y., & Siu, T. K. (2013). Stochastic differential game, Esscher transform and general equilibrium under a Markovian regime-switching Lévy model. *Insurance: Mathematics and Economics 53*(3), 757–768.
56. Tu, S., & Hao, W. (2014). Anticipated backward stochastic differential equations with jumps under the non-Lipschitz condition. *Statistics and Probabilty Letters, 92*, 215–225.
57. Yiu, K.-F. C., Liu, J., Siu, T. K., & Ching, W.-K. (2010). Optimal portfolios with regime switching and value-at-risk constraint. *Automatica, 46*(6), 979–989.
58. Zhang, Q. (2001). Stock trading: An optimal selling rule. *Siam Journal on Control and Optimization, 40*(1), 64–87.
59. Zhang, Q., & Yin, G. (2004). Nearly-optimal asset allocation in hybrid stock investment. *Journal of Optimization Theory and Applications, 121*(2), 121–419.
60. Zhang, X., Elliott, R. J., & Siu, T. K. (2012). A stochastic maximum principle for a Markov regime-switching jump-diffusion model and an application to finance. *SIAM Journal on Control and Optimization, 50*(2), 964–990.

The manufacturer's authorised representative in the EU is Springer Nature Customer Service Centre GmbH, Europaplatz 3, 69115 Heidelberg, Germany. If you have any concerns regarding our products, please contact ProductSafety@springernature.com

Printed and bound by CPI Group (UK) Ltd, Croydon, CR0 4YY
22/09/2026
02221059-0009